GEOMETRY AND LIGHT

The Science of Invisibility

Ulf Leonhardt
University of St. Andrews

Thomas Philbin
University of St. Andrews

Dover Publications, Inc.
Mineola, New York

Bibliographical Note

Geometry and Light: The Science of Invisibility is a new work,
first published by Dover Publications, Inc., in 2010.

International Standard Book Number

ISBN-13: 978-0-486-47693-3
ISBN-10: 0-486-47693-6

Manufactured in the United States by Courier Corporation
47693601
www.doverpublications.com

Contents

GEOMETRY AND LIGHT

The Science of Invisibility

Chapter 1

Prologue

Many mass–produced products of modern technology would have appeared completely magical two hundred years ago. Mobile phones and computers are obvious examples, but something as commonplace to us as electric light would perhaps be just as astonishing to an age of candles and oil lamps. It seems reasonable to assume that we are no more prescient than the children of the Enlightenment and that, as science and technology develop further, some things that appear impossible today will become ubiquitous in the future. As Arthur C. Clarke famously wrote, "Any sufficiently advanced technology is indistinguishable from magic". In this book we focus on optics and electromagnetism, an ancient subject so suffused with notions of magic that the word illusion is still used by its modern practitioners in their learned journals. We explain the science of the ultimate optical illusion, invisibility. The ingredients of invisibility can be used for other surprising optical effects, such as perfect imaging and laboratory analogues of black holes. Just as important as the particular applications discussed are the powerful ideas that underlie them, ideas that have a fascinating pedigree and that are far from exhausted. We hope to equip the reader with these versatile and fruitful tools of physics and mathematics.

Although invisibility may seem like magic, its roots are familiar to everyone with (literal) vision. Almost all we need to do is to wonder and ask questions. Take a simple observation from daily life and ask some questions: if a straw is placed in a glass of water it appears to be broken at the water's surface (Fig. 1.1). We know the straw is not really broken (and miraculously repaired when removed from the water), so what does the water change? It can only change our perception of the straw, its image carried by light. The water in the glass distorts our perception of space, and this perception is conveyed by light. We conclude that the water changes the measure of space for light, the way light "sees" distances—the geometry of space. Other transparent substances like glass or air, called optical materials or optical media, should not be qualitatively different from water in the way they distort geometry for light. So we are led to the hypothesis that media appear to light as geometries. In this book we take this geometrical perspective on light in media seriously and develop it to extremes. We also discuss its limitations and find the conditions when the geometry established by media is exact.

Taking some basic facts seriously, scrutinizing them and developing them to extremes is the way science generally develops. The tools for this development are sophisticated instrumentation for finding experimental facts and mathematical theory for refining the ideas; what seems like magic is a brew of applied mathematics.

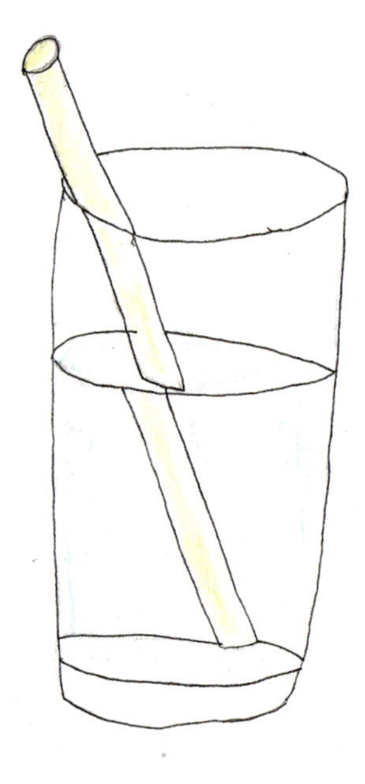

Figure 1.1: Refraction. The image of a straw in a glass of water appears refracted at the water surface. (Credit: Maria Leonhardt.)

But before going into mathematical detail, we can already deduce some aspects of the geometry of light by thinking about things we already know, encouraged by the saying that "research is to see what everybody has seen and to think what nobody has thought" (Jammer [1989]). We know, for example, that a convex lens focuses light (Fig. 1.2); parallel bundles of light rays are focused at one point, which suggests that in the geometry of light established by the lens parallel lines meet. The Greek mathematician Euclid, who developed geometry from five axioms, postulated that parallels never meet, but Euclid's geometry is the geometry of flat space. Euclid's parallel axiom is in fact the defining characteristic of flat space. The light rays focused by the lens do not seem to conform to Euclid's postulate; the geometry of light is non–Euclidean, light may perceive a medium as a curved space. Only in exceptional cases is the geometry established by an optical material that of flat space. One of the exceptional cases is obvious: imagine being completely immersed in a transparent substance, like a diver in water. In this situation space does not appear to be distorted at all, except when the diver looks from below at the water's surface where the flat space established by the water ends. We will prove that having two different media, say water and air, with an interface between them, is already sufficient to establish a curved geometry for light. The straw in the glass of water appears broken because the geometry of light is curved. We will deduce the conditions when the geometry made by media is flat and show that such media can make things disappear from view.

James Clerk Maxwell discovered that light is an electromagnetic wave. With his theory of electromagnetism he also laid the foundation for most of modern technology. The geometry of curved space, on the other hand, is normally encountered by physicists only in Albert Einstein's general relativity. To understand the geometry of light we need to combine aspects of both theories. Yet for most physics and engineering students, ordinary electromagnetism with its vector calculus is already a challenge. In this textbook we build up the required mathematics, differential geometry, step by step with many exercises designed to help the reader gain expertise and confidence in the mathematical machinery we set forth. We strongly recommend doing as many of the exercises as possible, because there is no easier path to

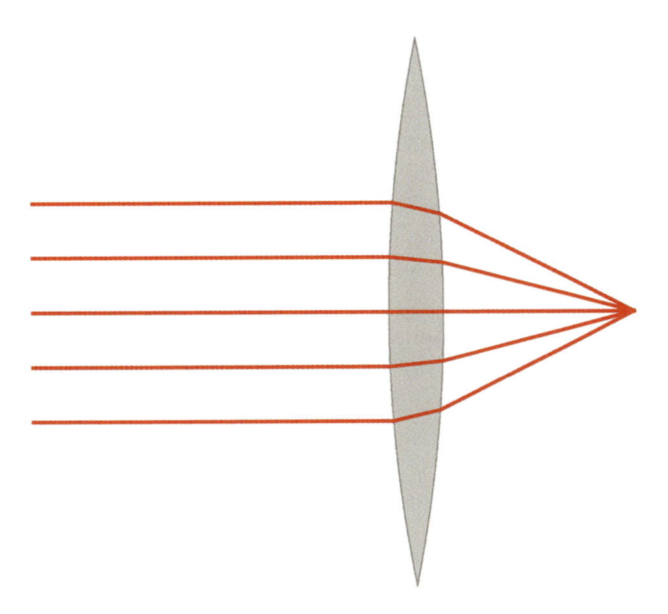

Figure 1.2: Parallel light rays (red) meet in the focus of a lens (grey).

the necessary geometry, no "royal road". We assume the reader knows basic calculus and analytic geometry, and has some acquaintance with Maxwell's equations. Differential geometry applied to electromagnetism gives insights into the nature of light and establishes the scientific foundations for the applications that follow.

We hope the applications and insights provide a strong enough incentive to work through this book. The profit for the reader is a working knowledge of differential geometry and other versatile tools, with a sense for the way in which physicists and engineers apply mathematics. Potential applications are not confined to optics and electromagnetism, but include waves in fluid mechanics and acoustics and the strange waves of quantum physics. The most difficult part of the book is probably the beginning, the Chapter on Fermat's principle, because there we introduce the main concepts with limited algebra, assisted by tailor–made arguments and visualizations. Concepts are the hardest part of science—one should always remember that their originators also struggled to master them.

One of the joys of this area of optics is that it makes use of a surprisingly wide range of classic physics and mathematics. The appearance of names such as Fermat, Newton, Hamilton, Maxwell, Riemann and Einstein shows that this book is built on old foundations. Indeed, one could describe the recent developments presented here as "new things in old things", to quote a phrase by Michael Berry. This illustrates the continuing importance of the old things, but also the gradual, hard–won shift in perspective that is required to see the new things. How else could it have taken so long before ideas for invisibility and perfect imaging appeared? As the reader will see, they are obvious with hindsight.

The materials required for cloaking and perfect imaging, metamaterials, are not new either; they date back to Ancient Rome. The Romans invented the first optical metamaterial—ruby glass. They probably did not know it, but their recipe for ruby glass contained a crucial ingredient: tiny gold droplets, typically 5–60 nm in size (Wagner et al. [2000]). These gold particles colour the glass in an extraordinary way, as demonstrated by the exquisite Lycurgus Cup (Fig. 1.3). In daylight the cup appears green, but when illuminated from the inside it glows with a ruby colour. The gold nanoparticles in the glass do not colour it golden, but red. One can also make other colours with metal particles; the brilliant colours of medieval stained–glass windows come from metal nanoparticles immersed in the glass. The sizes and shapes of the nanoparticles determine the colour. In a metamaterial, structures smaller than the wavelength of light control the optical properties of the material, their shapes and sizes matter more than their chemistry—metal nanostructures like the gold droplets in the Lycurgus Cup do not appear metallic. Thanks to advances in modern nanotechnology and the science behind it, engineers can now make carefully controlled subwavelength structures with designs based on accurate theoretical predictions, whereas Roman technology mostly relied on trial and error. Rome pioneered the technology of metamaterials and Greece, through geometry, the ideas to make use of them.

Figure 1.3: Lycurgus Cup. This Roman cup is made of ruby glass, the first optical metamaterial. When viewed in reflected light, for example in daylight, it appears green. However, when a light is shone into the cup and transmitted through the glass, it appears red. The cup illustrates the myth of King Lycurgus. He is seen being dragged into the underworld by the Greek nymph Ambrosia, who is disguised as a vine. (Credit: the Trustees of the British Museum.)

There are several excellent monographs on the science and technology of metamaterials (see the list in *Further Reading*), but this is the first textbook on the geometrical ideas behind some of their most exciting applications. We thus explore the Greek path rather than the Roman. Connections between general relativity and optics have been reviewed before (Schleich and Scully [1984]), but with different applications in mind and not in a textbook. The only other textbook that combines general relativity with electromagnetism in media is Post's "Formal Structure of Electromagnetics" (Post [1962]), but the book is, as the title says, formal. Here we hope to breathe life into formalism, to explain some "new things in old things", and to inspire the reader to discover others that, for now, are still magic.

Further reading

This book grew out of the review article Leonhardt and Philbin [2009]. We recommend Post [1962] and Schleich and Scully [1984] for getting a perspective on the geometry of light that complements our book.

On the practicalities and the underlying physics of metamaterials we recommend Milton [2002], Sarychev and Shalaev [2007], Cai and Shalaev [2009] and the monumental Metamaterials Handbook (Capolino [2009]). On numerical aspects we suggest to consult Hao and Mittra [2008]. Wave propagation in metamaterials is discussed in Solymar and Shamonina [2009].

The practical use of general relativity in electrical and optical engineering may seem surprisingly unorthodox: traditionally, relativity has been associated with the physics of gravitation (Misner, Thorne and Wheeler [1973]) and cosmology (Peacock [1999]) or, in engineering (Van Bladel [1984]) has been considered a complication, not a simplification. This situation changed with the advent of transformation optics (Chen, Chan and Sheng [2010]). Geometrical ideas have been applied to construct conductivities that are undetectable by static electric fields (Greenleaf, Lassas and Uhlmann [2003a,b]) which was the precursor of invisibility devices (Gbur [2003], Alu and Engheta [2005], Leonhardt [2006a,b], Milton and Nicorovici [2006], Pendry, Schurig and Smith [2006], Schurig, Pendry and Smith [2006]) based on optical implementations of coordinate transformations. From these developments grew the subject of transformation optics (Chen, Chan and Sheng [2010]).

In Chapter 2 we mention the fascinating history of ideas behind the geometrical perspective on optics and electromagnetism, a history that spans more than three centuries. More recently, in 1923 Gordon noticed that moving isotropic media appear to electromagnetic fields as certain effective space–time geometries. Bortolotti [1926] and Rytov [1938] pointed out that ordinary isotropic media establish non–Euclidean geometries for light. Tamm [1924, 1925] generalized the geometrical approach to anisotropic media and briefly applied this theory (Tamm [1925]) to the propagation of light in curved geometries. Plebanski [1960] formulated the electromagnetic effect of curved space–time or curved coordinates in concise constitutive equations. Dolin [1961] published an early precursor of transformation optics that, however, rather focuses on the construction of new solutions of Maxwell's equations than on the invention of new devices.

Acknowledgments

We are privileged to have benefited from many inspiring conversations about "geometry, light and a wee bit of magic". In particular, we would like to thank John Allen, Sir Michael Berry, Leda Boussiakou, Che Ting Chan, Huanyang Chen, Aaron Danner, Luciana Davila–Romero, Mark Dennis, Malcolm Dunn, Ildar Gabitov, Lucas Gabrielli, Greg Gbur, Andrew Green, Awatif Hendi, Julian Henn, Chris Hooley, Sir Peter Knight, Natalia Korolkova, Irina Leonhardt, Michal Lipson, Renaud Parentani, Harry Paul, Sir John Pendry, Ulf Peschel, Paul Piwnicki, Sahar Sahebdivan, Wolfgang Schleich, David Smith, Stig Stenholm, Arran Tamsett, Tomáš Tyc and Grigori Volovik. Our work has been supported by the Scottish Government, the Royal Society of Edinburgh and the Royal Society of London.

Chapter 2

Fermat's principle

The fact that optical materials change the geometry perceived by light appears in quantitative, mathematical form in Fermat's principle. The principle says that light rays follow extremal optical paths; in most cases these are the optically shortest paths. Pierre de Fermat introduced the crucial idea that in materials the measure of optical path length is not the geometrical measure of length, but the geometrical length multiplied by the refractive index of the material. If the index varies the optimal paths are no longer straight lines, but are curved, which may cause optical illusions including the ultimate illusion, invisibility. As the optical length is not the geometrical length, the optical medium defines a different geometry for light.

In this Chapter we take Fermat's principle as the starting point of a first preliminary expedition that explores the scenes and themes described later in this book in full detail. On our journey we will encounter some of the ideas that have shaped modern physics, in particular the connection between optics and mechanics that, via Hamilton's principle of least action, inspired quantum mechanics and field theory. Combined with conformal mapping and the mathematics of tomography, these ideas lead to the design of a range of optical instruments, from invisible spheres to perfect lenses. On our way we will also have glimpses into some fascinating mathematics, from complex analysis to the Escheresque Riemann sheets behind Fermat's Last Theorem. Let the journey begin.

§1. Letters from Pierre de Fermat

In this Section we introduce Fermat's principle that governs the propagation of light rays in optical materials. We derive Snell's law of refraction from Fermat's principle and, conversely, Fermat's principle from Snell's law. We introduce the idea that optical media alter the perception of space, establishing a geometry different from the familiar geometry of physical space, the geometry of light. Our story begins with a letter.

On January 1, 1662 Pierre de Fermat, parlamentarian in Toulouse and amateur scientist, sent a letter to his long–term acquaintance Marin Cureau de la Chambre, physician to Cardinal Mazarin and the King of France (Mahoney [1994]). In this letter, Fermat established a physical principle that was destined to shape geometrical optics, to give rise to Lagrangian and Hamiltonian dynamics, and to inspire Schrödinger's quantum mechanics and Feynman's form of quantum field theory and statistical physics. *Fermat's principle* is the principle of the shortest optical path: light rays passing between two spatial points A and B choose the optically shortest route (see Fig. 2.1). In some cases, however, light takes the longest path (see Fig.

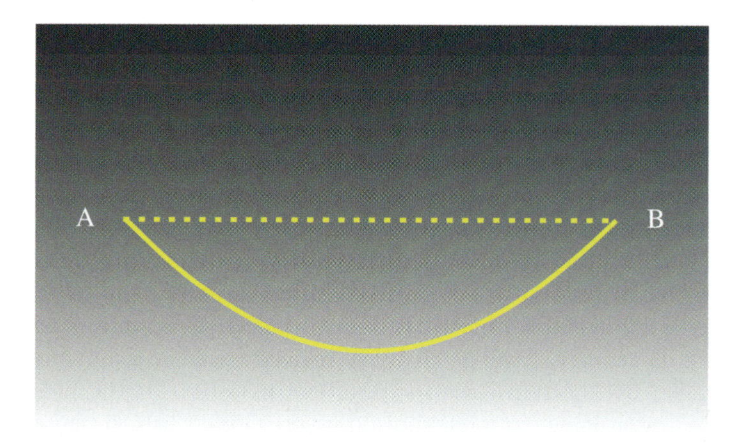

Figure 2.1: Fermat's principle. Light takes the shortest optical path from A to B (solid line) which is not a straight line (dotted line) in general. The optical path length is measured in terms of the refractive index n integrated along the trajectory. The greylevel of the background indicates the refractive index; darker tones correspond to higher refractive indices. The figure illustrates the ray trajectories involved in forming a mirage.

2.2); in any case, light follows extremal optical paths (stationary paths, to be absolutely precise). The *optical path length* is defined in terms of the refractive index n as

$$s = \int n \, \mathrm{d}l \,. \tag{1.1}$$

Here $\mathrm{d}l$ denotes the infinitesimal increment of the geometrical path length,

$$\mathrm{d}l = \sqrt{\mathrm{d}x^2 + \mathrm{d}y^2 + \mathrm{d}z^2} \tag{1.2}$$

in Cartesian coordinates x, y, z. The *refractive index* n is the ratio between the speed of light in vacuum c, and the phase velocity of light in the optical medium.* Fermat's principle is thus the principle of the shortest or longest time, because the traveling time of light between infinitesimally close points is $n \, \mathrm{d}l$ divided by the constant c. When the refractive index varies in space—for non-uniform media—the extremal optical path is not a straight line, but is curved. This bending of light is the cause of many optical illusions. For example, picture a mirage in the desert (Feynman,

*Typically, the phase velocity c/n is less than c and the refractive index n is larger than 1, but materials with $n < 1$ for some bands of the spectrum can be found or made.

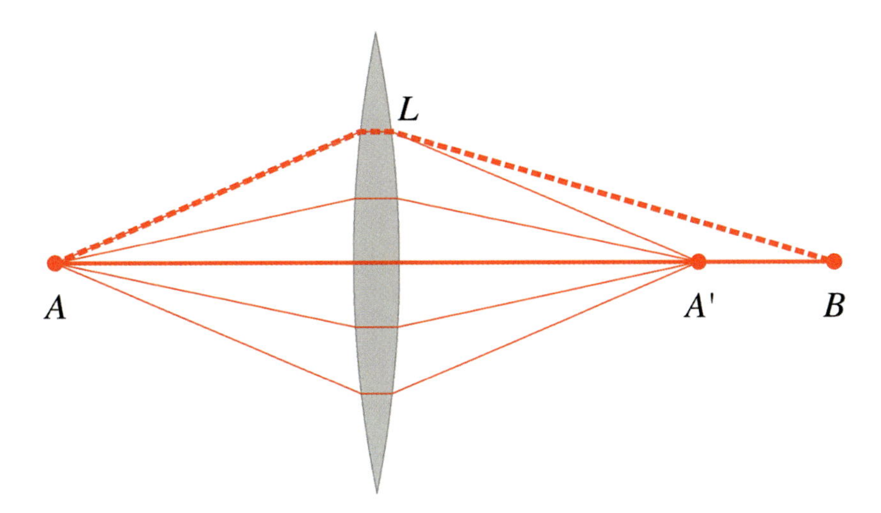

Figure 2.2: Fermat's principle: longest optical path. The figure shows a simple example where light takes the longest optical path, a lens. The lens focuses all light rays (red lines) from A at A′. According to Fermat's principle, this is only possible if their optical path lengths (1.1) are the same; differences in geometrical path length are thus compensated by the different propagation lengths in the material of the lens. All rays are focused at A′, but only one of them (thick red line) continues to B. The optical path length of this ray from A to B is the longest, not the shortest, for the following reason: compare the solid line of the actual path from A to B with an example of a virtual trajectory, the dashed path. As the optical path lengths from A to A′ are the same, the optical path length taken differs from the virtual path length by the difference between the two short sides and the long side of the triangle from L to A′ and B. The sum of the two short sides of a triangle is always longer than the long side: light has taken the locally longest optical path from A to B.

Leighton and Sands [1963]). The tremulous air above the hot sand conjures up images of water in the distance, but it would be foolish to follow these deceptions; they are not water, but images of the sky. The hot air above the sand bends light from the sky, because hot air is thin with a low refractive index and so light prefers to propagate there (see Fig. 2.1).

The simplest optical illusion is the *refraction* of light at the interface between two optical materials with different constant refractive indices n_1 and n_2. Figure 2.3 shows a fish in water that appears to the observer to be at a different position, because the light rays carrying the image of the fish are refracted at the water's surface. Refraction obeys *Snell's law*,

$$n_1 \sin \alpha_1 = n_2 \sin \alpha_2 \tag{1.3}$$

where α_i denote the angles of the light rays relative to the normal vector on the interface. How does Snell's law follow from Fermat's principle? Suppose we know that light travels between point (x_1, y_1) and (x_2, y_2), and the interface is located

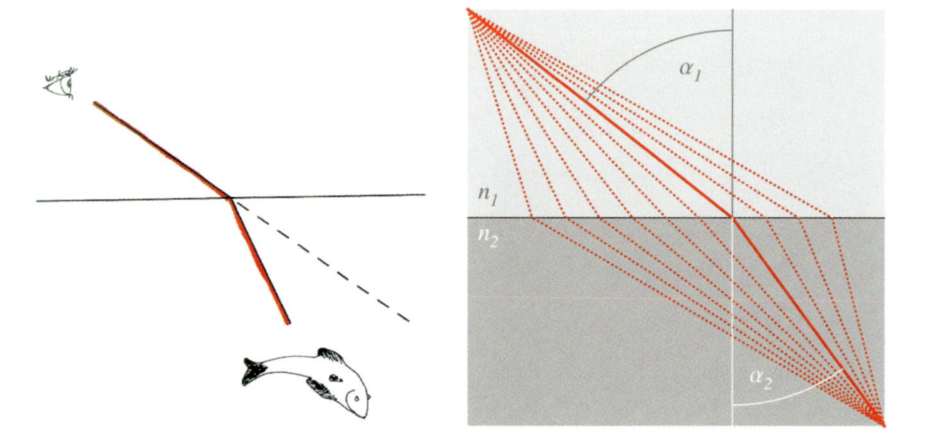

Figure 2.3: Refraction, Fermat's principle and Snell's law. Light is refracted at the interface between two optical media with different, but uniform, refractive indices n_1 and n_2, say air and water. The left picture (Credit: Erika Andersson) shows how light (red ray) coming from the fish in water is refracted at the water surface. For example, sunlight is reflected at the fish, carrying its image. Because of refraction, the fish appears to be at a different location than where it actually is. Optical media thus cause optical illusions. The right picture compares various possible ray trajectories (dotted lines) with the actual optical path (red line). The different greylevels indicate the different refractive indices of air ($n_1 = 1.00$) and water ($n_2 = 1.33$). Look at the dotted ray trajectories to the right of the red line. They pass through a short distance in water where the refractive index is high, which helps to reduce travel time, but they must then travel a rather long distance in air, which may cancel out the time saved in the water. By making the segment in water long enough, but not too long, the travel time becomes optimal at the red line. The dotted rays on the left spend too much time in water and so are no longer optimal. It turns out that the optimal path is the actual path, the path consistent with Snell's law (1.3) indicated by the angles α_1 and α_2.

at the line $y = 0$. Imagine light is able to explore the various options in traveling between (x_1, y_1) and (x_2, y_2) for finding the optimal path. In the two regions of uniform refractive index, the shortest optical path is obviously a straight line, but this line may be tilted—refracted—at the interface. The total optical path length is given by

$$s = n_1\sqrt{(x' - x_1)^2 + y_1^2} + n_2\sqrt{(x_2 - x')^2 + y_2^2}\,, \tag{1.4}$$

according to Pythagoras' theorem, where x' is the point on the x–axis where the ray crosses the interface $y = 0$.

Exercise 1.1

Express Snell's law (1.3) in terms of the coordinates of the points (x_1, y_1) and (x_2, y_2) and the point $(x', 0)$ at the interface. Draw examples of these points and the light ray in a diagram.

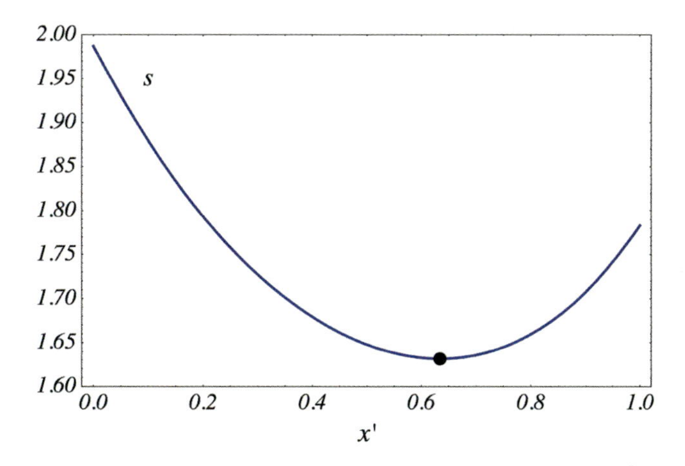

Figure 2.4: Minimum of the optical path length. Plot of the optical path length (1.4) as a function of refraction point x' for the ray trajectories shown on the right side of Fig. 2.3 (with $x_1 = 0$, $y_1 = 0.5$, $x_2 = 1.0$, $y_2 = -0.5$ in arbitrary units and $n_1 = 1.00$, $n_2 = 1.33$). The actual ray is refracted at the point x' of the curve's minimum (the dot).

Figure 2.4 shows s as a function of x' for fixed parameters x_i, y_i, n_i. The extremum of s lies at the zero of its first derivative with respect to the variable x',

$$\frac{\partial s}{\partial x'} = 0\,. \tag{1.5}$$

We differentiate s and obtain

$$\frac{n_1(x' - x_1)}{\sqrt{(x' - x_1)^2 + y_1^2}} = \frac{n_2(x_2 - x')}{\sqrt{(x_2 - x')^2 + y_2^2}} \tag{1.6}$$

which gives Snell's law (1.3) according to Exercise 1.1. Conversely, if we know Snell's law to be true we can derive Fermat's principle by running the argument the other way round. We express Snell's formula (1.3) in the form (1.6) that corresponds to condition (1.5) for the extremal optical path, which proves Fermat's principle for the refraction of light at the interface between spatial regions with different, but uniform, refractive indices. Next we generalize Fermat's principle to non–uniform media. When $n(x, y, z)$ varies across space we can assume n to be uniform in infinitesimal patches; the optical medium is a patchwork of infinitesimally uniform pieces. A light ray passing from piece to piece is refracted at their boundaries, however slightly. As each infinitesimal refraction obeys Snell's law (1.3), Fermat's principle holds piece by piece. As the total optical path length (1.1) is the sum of the infinitesimal path lengths, Fermat's principle holds globally.

Let us return to refraction for a moment. We have seen that if the optical path length is stationary, the first derivative with respect to x' vanishes. Is it minimal or maximal (or neither: flat)? This issue is a matter of the second derivative. The reader is asked to verify that the second derivative of the optical path length (1.4) is positive at the extremum (1.6),

$$\left.\frac{\partial^2 s}{\partial x'^2}\right|_{\text{Snell}} > 0\,. \tag{1.7}$$

Exercise 1.2
Calculate the second derivative of the optical path length (1.4) and argue why it is positive.

As the second derivative is positive, the extremum is a minimum. We argued that Snell's law describes the bending of light from one infinitesimal patch of a non–uniform medium to the next. Consequently, for short distances the optical path length (1.1) is always a minimum. But this conclusion concerning short distances is not universally true for long distances; Fig. 2.2 shows a case where the actual optical path taken is a local maximum, not a minimum. So light may continuously follow the short–term goal of the shortest travel time, only to end up with the longest.

Fermat's principle has profoundly influenced modern physics, and like most if not all profound discoveries, it has deep roots in the history of science. Cureau, the receiver of Fermat's letter, was not only a physician, but also an amateur physicist. He had sent his treatise *Light* to Fermat in the summer of 1657. In this work, Cureau explained the law of reflection by a mirror according to a principle of the shortest path (see Fig. 2.5), but Cureau's path length was the geometrical length (1.2) rather than the optical length (1.1), both of which give the same shortest paths in a uniform medium.

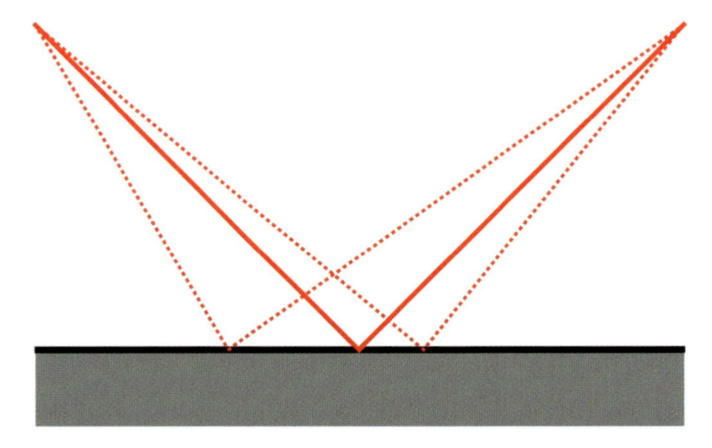

Figure 2.5: Heron's principle: a mirror reflects light such that the geometrical path is the shortest. The geometrical length of the actual light path (red line) is shorter than any other trajectories (dotted red lines) that touch the mirror. It turns out that this condition is equivalent to the familiar law of reflection at a mirror (the angle of incidence equals the angle of reflection).

Exercise 1.3

Prove Heron's principle (Fig. 2.5). Write down the geometrical path length l of a ray between the points (x_1, y_1) and (x_2, y_2) that is reflected at the point $(x', 0)$ on the surface of the mirror. Differentiate l with respect to x' and put $\mathrm{d}l/\mathrm{d}x'$ to zero. Show that from this follows that the incident angle is equal to the reflected angle, the law of reflection at a mirror.

Prior to Cureau, the Greek polymath Hero of Alexandria (Heron) had come to the same conclusion in ancient times. The great Arab scientist Ibn al–Haytam (Alhazen) anticipated Fermat's principle in his *Book of Optics* (written during his house arrest in Cairo from 1011 to 1021). Snell's law is not Snell's law either, but was first published by Ibn Sahl in 984 (Rashed [1990]). Fermat's main innovation in optics was the introduction of the refractive index in the optical path length, or, as he put it, the resistance of the medium to light. Fermat outlined his idea in a letter of thanks to Cureau written in August 1657, but was not able to deduce Snell's law at the time. Calculus was not developed yet, and so Fermat lacked the elegant mathematical machinery we applied in our brief derivation. Pressed by friends to settle an old score with the rival Cartesians on optics, Fermat finally derived Snell's law from his principle in the letter to Cureau from 1662 mentioned at the beginning of this Section, thus putting the principle of the shortest optical path on solid ground—or so he thought.

Although Fermat's principle was destined for great things, it was immediately greeted with objections after his letter began circulating, because it is a principle about destiny itself. It says that when light travels from A to B it takes the extremal optical path, but how does it know that it will arrive at B? We assume that it is the destiny of a particular light ray to travel from A to B, and Fermat's principle describes how that destiny is fulfilled. The principle apparently violates local causality, being a statement about destinies or teleology. It thus appears to be more of a moral principle, rather than a law of physics—said the Cartesians. Fermat's principle fares better in wave optics, because a wave emitted at A propagates in all directions and may reach all positions B (at least with some amplitude, however small). The path of extremal optical length (1.1) is the place of constructive interference between all possible paths.

Philosophical ideas about destiny were en vogue in the 19th century and may have inspired William Rowan Hamilton to formulate the classical mechanics of point particles in a single principle, the principle of least action: Hamilton's principle. The relationship between ray optics and wave optics appeared to Louis de Broglie similar to the relationship between classical mechanics and the hotly debated quantum mechanics of the 1920's. Erwin Schrödinger put de Broglie's idea into precise form in the equation that carries his name. Max Born understood that Schrödinger's waves describe probability amplitudes, they are waves of possibilities interfering with each other. Richard Feynman vividly remembered the shock he felt when he first heard about the principle of least action in school. This early encounter with Hamilton's principle set Feynman on the course of developing quantum mechanics as the interference of all possible paths with a phase given by the action in units of $\hbar$. Nevertheless, how probability waves materialize to solid facts has still remained a mystery. In classical optics, the interference of paths is conceptually easier, because the waves are physical things themselves—electromagnetic waves, not probability amplitudes. We briefly discuss Feynman's interference of paths in §4.

In Fermat's principle, the optical path length (1.1) differs from the geometrical path length (1.2), as if the refractive index of the optical material defined a different measure of space for light. Regions of high refractive index appear larger than regions of small index. An optical medium seems to distort the geometry of space, or, equivalently, it establishes a geometry of its own, the geometry of light, the theme of this book. Objects inside the medium appear to have moved to different positions than where they actually are, as anyone who has looked into an aquarium will confirm. Walking around the aquarium one notices that the objects inside appear at different locations, depending on the viewpoint. Under exceptional circumstances, not realized in normal optical materials like water, the apparent positions are independent of the viewpoint. In such a case (Fig. 2.6), the optical medium simply shifts the points of physical space to new positions. The medium performs a transformation of space: physical space is transformed into a *virtual space* in which the effect of the medium has been fully accounted for by the shifting of the points. Virtual space is empty, and so light rays follow straight lines here. In physical space, the virtual ray trajectories are transformed back from virtual space, the straight virtual ray trajectories are curved in physical space. We can view the transformation of space as a coordinate transformation (Fig. 2.6). Suppose we employ an orthogonal grid of light rays in virtual space as Cartesian coordinates. In physical space, they define a curved coordinate grid: transformation media implement coordinate transformations. These ideas inspired the research area of *transformation optics* (Leonhardt [2006a], Pendry, Schurig and Smith [2006]). Imagine that the transformation compresses an extended region of physical space into a single point (Fig. 2.6). Anything inside this region vanishes from view. In short, the optical material acts as a cloaking device, creating the ultimate optical illusion: *invisibility*.

Exercise 1.4
Consider a uniform optical material with refractive index n. Write down the coordinate transformation from the points $\boldsymbol{r}$ of physical space to the points $\boldsymbol{r}'$ of virtual space performed by this medium.
Solution
According to Fermat's principle, $\mathrm{d}s$ represents the line element $\mathrm{d}l'$ in virtual space, and $\mathrm{d}l$ is the line element in physical space. Their relationship (1.1) suggests the transformation formula $\boldsymbol{r}' = n\boldsymbol{r}$.

Exercise 1.5
Develop an argument why two ordinary uniform materials attached to each other do not simply perform a coordinate transformation. Consider the refraction of light at the interface between two uniform optical materials with refractive indices n_1 and n_2 (Fig. 2.3).
Solution
In the half spaces of each medium we must use the transformations $\boldsymbol{r}' = n_1\boldsymbol{r}$ and $\boldsymbol{r}' = n_2\boldsymbol{r}$, respectively (Exercise 1.4), but the two expressions are not consistent at the interface, there they contradict each other.

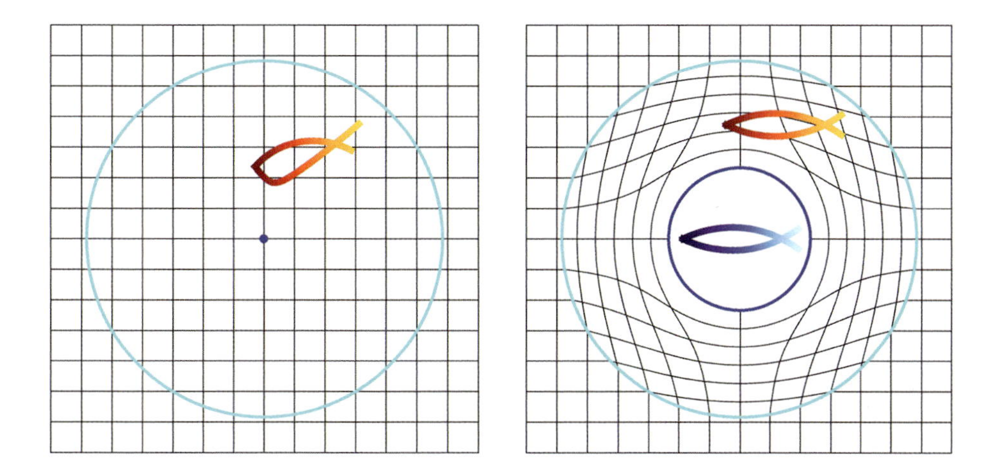

Figure 2.6: Transformation of space. Optical materials appear to change the perception of space; objects (fish) in physical space (right picture) appear at positions (left picture) different from where they actually are, as we have already seen in Fig. 2.3. If, as shown here, their virtual positions do not depend on the viewpoint, but only on their actual positions, the medium performs a coordinate transformation from physical space (right) to virtual space (left) and vice versa. Virtual space is empty and so light propagates along straight lines. In virtual space, we may draw a coordinate system as a rectangular grid of light rays (left grid). In physical space, the light rays are curved; the coordinate grid of virtual space is transformed into a curved coordinate system in physical space (right grid). As the coordinate transformation only changes space within the light–blue circle, this circle marks the boundary of the optical material used to transform space. We see that the images of the fish are distorted in virtual space, because the coordinate transformation illustrated here is not uniform. Moreover, the blue fish has completely disappeared, because it was swimming within a region of physical space (blue circle) that, in virtual space, is contracted to a single, invisible point. Such an optical material makes an invisibility device.

§2. VARIATIONAL CALCULUS

Calculus allowed us to derive Snell's law from Fermat's principle in a few lines. As we have argued, we could apply infinitesimal Snell refractions infinitely many times to calculate ray trajectories in non–uniform media, but it is infinitely better to develop the mathematical machinery for solving optimization problems *à la* Fermat in one stroke: variational calculus.

It is wise to consider a rather general optimization problem, not only Fermat's principle. Consider d real functions q_i of a real parameter ξ. In our case, the q_i are the Cartesian coordinates x, y, z and the parameter ξ draws the ray trajectory. In the following we use q as a shorthand notation for the components $(q_1, \ldots, q_d)$ and $\dot{q}$ for the derivative of q with respect to ξ. Suppose S is an accumulative quantity like the optical path length:

$$S = \int_{\xi_A}^{\xi_B} L \, \mathrm{d}\xi \,. \tag{2.1}$$

The function L describes the derivative $\mathrm{d}S/\mathrm{d}\xi$, it characterizes the increment $\mathrm{d}S$ to S. In classical mechanics, L is called the *Lagrangian* and S the *action*. In optics, S is s, Fermat's optical path length (1.1). We consider the specific case of optical paths in the next Section. Suppose further that the increment to S depends only on $q, \dot{q}$ and ξ,

$$L = L(q, \dot{q}, \xi) \,. \tag{2.2}$$

We wish to find the extremum of S subject to the constraint that $q(\xi_A)$ and $q(\xi_B)$ are fixed. If S depended only on one unknown variable, similar to the optical path length (1.4) in our derivation of Snell's law (Fig. 2.4), we could simply seek the zero of the first derivative of S with respect to that variable. However, S depends on all possible trajectories between $q(\xi_A)$ and $q(\xi_B)$; we can characterize these trajectories by an increasingly finer mesh of variables (Fig. 2.7).

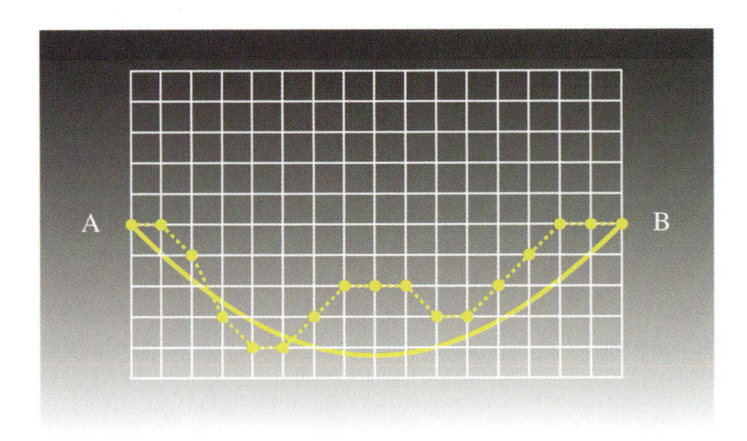

Figure 2.7: Mesh of trajectories. The picture compares the light ray (line) of Fig. 2.1 (in a mirage) with alternative ray trajectories (dotted line). Each alternative trajectory is characterized by the points it passes through in a coordinate mesh. By making the mesh finer and finer we can describe all possible trajectories. For finding the optimum of the optical path (1.1) we should vary all mesh points. Variational calculus does this for us.

In variational calculus we differentiate with respect to these infinitely many variables. We introduce the variational derivative similar to the ordinary derivative of a function: consider the trajectory $q(\xi)$ and an infinitesimally close neighbouring path $q(\xi) + \delta q(\xi)$. The difference $\delta q(\xi)$ between the two trajectories depends on infinitely many variables labeled by the parameter ξ. In the following we express the difference δS in S between the two trajectories as a differential in terms of the path difference δq similar to the ordinary differential $\mathrm{d}s$ of a function $s(x')$. As the two trajectories are infinitesimally close to each other, the difference between their Lagrangians is given by a first–order Taylor expansion,

$$\delta L = L(q + \delta q, \dot{q} + \delta\dot{q}, \xi) - L(q, \dot{q}, \xi) = \frac{\partial L}{\partial q} \cdot \delta q + \frac{\partial L}{\partial \dot{q}} \cdot \delta \dot{q} \,. \tag{2.3}$$

The difference $\delta\dot{q}$ between the derivatives $\dot{q}(\xi)$ and $\dot{q}(\xi) + \delta\dot{q}$ of the two trajectories is the derivative $\mathrm{d}\delta q(\xi)/\mathrm{d}\xi$ of the difference between the trajectories. We thus obtain for δS the expressions

$$\begin{aligned}
\delta S &= \int_{\xi_A}^{\xi_B} \delta L \, \mathrm{d}\xi = \int_{\xi_A}^{\xi_B} \left(\frac{\partial L}{\partial q} \cdot \delta q + \frac{\partial L}{\partial \dot{q}} \cdot \delta \dot{q} \right) \mathrm{d}\xi \\
&= \int_{\xi_A}^{\xi_B} \left[\frac{\partial L}{\partial q} \cdot \delta q + \frac{\mathrm{d}}{\mathrm{d}\xi} \left(\frac{\partial L}{\partial \dot{q}} \cdot \delta q \right) - \delta q \cdot \frac{\mathrm{d}}{\mathrm{d}\xi} \frac{\partial L}{\partial \dot{q}} \right] \mathrm{d}\xi \\
&= \int_{\xi_A}^{\xi_B} \left(\frac{\partial L}{\partial q} - \frac{\mathrm{d}}{\mathrm{d}\xi} \frac{\partial L}{\partial \dot{q}} \right) \cdot \delta q \, \mathrm{d}\xi + \left. \frac{\partial L}{\partial \dot{q}} \cdot \delta q \right|_{\xi_A}^{\xi_B} \,.
\end{aligned} \tag{2.4}$$

The last term in Eq. (2.4) vanishes, because the two trajectories match at the end points such that $\delta q(\xi_A) = \delta q(\xi_B) = 0$. In this way we obtain the desired variational differential

$$\delta S = \int_{\xi_A}^{\xi_B} \left(\frac{\partial L}{\partial q} - \frac{\mathrm{d}}{\mathrm{d}\xi} \frac{\partial L}{\partial \dot{q}} \right) \cdot \delta q \, \mathrm{d}\xi \,. \tag{2.5}$$

At the extremum of S the differential (2.5) is zero for all variations δq, therefore

$$\frac{\mathrm{d}}{\mathrm{d}\xi} \frac{\partial L}{\partial \dot{q}} = \frac{\partial L}{\partial q} \,. \tag{2.6}$$

This is the *Euler–Lagrange equation* for the extremal trajectory. As L does not depend on higher derivatives of q (such as $\ddot{q}$ etc), the Euler–Lagrange equation constitutes a coupled system of ordinary second–order differential equations. The solution $q(\xi)$ is uniquely determined by the initial conditions $q(\xi_0)$ and $\dot{q}(\xi_0)$ at the same ξ_0. Hence, although we fixed both end points in the optimization problem, the solution depends on only one point and the rate of change there. We may have formulated an apparently acausal optimization problem, such as Fermat's principle, but the actual equation of motion, the Euler–Lagrange equation (2.6) is perfectly causal. Moreover, the numerical solution of ordinary differential equations is straightforward and in some important cases one can solve the Euler–Lagrange equation analytically. Variational calculus has given us the tools to handle Fermat's principle.

Exercise 2.1

Consider a mechanical particle at position $\boldsymbol{r}$ moving with velocity $\boldsymbol{v}$ in a potential $U(\boldsymbol{r})$. Assume the Lagrangian $L = mv^2/2 - U(\boldsymbol{r})$, write down the Euler–Lagrange equation for $\xi = t$ and show that it agrees with Newton's law:

$$L = \frac{m}{2} v^2 - U(\boldsymbol{r}) \quad \Rightarrow \quad m \frac{\mathrm{d}^2 \boldsymbol{r}}{\mathrm{d}t^2} = -\boldsymbol{\nabla} U \,. \tag{2.7}$$

This calculation proves the principle of least action in classical mechanics.

Solution
Here $q(\xi)$ is the position $\boldsymbol{r}(t)$, parameterized by the time t, and $\dot{q}$ is the velocity $\boldsymbol{v} = \mathrm{d}\boldsymbol{r}/\mathrm{d}t$. We obtain from the Euler–Lagrange equation (2.6):

$$\frac{\partial L}{\partial \dot{q}} = \frac{\partial L}{\partial \boldsymbol{v}} = m\boldsymbol{v}\,, \quad \frac{\partial L}{\partial q} = \frac{\partial L}{\partial \boldsymbol{r}} = -\boldsymbol{\nabla} U \quad \Rightarrow \quad m\frac{\mathrm{d}^2\boldsymbol{r}}{\mathrm{d}t^2} = m\frac{\mathrm{d}\boldsymbol{v}}{\mathrm{d}t} = -\boldsymbol{\nabla} U\,. \tag{2.8}$$

Exercise 2.2
Some of the pioneering experiments in mechanics consisted of measuring the time taken for bodies to roll or slide down inclined planes. This kind of experiment raises an obvious question: given the starting point A and the (lower) end point B, what curve joining A and B should the body slide along in order for the time taken to be a minimum (Fig. 2.8).

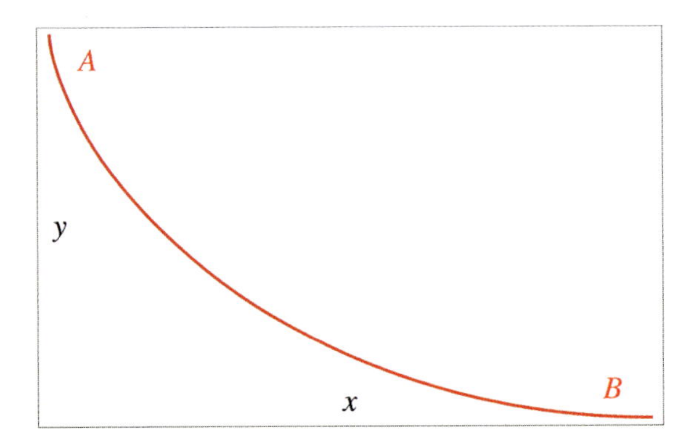

Figure 2.8: Brachistochrone problem. Optimal path for an object to slide down from A to B in the shortest possible time.

Such a curve would give the optimal shape of an aircraft evacuation slide. Also the slopes of traditional Japanese roofs follow this optimal curve, for letting snow slide down as fast as possible. Consideration of this problem quickly leads to the suspicion that the extremal curve is not the straight line of the inclined plane. If the curve initially has a steep vertical drop from A, the body will pick up speed and so will move quickly along the remaining shallow path to B. On the other hand, if the initial drop is too steep for too long the body may have too far to travel along the shallow part of the curve to make the extra speed advantageous. This is the famous problem of least (βραχίστος) time (χρόνος): the *brachistochrone problem.* It is clearly a problem of variational calculus; indeed, solutions of the brachistochrone problem by various mathematicians played an important part in the development of the calculus of variations.* Use the machinery of variational calculus to solve the brachistochrone problem.
Solution
Let the points A and B lie in the xy–plane, with the x–axis horizontal and the y–axis vertical (Fig. 2.8). Let A have coordinates $(0,0)$ and let B lie to the right of A, so that B has positive x–coordinate and negative y–coordinate. We require an expression for the time taken for a particle to slide without friction along an arbitrary curve joining A and B; we can then extract the Lagrangian

*In 1696 the Swiss mathematician Jacob Bernoulli proposed the brachistochrone problem and offered a prize for its solution. He then solved the problem, relating it to an appropriate refractive–index profile and applying Fermat's principle, and awarded the prize to himself, disregarding the efforts of his brother Jean who also solved the brachistochrone problem by inventing the beginnings of variational calculus (Stavroudis [2006]). In 1766 Leonard Euler refined and generalized variational calculus.

of the problem from this expression and use the Euler–Lagrange equation (2.6) to find the extremal curve that minimizes the time. The time taken to travel an infinitesimal distance ds along the curve is ds/v, where v is the speed of the particle as it traverses ds; the total time from A to B is therefore

$$T = \int_A^B \frac{\mathrm{d}s}{v} . \tag{2.9}$$

We now require an expression for the speed v. Although v varies along the curve, the total energy of the particle, given by the sum of kinetic energy $mv^2/2$ and gravitational potential energy mgy is conserved; at point A the particle is at rest ($v = 0$) and $y = 0$, so the total energy is zero:

$$\frac{1}{2}mv^2 + mgy = 0 , \tag{2.10}$$

which shows that v is

$$v = \sqrt{-2gy} . \tag{2.11}$$

(Note that on the curve $y < 0$, so v is real.) Inserting the speed (2.11) into Eq. (2.9) and rearranging we obtain

$$T = \int_A^B \frac{\mathrm{d}s}{\sqrt{-2gy}} = \int_A^B \frac{\sqrt{\mathrm{d}x^2 + \mathrm{d}y^2}}{\sqrt{-2gy}} = \int_A^B \mathrm{d}y \sqrt{\frac{1}{-2gy}\left[\left(\frac{\mathrm{d}x}{\mathrm{d}y}\right)^2 + 1\right]} , \tag{2.12}$$

where in the last expression the curve is regarded as a function $x(y)$. Equation (2.12) presents a variational–calculus problem for the curve $x(y)$ joining A and B that minimizes the time T (there is clearly no extremal curve that gives a *maximum* time as we can make T arbitrarily large by starting the curve at A with an arbitrarily shallow incline). The Lagrangian is

$$L = \sqrt{\frac{\dot{x}^2 + 1}{-2gy}} , \qquad \dot{x} \equiv \frac{\mathrm{d}x}{\mathrm{d}y} , \tag{2.13}$$

and the curve $x(y)$ of minimum time satisfies the Euler–Lagrange equation

$$\frac{\mathrm{d}}{\mathrm{d}y}\left(\frac{\partial L}{\partial \dot{x}}\right) - \frac{\partial L}{\partial x} = 0 . \tag{2.14}$$

The Lagrangian (2.13) has no explicit dependence on x, so Eq. (2.14) is

$$\frac{\mathrm{d}}{\mathrm{d}y}\left(\frac{\dot{x}}{\sqrt{-2gy(\dot{x}^2 + 1)}}\right) = 0 , \tag{2.15}$$

which can be integrated to give

$$\frac{\dot{x}}{\sqrt{-2gy(\dot{x}^2 + 1)}} = b \qquad \Longrightarrow \qquad x = \int \mathrm{d}y \sqrt{\frac{-y}{a + y}} , \qquad a \equiv \frac{1}{2gb^2} , \tag{2.16}$$

where b and a are constants. The integral in the formula for x can be evaluated using the substitution

$$y = -a \sin^2\left(\frac{\theta}{2}\right) \tag{2.17}$$

(recall again that $y < 0$ on the curve) and we find

$$x = a \int \mathrm{d}\theta \, \sin^2\left(\frac{\theta}{2}\right) = \frac{a}{2}\left(\theta - \sin\theta\right) + d , \tag{2.18}$$

where d is a constant. Equations (2.17) and (2.18) give the solution for the brachistochrone curve, parametrized by θ. It is convenient to choose the parameter value $\theta = 0$ to correspond to the starting point A ($x = 0, y = 0$); this requires $d = 0$ in Eq. (2.18) and, with a rewriting of Eq. (2.17), we have the final expression for the curve:

$$x = \frac{a}{2}(\theta - \sin\theta), \qquad y = -\frac{a}{2}(1 - \cos\theta). \tag{2.19}$$

The condition that the curve passes through the specified endpoint B determines both the constant a and the parameter value at B. A simple construction produces the curve (2.19): it is the path traced out by a point on a circle of radius $a/2$ as it rolls along the x–axis with its centre below the x–axis at $y = -a/2$; θ is the angle of rotation of the rolling circle. The full curve is called a *cycloid*; the brachistochrone curve is thus a section of a cycloid (Fig. 2.9). Note that the brachistochrone curve is independent not only of the mass of the particle but also of the strength g of the gravitational field. It is straightforward to show the following remarkable fact: if the endpoint B has coordinates (x, y) such that $x \leq |y|(\pi/2)$ then the curve of least time descends all the way from A to B, whereas if $x > |y|(\pi/2)$ the curve descends to a point lower than B and then moves upward to reach B.

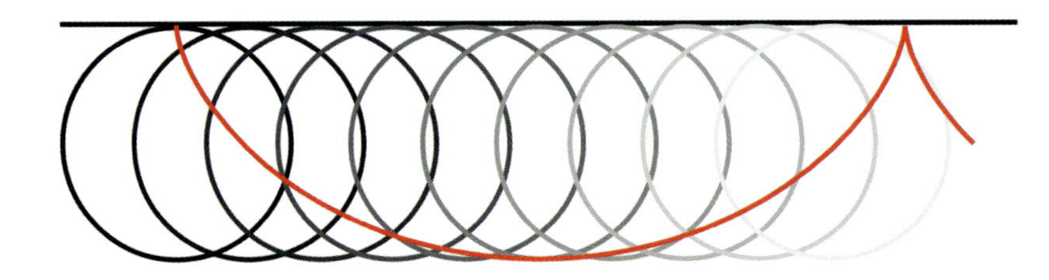

Figure 2.9: Cycloid. Trajectory (red curve) of a point on a circle as it rolls along a line. The brachiostochrone curve (Fig. 2.8) consists of a cycloid segment.

---

§3. Newtonian analogy

We have developed variational calculus in a general form. In this Section we apply this mathematical machinery to Fermat's principle. We deduce a useful correspondence between optics and classical mechanics and design our first nontrivial optical instrument, the Luneburg lens.

Let us derive the Euler–Lagrange equation for Fermat's principle. We denote the points along the ray trajectory by $\boldsymbol{r}$ as a function of the parameter ξ. We obtain from formula (1.1) of the optical path length and the expression (1.2) for the geometric line element

$$s = \int n\sqrt{\left(\frac{\mathrm{d}x}{\mathrm{d}\xi}\right)^2 + \left(\frac{\mathrm{d}y}{\mathrm{d}\xi}\right)^2 + \left(\frac{\mathrm{d}z}{\mathrm{d}\xi}\right)^2}\,\mathrm{d}\xi = \int \sqrt{n^2v^2}\,\mathrm{d}\xi \quad \text{with} \quad \boldsymbol{v} = \frac{\mathrm{d}\boldsymbol{r}}{\mathrm{d}\xi}. \tag{3.1}$$

Comparing this expression with Eqs. (2.1) and (2.2), we see that the Lagrangian in this case is

$$L = \sqrt{n^2v^2}, \tag{3.2}$$

where q and $\dot{q}$ correspond to $\boldsymbol{r}$ and $\boldsymbol{v}$. The Euler–Lagrange equation (2.6) is therefore

$$\frac{\mathrm{d}}{\mathrm{d}\xi}\frac{n^2\boldsymbol{v}}{L} = \frac{v^2\boldsymbol{\nabla}n^2}{2L} \tag{3.3}$$

where $\boldsymbol{\nabla}$ denotes the vector of the partial derivatives $\partial/\partial x^i$ with respect to the components of $\boldsymbol{r}$. Equation (3.3) determines the extremal trajectory followed by the light ray. Suppose that for this trajectory we choose a parameter ξ given by the relation

$$\mathrm{d}\xi = \frac{\mathrm{d}l}{n}\,. \tag{3.4}$$

In this case, we get for the "speed"

$$v = \left|\frac{\mathrm{d}\boldsymbol{r}}{\mathrm{d}\xi}\right| = n\left|\frac{\mathrm{d}\boldsymbol{r}}{\mathrm{d}l}\right| = n\,, \tag{3.5}$$

where in the last step we used definition (1.2) for $\mathrm{d}l$. The parameter ξ obviously does not correspond to physical time t, for otherwise the speed of light would be inversely proportional to the refractive index n. Yet with this parameterization the Euler–Lagrange equation (3.3) reduces to the simple equation of motion

$$\frac{\mathrm{d}^2\boldsymbol{r}}{\mathrm{d}\xi^2} = \frac{\boldsymbol{\nabla}n^2}{2}\,. \tag{3.6}$$

This is Newton's law (2.7) for a mechanical particle with unit mass moving in "time" ξ under the influence of the potential

$$U = -\frac{n^2}{2} + E \tag{3.7}$$

where E is an arbitrary constant, as if the light ray were drawn by a tracer particle that responds to a mechanical force. We see from formula (3.5) that

$$E = \frac{v^2}{2} + U\,, \tag{3.8}$$

which implies that E is the sum of kinetic and potential energy, the total energy of the particle. Both U and E are dimensionless, because ξ in the equation of motion (3.6) has the dimension of a length, see Eq. (3.4). Requiring that U vanishes in empty space where n is unity, we put

$$E = \frac{1}{2}\,, \tag{3.9}$$

but of course we can set the zero of the potential to any other value of n, depending on convenience. Ray optics is reduced to Newtonian mechanics (and vice versa).

Note that we can multiply the refractive index profile n by an arbitrary constant factor η and get the same trajectories as for the original profile n, because this multiplication does not affect Fermat's principle: the extremum of the optical path is still the same, whether we multiply n by η or not, only the actual value of the extremal path length is different. Translated into the Newtonian analogy of ray optics, we can multiply both the potential U and the energy E by a constant η^2 without changing the trajectories.

Exercise 3.1
Take the optical path length s as the parameter of the ray trajectory. Deduce the Euler–Lagrange equation for this case.
Solution
From $\mathrm{d}s = n\,\mathrm{d}l$ follows $v = |\mathrm{d}\boldsymbol{r}/\mathrm{d}s| = 1/n$ and hence $L = 1$, which gives from the Euler–Lagrange equation (3.3) the equation of motion (4.12) with $\xi = ct$.

Consider an important special case for mechanics and ray optics alike, a radially symmetric refractive–index profile $n(r)$ with

$$r = \sqrt{x^2 + y^2 + z^2}\,. \tag{3.10}$$

In Newtonian mechanics, this case corresponds to motion under the influence of a central force, because the gradient of n^2 points in the radial direction. Defining the angular momentum as the vector product

$$\boldsymbol{L} = \boldsymbol{r} \times \frac{\mathrm{d}\boldsymbol{r}}{\mathrm{d}\xi}\,, \tag{3.11}$$

carrying here the physical unit of a length, we see that $\boldsymbol{L}$ is a conserved quantity, as

$$\frac{\mathrm{d}\boldsymbol{L}}{\mathrm{d}\xi} = \boldsymbol{r} \times \frac{\mathrm{d}^2\boldsymbol{r}}{\mathrm{d}\xi^2} = \boldsymbol{r} \times \frac{\boldsymbol{\nabla} n^2(r)}{2} = 0\,. \tag{3.12}$$

Consequently, the entire ray trajectory $\boldsymbol{r}(\xi)$ is orthogonal to the constant vector $\boldsymbol{L}$; the ray lies in a plane, unless $\boldsymbol{L} = \boldsymbol{0}$. For vanishing angular momentum (3.11) the position $\boldsymbol{r}$ and the velocity $\mathrm{d}\boldsymbol{r}/\mathrm{d}\xi$ must be parallel, and, as $\boldsymbol{L}$ is a conserved quantity, $\boldsymbol{r}$ and $\mathrm{d}\boldsymbol{r}/\mathrm{d}\xi$ must remain parallel for all ξ, which is only possible if the trajectory is a straight line through the origin (the trajectory of a direct hit). Consider for the case $\boldsymbol{L} \neq \boldsymbol{0}$ light propagation in the plane orthogonal to $\boldsymbol{L}$. We orient the coordinate system such that x and y are the Cartesian coordinates of the propagation plane and $z = 0$. It is convenient to combine the coordinates x and y in the complex numbers*

$$z = x + \mathrm{i}y\,, \tag{3.13}$$

because a complex number has geometrical meaning: the modulus $|z| = \sqrt{z^*z}$ gives the radius $r = \sqrt{x^2 + y^2}$ while the argument $\arg z$ describes the angle of the point z from the positive x axis (we use the asterisk to symbolize complex conjugation). Many operations with complex numbers have geometrical meanings as well. For instance, the multiplication of z with $\exp(\mathrm{i}\alpha)$ describes a rotation by the angle α. Throughout this entire Chapter we will frequently use complex numbers for combining analysis with geometrical visualization.

---

*It is understood that z is no longer the z coordinate in this context.

Exercise 3.2
Show that the multiplication of z by an arbitrary complex number a correspond to a rotation and a scaling of space.
Solution
Write $z = |z| \exp(\mathrm{i} \arg z)$ and $a = |a| \exp(\mathrm{i} \arg a)$. Multiplication gives $z' = |a||z| \exp[\mathrm{i}(\arg a + \arg z)]$. As $\arg a$ is added to the angle of z, the vector z is rotated by $\arg a$; as the modulus is multiplied by $|a|$ the length of z is changed. Complex multiplication thus corresponds to a rotation and a scaling.

Arguably the most important nontrivial example of Newtonian motion is harmonic oscillation, the case of a linear restoring force according to Hooke's law. Such a linear force is described by a potential quadratic in r. In optics, Hooke's case corresponds to the refractive–index profile

$$n = \sqrt{2 - r^2}\,, \tag{3.14}$$

because, substituted into relation (3.7), this gives

$$U = \frac{r^2}{2}\,, \qquad E = 1\,. \tag{3.15}$$

The spatial coordinates are described in units of the spatial extension of the device implementing the profile (3.14); this profile is only valid for $r \leq \sqrt{2}$ where n is real. In complex notation (3.13), we obtain from Newton's law (3.6) with the Hooke profile (3.14) the equation of motion

$$\frac{\mathrm{d}^2 z}{\mathrm{d}\xi^2} + z = 0 \tag{3.16}$$

with the general solution written as

$$z = \mathrm{e}^{\mathrm{i}\alpha}\left(\frac{a+b}{2}\,\mathrm{e}^{\mathrm{i}\xi} + \frac{a-b}{2}\,\mathrm{e}^{-\mathrm{i}\xi}\right) = \mathrm{e}^{\mathrm{i}\alpha}\left(a\cos\xi + \mathrm{i}\,b\sin\xi\right)\,. \tag{3.17}$$

The real constants a, b and α have geometrical meaning: α describes the rotation of the trajectory relative to the axes of the coordinate system, whereas a and b define the extensions of the curve. For $\alpha = 0$, the trajectory (3.17) with the definition (3.13) obeys the equation of an ellipse,

$$\frac{x^2}{a^2} + \frac{y^2}{b^2} = 1\,. \tag{3.18}$$

Exercise 3.3
Prove from expression (3.17) the equation of the ellipse (3.18) in the case $\alpha = 0$.

The a and b define the semimajor and semiminor axes of an ellipse (Fig. 2.10) with half the focal length

$$f = \sqrt{a^2 - b^2}\,. \tag{3.19}$$

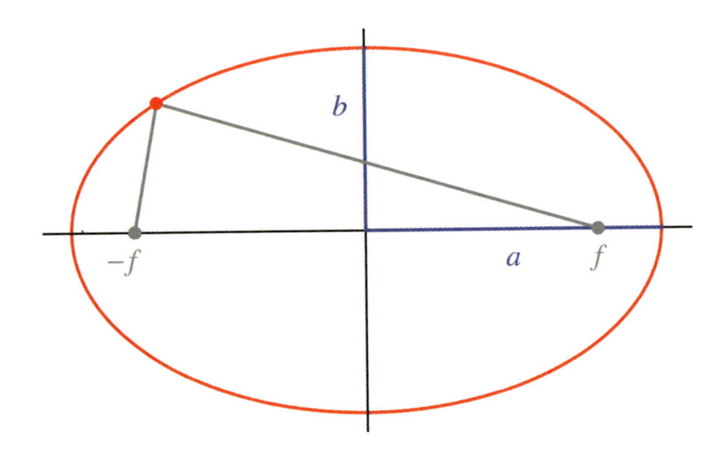

Figure 2.10: Ellipse. The parameters a and b in Eq. (3.18) describe the lengths of the semimajor and semiminor axes; f is half the focal length.

Exercise 3.4
You can draw an ellipse with two pins, a piece of string and a pencil (Fig. 2.10). Stick the pins into the paper at two points, which will become the focal points. Tie each end of the string to one of the pins. Pull the string taut with the pencil's tip, so as to form a triangle. Move the pencil around, while keeping the string taut, and its tip will trace out an ellipse.

Exercise 3.5
Does the geometric construction of Exercise 3.4 agree with the equation (3.18) of an ellipse? First, think how the parameters a and b in the equation are related to the parameters of the construction (Fig. 2.10), the distance $2f$ between the pins and the length l of the string, assuming that the construction agrees with formula (3.18). Show that f is given by Eq. (3.19) and $l = 2a$.

Exercise 3.6
Now prove that your construction agrees with formula (3.18). The length of the string is the sum of the distances from the point on the ellipse to the two foci,

$$l = \sqrt{(x-f)^2 + y^2} + \sqrt{(x+f)^2 + y^2} = 2a\,. \tag{3.20}$$

All you need to show is that this expression is consistent with Eqs. (3.18) and (3.19).
Solution
It is wise to square expression (3.20):

$$2a^2 = x^2 + y^2 + f^2 + \sqrt{(x^2+y^2+f^2)^2 - 4f^2x^2}\,. \tag{3.21}$$

This is the relationship we need to deduce from the equation (3.18) of the ellipse. For this purpose, consider $x^2 + y^2 + f^2 - 2a^2$ and square it:

$$\begin{aligned} (x^2+y^2+f^2-2a^2)^2 &= (x^2+y^2+f^2)^2 - 4a^2(x^2+y^2+f^2-a^2) \\ \Rightarrow\quad a^2(x^2+y^2+f^2-a^2) &= a^2x^2 + b^2(a^2-x^2) + a^2(f^2-a^2) = f^2x^2 \end{aligned} \tag{3.22}$$

using Eqs. (3.18) and (3.19) in the last step. Equation (3.22) proves that

$$(x^2+y^2+f^2-2a^2)^2 = (x^2+y^2+f^2)^2 - 4f^2x^2\,, \tag{3.23}$$

which gives relation (3.21): the geometrical construction of Exercise 3.4 agrees with the equation (3.18) of the ellipse.

For $\alpha \neq 0$ the ray trajectories (3.17) in Hooke's profile (3.14) are rotated ellipses (Fig. 2.11). The constants a and b are not entirely independent, as we see as follows. From the Newtonian speed (3.5) and the Hooke profile (3.14) we obtain:

$$v^2 = n^2 = 2 - r^2 \,. \tag{3.24}$$

The solution (3.17) gives

$$r^2 = |z|^2 = a^2 \cos^2 \xi + b^2 \sin^2 \xi \,, \quad v^2 = \left|\frac{\mathrm{d}z}{\mathrm{d}\xi}\right|^2 = a^2 \sin^2 \xi + b^2 \cos^2 \xi \,, \tag{3.25}$$

and substituting these into relation (3.24) we arrive at the condition

$$a^2 + b^2 = 2 \quad \text{or} \quad a + \mathrm{i}\, b = \sqrt{2}\, \mathrm{e}^{\mathrm{i}\gamma} \,, \tag{3.26}$$

for some real constant γ. This being the case, we obtain from Eq. (3.25):

$$|z|^2 = 1 + (a^2 - 1) \cos(2\xi) \,. \tag{3.27}$$

We see from this expression that all trajectories reach the unit circle at the zeros of $\cos(2\xi)$, irrespective of the parameters of the ellipse. The zeros of $\cos(2\xi)$ lie at

$$\xi_m = \left(m - \frac{1}{2}\right) \frac{\pi}{2} \quad \text{for integer } m. \tag{3.28}$$

We obtain from Eqs. (3.17) and (3.26)

$$z\big|_{\xi=\xi_1} = \mathrm{e}^{\mathrm{i}(\alpha+\gamma)} = \left.\frac{\mathrm{d}z}{\mathrm{d}\xi}\right|_{\xi=\xi_0} \,. \tag{3.29}$$

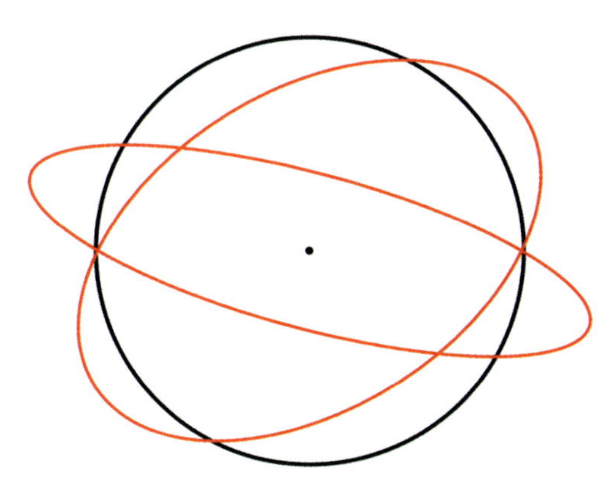

Figure 2.11: Hooke's ellipses. Light rays (red) in the refractive–index profile (3.14). In Newtonian mechanics, this profile corresponds to a harmonic oscillator with linear restoring force according to Hooke's law. The black circle indicates the characteristic scale of the device, the circle where $n = 1$. The light rays form rotated ellipses centred at the origin (dot).

Exercise 3.7
Derive property (3.29).

We can draw an interesting conclusion from this remarkable mathematical property of the Hooke profile (3.14) that leads to a useful optical instrument, the *Luneburg lens* (Luneburg [1964]). This lens is a sphere of radius 1 in our units, filled with a material that establishes the Hooke profile (3.14) for $r \leq 1$. Outside of the Luneburg lens, for $r > 1$, light propagates through empty space with $n = 1$. Let us find out what the Luneburg lens does. As we argued before, it is sufficient to discuss ray propagation in the x, y plane. Picture a family of light rays incident from one direction (Fig. 2.12). Initially, the incident rays follow straight lines with the fixed angle ϕ_0, the angle of incidence. A straight line is described by the equation

$$z = \mathrm{e}^{\mathrm{i}\phi_0}(\xi - \xi_0) + z_0 \tag{3.30}$$

with real constants ϕ_0 and ξ_0 and the complex constant z_0, because the derivative of this trajectory, $\mathrm{d}z/\mathrm{d}\xi$, is a constant, $\exp(\mathrm{i}\phi_0)$, which defines a straight line. As the derivative of the trajectory (3.30) gives the Newtonian velocity $\boldsymbol{v}$, we see that the velocity vectors of the family of incident rays are all the same. We also see that the modulus of the Newtonian velocity, the speed v, is 1, as required by Eq. (3.5), which proves that Eq. (3.30) correctly describes the incident light rays. When the rays strike the boundary of the device they are not refracted, but enter with velocity $\exp(\mathrm{i}\phi_0)$, because the refractive index at the boundary matches the index of empty space, $n(1) = 1$. Inside of the Luneburg lens, where the refractive index obeys the Hooke profile (3.14), the rays no longer follow straight lines, but segments of the Hooke ellipses we discussed. In order to describe the complete ray trajectories, we simply match the family (3.30) of incident straight lines to suitable ellipses (3.17) at the boundary of the device at $r = 1$. For this we only need to adjust the

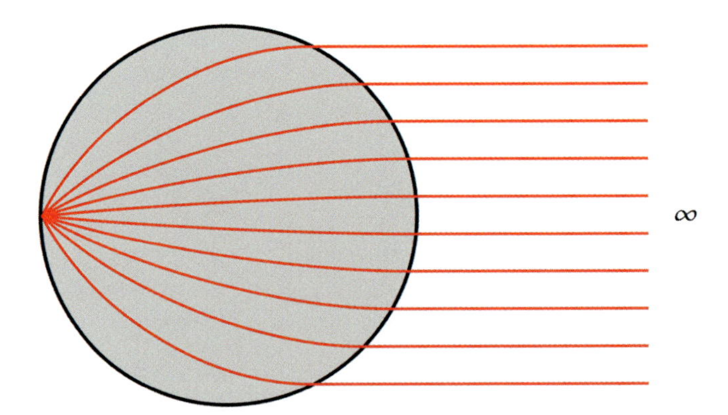

Figure 2.12: Luneburg lens. Light rays coming from ∞ enter the refractive–index profile (3.14) for $r \leq 1$. The rays are focused on the surface of the Luneburg lens.

constants z_0 and ξ_0 such that both the lines (3.30) and the ellipses (3.17) intersect the unit circle at the same parameter (3.28), say at $\xi = \xi_0$. Now, formula (3.29) for Hooke ellipses says that all these rays with velocity $\exp(\mathrm{i}\phi_0)$ meet at the point $z = x + \mathrm{i}y = \exp(\mathrm{i}\phi_0)$; they are focused there. For example, rays incident in the negative x direction focus at the point $(x, y) = (-1, 0)$. The Luneburg lens thus focuses light rays or other electromagnetic radiation on points on the surface of its sphere that lie in the direction of incidence. Luneburg lenses are useful in microwave technology, for example for satellite tracking. A single Luneburg lens can focus signals from several satellites on its surface. Movable detectors placed at the radio images can follow the signal by feedback—they track the satellites on the Luneburg sphere.

§4. Hamilton's equations

In this Section we prove Fermat's principle from Feynman's interference of paths. We begin by discussing linear waves in general. Then we turn to Feynman's version of Huygens' principle—the idea that in a wave all paths interfere with each other, which gives Hamilton's equations for rays and Fermat's principle. Finally we deduce the wave equation for scalar waves.

In wave optics, Fermat's principle poses less of a conceptual problem than in ray optics where it appears to violate causality. A *wave* is an oscillation that propagates in time across space. Each spatial point is equipped with an oscillator. For electromagnetic waves the oscillator is the electromagnetic field with electric and magnetic field strengths $\boldsymbol{E}$ and $\boldsymbol{H}$. Each oscillator is in contact with its neighbours—the magnetic flux through an infinitesimal area induces an electric field at its boundaries and vice versa: waves are made by coupled oscillators. At each position $\boldsymbol{r}$ and time t the oscillation is characterized by the *phase* φ, the oscillation angle. As the oscillators are coupled, the oscillation propagates through space. For example, in a plane wave propagating in the x–direction the phase evolves as $\varphi = 2\pi(x/\lambda - t/T)$ where T is the *period* and λ the *wavelength*, because φ increases by 2π over T in time and λ in space. It is convenient to express the phase in terms of the *wave number* k and the *angular frequency* ω as $\varphi = kx - \omega t$ with

$$\lambda = \frac{2\pi}{k}, \quad T = \frac{2\pi}{\omega}. \tag{4.1}$$

Throughout this book we call ω the frequency. It is even more convenient to introduce the *wave vector* $\boldsymbol{k}$, with magnitude equal to the wave number k and pointing in the direction in which the oscillation propagates, because then we can write the phase as $\varphi = \boldsymbol{k} \cdot \boldsymbol{r} - \omega t$, regardless of the direction of propagation. A given value of the phase propagates with the velocity ω/k in the direction of $\boldsymbol{k}$, the *phase velocity*. In optically isotropic materials such as glass the magnitude of the phase velocity is determined by the refractive index n in terms of the *dispersion relation*

$$\omega = \frac{ck}{n}. \tag{4.2}$$

The refractive index is in general different for waves of different frequencies, but in this book we are mostly concerned with stationary waves of fixed frequency ω. In a non–uniform medium, n varies across space, so for fixed ω the wave vector must vary accordingly. Let us simply define the wave vector and the frequency as the spatial and temporal gradient of the phase,

$$\boldsymbol{k} = \boldsymbol{\nabla}\varphi\,, \quad \omega = -\frac{\partial \varphi}{\partial t}\,, \tag{4.3}$$

which includes and generalizes the wave vector and frequency of a plane wave. Then φ is given by the *phase integral*

$$\varphi = \int (\boldsymbol{k}\cdot \mathrm{d}\boldsymbol{r} - \omega\,\mathrm{d}t)\ . \tag{4.4}$$

Linear waves obey the *superposition principle*: wave amplitudes simply add up. They may do so by *constructive interference* where two amplitudes are in–phase and enhance each other, or by *destructive interference* where the waves are out–of–phase and suppress each other. Electromagnetic waves in vacuum are linear waves to an extremely good approximation (unless they are so intense that quantum pair production plays a role). However, when electromagnetic waves interact with electric charges and currents, electromagnetism becomes nonlinear. Here we assume that the electromagnetic waves we are considering in optical media are still linear waves.*

Exercise 4.1
Consider a light wave emitted at the point $\boldsymbol{r}_0$ in a uniform medium. Assume that the wave vectors point in the directions of $\boldsymbol{r} - \boldsymbol{r}_0$. Calculate the phase profile $\varphi(\boldsymbol{r}, t)$ of this wave. Draw the phase fronts (surfaces of equal phase) and wave vectors for such a wave.
Solution
Set the coordinate system such that $\boldsymbol{r}_0 = \boldsymbol{0}$. From the dispersion relation follows that $\boldsymbol{k} = k_0\boldsymbol{r}/r$ with $k_0 = n\omega/c = \text{const}$. Integration along the r–direction with integration constant $-\omega t$ gives $\varphi = k_0 r - \omega t$ and so, in general,

$$\varphi = k_0|\boldsymbol{r} - \boldsymbol{r}_0| - \omega t\,. \tag{4.5}$$

The phase fronts are spherical surfaces around $\boldsymbol{r}_0$ and the wave vectors are the normal vectors on these spheres.

Consider the propagation of a linear wave from the source point A to the spectator point B. According to the superposition principle, all paths from A to B may interfere with each other. Imagine that this actually happens; the wave oscillates along all the paths with the same magnitude, but with phases that depend on the path (Fig. 2.13). This is Feynman's version (Feynman [1948]) of Huygens' principle—any wave consists of infinitely many elementary waves, one for each path, that interfere with each other. Most of this massive interference is destructive, for the following reason. Suppose the phase varies rapidly in comparison with the variation of the medium. In this case one can find neighbouring paths with a phase difference of π and hence opposite amplitude that cancel each other (Fig. 2.14). Only paths

*The physics holding the optical material together, creating atomic and molecular forces, is highly nonlinear however.

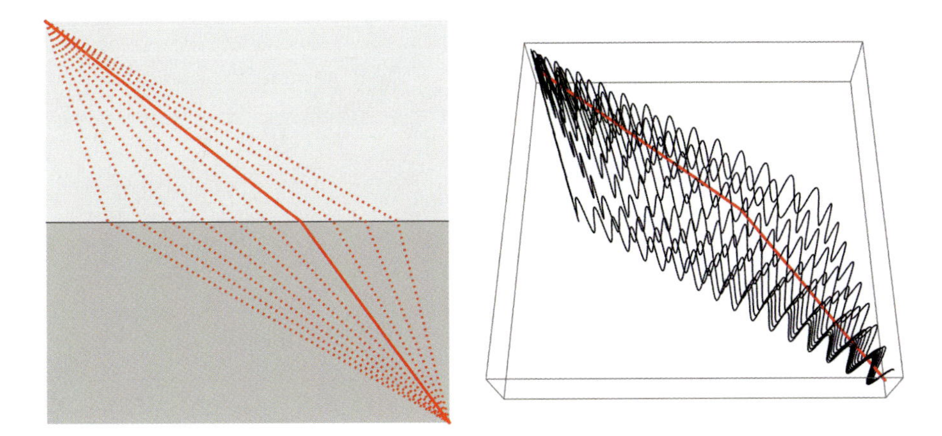

Figure 2.13: Feynman's interference of paths. Consider the example of refraction illustrated in Fig. 2.3. The left picture shows several paths of light rays, the actual path (red line) and virtual paths (dotted lines), travelling from the bottom of the picture to the top. In the right picture all these paths carry oscillating amplitudes with a phase that is proportional to the optical path length (1.1). The virtual paths interfere destructively with each other (Fig. 2.14); only the actual path where the phase is stationary manifests itself as a ray trajectory. As the phase is stationary when the optical path length is stationary, the light ray appears to follow Fermat's principle.

without the destructive interference of their neighbours make a contribution. Near them the phase does not change,

$$\delta\varphi = 0\,. \tag{4.6}$$

Similar to the variational calculus of §2, imagine we vary the path by the increment $\delta\boldsymbol{r}$ around the optimal trajectory with stationary phase (4.6), but keep the end points fixed. Assume in addition that $\boldsymbol{k}$ is another independent variable that can be varied in the phase. The frequency ω depends on both $\boldsymbol{r}$ and $\boldsymbol{k}$ according to the dispersion relation (4.2). We obtain by Taylor expansion from the phase integral (4.4):

$$\begin{aligned}
\delta\varphi &= \int\left(\delta\boldsymbol{k}\cdot\mathrm{d}\boldsymbol{r} + \boldsymbol{k}\cdot\mathrm{d}\delta\boldsymbol{r} - \frac{\partial\omega}{\partial\boldsymbol{k}}\cdot\delta\boldsymbol{k}\,\mathrm{d}t - \frac{\partial\omega}{\partial\boldsymbol{r}}\cdot\delta\boldsymbol{r}\,\mathrm{d}t\right) \\
&= \int\left[\delta\boldsymbol{k}\cdot\mathrm{d}\boldsymbol{r} - \delta\boldsymbol{r}\cdot\mathrm{d}\boldsymbol{k} - \frac{\partial\omega}{\partial\boldsymbol{k}}\cdot\delta\boldsymbol{k}\,\mathrm{d}t - \frac{\partial\omega}{\partial\boldsymbol{r}}\cdot\delta\boldsymbol{r}\,\mathrm{d}t + \mathrm{d}\Big(\boldsymbol{k}\cdot\delta\boldsymbol{r}\Big)\right] \\
&= \int\left[\left(\mathrm{d}\boldsymbol{r} - \frac{\partial\omega}{\partial\boldsymbol{k}}\mathrm{d}t\right)\cdot\delta\boldsymbol{k} - \left(\mathrm{d}\boldsymbol{k} + \frac{\partial\omega}{\partial\boldsymbol{r}}\mathrm{d}t\right)\cdot\delta\boldsymbol{r}\right] + \Big(\boldsymbol{k}\cdot\delta\boldsymbol{r}\Big)\Big|_{\boldsymbol{r}_A}^{\boldsymbol{r}_B} \\
&= \int\left[\left(\mathrm{d}\boldsymbol{r} - \frac{\partial\omega}{\partial\boldsymbol{k}}\mathrm{d}t\right)\cdot\delta\boldsymbol{k} - \left(\mathrm{d}\boldsymbol{k} + \frac{\partial\omega}{\partial\boldsymbol{r}}\mathrm{d}t\right)\cdot\delta\boldsymbol{r}\right],
\end{aligned} \tag{4.7}$$

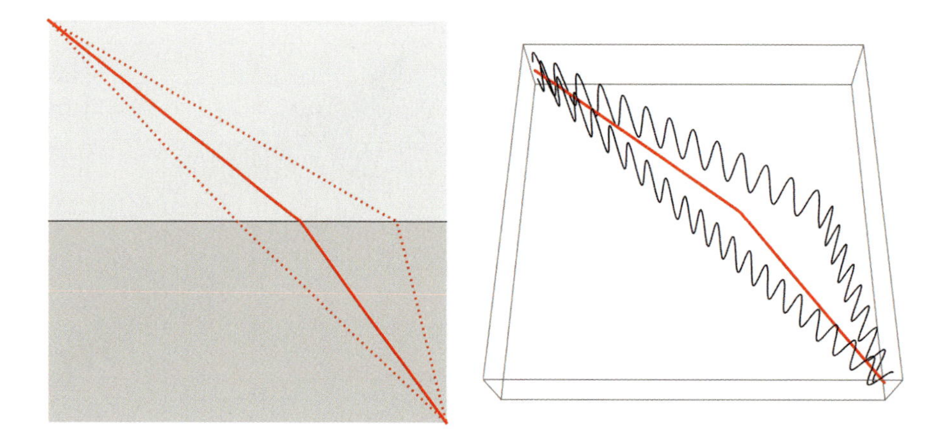

Figure 2.14: Destructive interference of paths. The figure focuses on two of the interfering paths of Fig. 2.13. In the left picture, the two virtual paths (dotted lines) around the actual path (red line) are exactly out of phase, as the right picture shows; they start from the same amplitude in the bottom right corner, but their final amplitudes at the top left corner are opposite to each other. According to the superposition principle, their individual amplitudes add, but as they carry opposite signs, they cancel each other. One can always find a pair of such cancelling trajectories, except for the path where the optical path length and hence the phase reaches an extremum. Consequently, this extremal path survives interference.

because $\delta \boldsymbol{r}_A = \delta \boldsymbol{r}_B = 0$. The variation of φ vanishes for all $\delta \boldsymbol{r}$ and $\delta \boldsymbol{k}$ if and only if

$$\frac{\mathrm{d}\boldsymbol{r}}{\mathrm{d}t} = \frac{\partial \omega}{\partial \boldsymbol{k}}, \quad \frac{\mathrm{d}\boldsymbol{k}}{\mathrm{d}t} = -\frac{\partial \omega}{\partial \boldsymbol{r}}. \tag{4.8}$$

These are Hamilton's equations.* If the medium does not change in time the refractive index does not depend on t and so the frequency ω given by the dispersion relation (4.2) does not explicitly depend on t either; $\partial\omega/\partial t = 0$. The frequency might dynamically depend on time, because $\boldsymbol{r}$ and $\boldsymbol{k}$ vary in time, but we obtain from Hamilton's equations (4.8) that ω is a conserved quantity:

$$\frac{\mathrm{d}\omega}{\mathrm{d}t} = \frac{\partial \omega}{\partial \boldsymbol{r}} \cdot \frac{\mathrm{d}\boldsymbol{r}}{\mathrm{d}t} + \frac{\partial \omega}{\partial \boldsymbol{k}} \cdot \frac{\mathrm{d}\boldsymbol{k}}{\mathrm{d}t} + \frac{\partial \omega}{\partial t} = 0\,. \tag{4.9}$$

The conservation of the frequency ω is related to the conservation of energy, because the energy of a light quantum is $\hbar\omega$. Note that the frequency is generally not conserved when the medium changes in time. This may be the case in nonlinear

*In mechanics, for a particle with position $\boldsymbol{r}$, momentum $\boldsymbol{p} = \hbar\boldsymbol{k}$ and Hamiltonian $H = \hbar\omega$, Hamilton's equations are

$$\frac{\mathrm{d}\boldsymbol{r}}{\mathrm{d}t} = \frac{\partial H}{\partial \boldsymbol{p}}, \quad \frac{\mathrm{d}\boldsymbol{p}}{\mathrm{d}t} = -\frac{\partial H}{\partial \boldsymbol{r}}.$$

optics (Agrawal [2001], Boyd [1992], Shen [1984]). For example, the oscillating electric field of a strong pump beam may create oscillating electric dipoles in a suitable optical material that, to another light beam, appears as an oscillating medium. As a consequence, the second beam may change frequency—colour. Let us return to the linear optics of passive materials. Here we obtain from the dispersion relation (4.2) and Hamilton's equations (4.8) the equations of motion

$$\frac{\mathrm{d}\boldsymbol{r}}{\mathrm{d}t} = \frac{c}{n}\frac{\boldsymbol{k}}{|\boldsymbol{k}|}\,, \quad \frac{\mathrm{d}\boldsymbol{k}}{\mathrm{d}t} = \frac{c|\boldsymbol{k}|}{n^2}\,\boldsymbol{\nabla} n\,. \tag{4.10}$$

Furthermore, using the dispersion relation (4.2) we get

$$\frac{\mathrm{d}\boldsymbol{r}}{\mathrm{d}t} = \frac{c^2}{n^2\omega}\,\boldsymbol{k}\,, \quad \frac{\mathrm{d}\boldsymbol{k}}{\mathrm{d}t} = \frac{\omega}{n^2}\,\frac{\boldsymbol{\nabla} n^2}{2} \tag{4.11}$$

and, since ω is time–independent:

$$\frac{n^2}{c}\frac{\mathrm{d}}{\mathrm{d}t}\frac{n^2}{c}\frac{\mathrm{d}\boldsymbol{r}}{\mathrm{d}t} = \frac{\boldsymbol{\nabla} n^2}{2}\,. \tag{4.12}$$

Introducing the propagation parameter ξ with

$$\mathrm{d}\xi = \frac{c}{n^2}\,\mathrm{d}t \tag{4.13}$$

we see that the equation of motion (4.12) agrees with the Newtonian Euler–Lagrange equation (3.6) for ray trajectories with extremal optical path (1.1). Feynman's interference of paths thus explains Fermat's principle: the path taken is the path of constructive interference where the optical path length varies the least (Fig. 2.13).

Exercise 4.2

One may parameterize ray trajectories by the Hamiltonian time t or the Newtonian parameter ξ. Consider the Hooke profile (3.14). How are t and ξ related in this case?

Solution

Integrate formula (4.13) with profile (3.14) and relationship (3.27). Expressed in terms of condition (3.26) the result is

$$t = t_0 - \frac{\cos(2\gamma)\sin(2\xi)}{2c} + \frac{\xi}{c}\,. \tag{4.14}$$

For *ray tracing*, the numerical solution of Hamilton's equations of the form (4.10) is highly recommended, because they tend to be numerically stable. In addition, the conservation of frequency (4.2) can be used for assessing the numerical accuracy.

Exercise 4.3

Numerically solve Hamilton's equations and plot ray trajectories. For instance, take the Luneburg lens shown in Fig. 2.12 and play with various initial conditions. You may use one of the commercial mathematics packages where the numerical solution of ordinary differential equations and decent graphics is part of the package.

The Newtonian form (3.6) of Hamilton's equations is often more suitable for finding analytic solutions and it reveals a curious duality of position and momentum space for spherically symmetric index profiles $n(r)$ (Miñano, Bentez and Santamaria [2006]).

Exercise 4.4
For simplicity, let us measure time in units such that $\omega/c = 1$. In these units, we obtain from the dispersion relation (4.2):

$$k = n(r)\,. \tag{4.15}$$

Suppose we invert this relationship by solving the equation for r,

$$r = \widetilde{n}(k)\,. \tag{4.16}$$

Show that the ray trajectory in momentum space (k space) is the trajectory in a position space with the inverted index profile (4.16).
Solution
For spherically symmetric profiles $n(r)$ we obtain from Hamilton's equations (4.11) with the parameterization (4.13)

$$\boldsymbol{k} = \frac{\mathrm{d}\boldsymbol{r}}{\mathrm{d}\xi}\,, \quad \frac{\mathrm{d}\boldsymbol{k}}{\mathrm{d}\xi} = n\,\frac{\mathrm{d}n}{\mathrm{d}r}\,\frac{\boldsymbol{r}}{r}\,. \tag{4.17}$$

We introduce a new parameter ξ' such that $\mathrm{d}\boldsymbol{k}/\mathrm{d}\xi' = (\mathrm{d}\xi/\mathrm{d}\xi')\mathrm{d}\boldsymbol{k}/\mathrm{d}\xi$ with $\mathrm{d}\boldsymbol{k}/\mathrm{d}\xi$ given by Eq. (4.17) reduces to $\boldsymbol{r}$ and use Eqs. (4.15) and (4.16):

$$\mathrm{d}\xi' = \frac{n}{r}\,\frac{\mathrm{d}n}{\mathrm{d}r}\,\mathrm{d}\xi = \frac{k}{\tilde{n}}\,\frac{\mathrm{d}k}{\mathrm{d}\tilde{n}}\,\mathrm{d}\xi \quad \Rightarrow \quad \boldsymbol{r} = \frac{\mathrm{d}\boldsymbol{k}}{\mathrm{d}\xi'}\,, \quad \frac{\mathrm{d}\boldsymbol{r}}{\mathrm{d}\xi'} = \tilde{n}\,\frac{\mathrm{d}\tilde{n}}{\mathrm{d}k}\,\frac{\boldsymbol{k}}{k}\,. \tag{4.18}$$

In k space—momentum space—rays thus propagate in the inverted index profile (4.16).

Exercise 4.5
Show that the Luneburg lens has the same profile in momentum space as in position space.

Let us return to Feynman's interference of paths. We can use this physical picture to deduce the *wave equation*, a partial differential equation that describes how the amplitude ψ of the wave propagates across infinitesimal distances. In practice, the solution of a wave equation is often infinitely easier than the summation of infinitely many paths. We write ψ as an integral of the phase factor $\exp(\mathrm{i}\varphi)$ over all possible paths, symbolically expressed by Feynman's path integral

$$\psi = \int \mathrm{e}^{\mathrm{i}\varphi}\,\mathrm{D}\boldsymbol{r}\,. \tag{4.19}$$

The "differential" $\mathrm{D}\boldsymbol{r}$ should indicate that the integral is to be performed over all possible paths from the source $\boldsymbol{r}_0$ to the spectator point $\boldsymbol{r}$, a bewildering manifold of paths. This book is not the place to give a mathematically precise definition of path integrals, nor are they mathematically well–established in any case. Nevertheless, we can draw some physical conclusions from Feynman's formula (4.19). Differentiating with respect to $\boldsymbol{r}$ and using the definition (4.3) of $\boldsymbol{k}$ and the dispersion relation (4.2) we obtain

$$\boldsymbol{\nabla}^2\psi = \int \left(\boldsymbol{\nabla}^2\mathrm{e}^{\mathrm{i}\varphi}\right)\mathrm{D}\boldsymbol{r} = -\int k^2\mathrm{e}^{\mathrm{i}\varphi}\,\mathrm{D}\boldsymbol{r} = -\int \frac{n^2\omega^2}{c^2}\mathrm{e}^{\mathrm{i}\varphi}\,\mathrm{D}\boldsymbol{r} = -\frac{n^2\omega^2}{c^2}\,\psi\,, \tag{4.20}$$

which gives the *Helmholtz equation*

$$\left(\nabla^2 + n^2\frac{\omega^2}{c^2}\right)\psi = 0\,. \tag{4.21}$$

In our derivation we made a tacid assumption, however; we assumed that the wave is solely characterized by a scalar amplitude ψ. But electromagnetic waves are carried by the electric and magnetic field strengths $\boldsymbol{E}$ and $\boldsymbol{H}$ that are vectors. Electromagnetic waves are polarized vector waves. The different vector components are not always proportional to a common amplitude ψ; it turns out they mix at sharp boundaries (Exercise 29.3). More generally and more precisely, the field components mix when the refractive–index profile varies over scales comparable with the wavelength. The Helmholtz equation (4.21) thus is a simplification and approximation valid in the regime of *geometrical optics* where the wavelength (4.1) with dispersion relation (4.2) varies slowly with distance,

$$|\nabla\lambda| \ll 1\,. \tag{4.22}$$

In this regime, the phase of ψ rather accurately describes the phase φ of the electromagnetic wave, but the amplitude of ψ cannot fully account for the polarization transport. In geometrical optics, the phase generates the ray trajectories according to relations (4.3) and (4.10); $\boldsymbol{r}(t)$ is given by the differential equation

$$\frac{\mathrm{d}\boldsymbol{r}}{\mathrm{d}t} = \frac{c}{n(\boldsymbol{r})}\frac{\boldsymbol{k}(\boldsymbol{r})}{|\boldsymbol{k}(\boldsymbol{r})|} \quad \text{with} \quad \boldsymbol{k}(\boldsymbol{r}) = \nabla\varphi\,. \tag{4.23}$$

In §30 we show that the field strength of the electromagnetic wave is transported along the ray trajectory given by the phase, parallel–transported to be precise (a term we define in §20). The phase rules the waves.

When the electromagnetic wave depends only on one field component the Helmholtz equation (4.21) turns out to be exact. This is the case in planar integrated optics (§27). Suppose light is confined to a layer with an effective refractive index $n(x, y)$ that varies only in the two directions x and y. If the polarization is chosen such that the electric field $\boldsymbol{E}$ always points orthogonal to x and y only this component E is different from zero. The electromagnetic wave is effectively a scalar wave (provided the optical material of the layer only responds to the electric field). We show in §27 that in this case the Helmholtz equation (4.21) for $\psi = E$ is exact.

§5. Optical conformal mapping

In this Section we develop the concept of transformation media for the physically simplest case: isotropic materials in 2D. Such optical media perform conformal mappings of the plane. We also introduce some aspects of complex analysis, because analytic functions turn out to generate conformal maps.

Consider waves subject to the Helmholtz equation (4.21) in two dimensions. They may be light waves in planar integrated optics or light rays propagating in a plane in 3D, for example in spherically symmetric refractive–index profiles. As before, we combine the Cartesian coordinates x and y in complex numbers. We denote the complex conjugate by an asterisk such that

$$z = x + \mathrm{i}y\,, \quad z^* = x - \mathrm{i}y\,. \tag{5.1}$$

Since $x = (z + z^*)/2$ and $y = (z - z^*)/(2\mathrm{i})$ a function in the plane is a function of z and z^*. Abbreviating partial derivatives $\partial/\partial w$ by ∂_w we obtain from definition (5.1) by the chain rule of differentiation

$$\begin{aligned} \partial_x &= \frac{\partial z}{\partial x}\partial_z + \frac{\partial z^*}{\partial x}\partial_{z^*} = \partial_z + \partial_{z^*}\,, \\ -\mathrm{i}\,\partial_y &= -\mathrm{i}\,\frac{\partial z}{\partial y}\partial_z - \mathrm{i}\,\frac{\partial z^*}{\partial y}\partial_{z^*} = \partial_z - \partial_{z^*}\,. \end{aligned} \tag{5.2}$$

Consequently,

$$\boldsymbol{\nabla}^2 = \partial_x^2 + \partial_y^2 = (\partial_z + \partial_{z^*})^2 - (\partial_z - \partial_{z^*})^2 = 4\,\partial_z\partial_{z^*}\,, \tag{5.3}$$

and so we can write the Helmholtz equation (4.21) as

$$\left(4\,\partial_z\partial_{z^*} + n^2\frac{\omega^2}{c^2}\right)\psi = 0\,. \tag{5.4}$$

Imagine we transform the Cartesian coordinates x and y to the coordinates u and v with

$$w = u + \mathrm{i}v \tag{5.5}$$

such that w only depends on z, but not on z^*,

$$w = w(z) \quad \text{or, equivalently,} \quad \partial_{z^*}w = 0\,. \tag{5.6}$$

The complex conjugate w^* is then a function of z^*, but not of z. We obtain from the chain rule

$$\partial_z = \frac{\mathrm{d}w}{\mathrm{d}z}\,\partial_w\,, \quad \partial_{z^*} = \frac{\mathrm{d}w^*}{\mathrm{d}z^*}\,\partial_{w^*}\,, \tag{5.7}$$

and therefore from the Helmholtz equation (5.4):

$$\left(4\,\partial_w\partial_{w^*} + n'^2\frac{\omega^2}{c^2}\right)\psi = 0 \tag{5.8}$$

where the original refractive index n is replaced by

$$n' = \left|\frac{\mathrm{d}z}{\mathrm{d}w}\right| n \quad \text{or} \quad n = \left|\frac{\mathrm{d}w}{\mathrm{d}z}\right| n'\,. \tag{5.9}$$

Now, the wave equation (5.8) is exactly the Helmholtz equation (5.4) with z replaced by the new Cartesian coordinates in complex representation (5.5). Transformations of the type (5.6) therefore map light propagation in physical space to light propagation in a different space filled with the transformed refractive–index profile n' (Fig.

2.15). We also see this in Fermat's principle where the squared optical line element appears as

$$dl^2 = n^2(dx^2 + dy^2) = n^2\, dz\, dz^* = n'^2\, dw\, dw^* . \tag{5.10}$$

Fermat's principle thus holds both in physical and transformed space, but with different refractive–index profiles (5.9).*

Exercise 5.1
Suppose w is a function of both z and z^*. Why do we not get a transformed refractive–index profile in this case?

Imagine the transformed refractive index n' is constant, say $n' = 1$, such that

$$n = |g(z)| , \quad g(z) = \frac{dw}{dz} . \tag{5.11}$$

In this case the transformed space appears to be empty—by definition, because $n' = 1$. The medium creates the illusion that light propagates through empty space, whereas in reality light waves are bent, following the coordinate transformation. The transformation $w(z)$ establishes a virtual space where physical objects appear at new positions assigned by the transformation. The optical material acts as a *transformation medium* (Leonhardt [2006a]).

Exercise 5.2
Suppose physical space is filled with a uniform medium of constant refractive index n. Write down the transformation to virtual space this medium performs.

Clearly, not all optical materials are transformation media. Although materials such as glass or water seem to alter the geometry of space such that objects appear at different positions than where they actually are, the virtual positions depend on the viewpoint. Transformation media are exceptional. Given a refractive–index profile $n(x, y)$, how can we find out whether n acts as a transformation medium? Consider the quantity

$$R = \frac{2(\nabla n)^2}{n^4} - \frac{2\nabla^2 n}{n^3} \tag{5.12}$$

or, in terms of z and z^*,

$$R = \frac{8(\partial_z n)(\partial_{z^*} n)}{n^4} - \frac{8\,\partial_z \partial_{z^*} n}{n^3} . \tag{5.13}$$

For a transformation medium (5.11), $n = \sqrt{gg^*}$ where g depends only on z and g^* only on z^*. Substituting $\sqrt{gg^*}$ for n in formula (5.13) we obtain

$$R = 0 . \tag{5.14}$$

This is the acid test for transformation media in two dimensions. Note that R has an important geometrical meaning: R is the *curvature scalar*, as we show in §23.

*Note that we could get the same feature for so–called anti–analytic transformations $w(z^*)$.

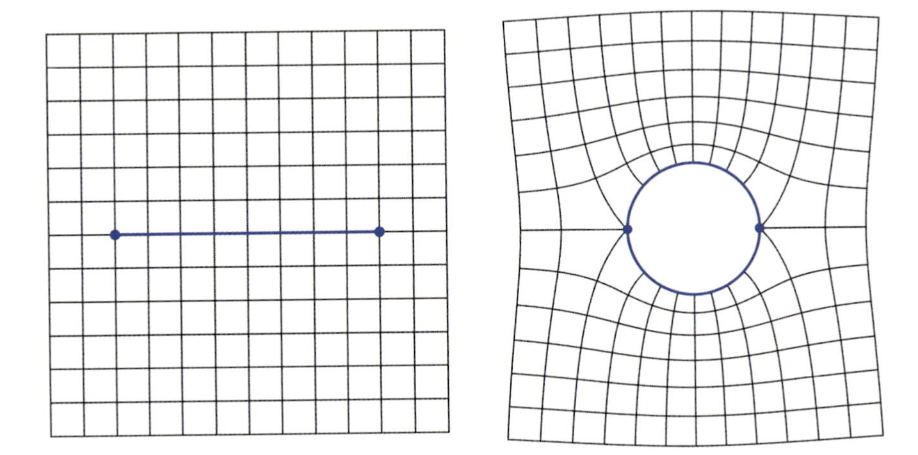

Figure 2.15: Optical conformal mapping. The medium performs a coordinate transformation of physical space; it creates the illusion that light propagates in a virtual space. Left: empty virtual space (w–space); light propagates along straight lines, such as the lines of the coordinate grid. Right: physical space (z–space); light rays are the transformed rays of virtual space. Virtual space is the space perceived by light; physical space appears to be distorted. For example, the blue circle in the right picture has been flattened, indicated by the blue line in the left picture. However, both physical space and the virtual space of light are flat spaces, because the material only performs a coordinate transformation and does not establish a curved geometry (§18); the effective curvature (5.12) of the medium vanishes. The figure corresponds to the Zhukovski transform (5.24).

$R = 0$ is a test for flatness. The transformation medium shifts the points of physical space to new locations in virtual space, but it does not fundamentally alter the geometry: space is still flat with $R = 0$. The converse is also true, when $R = 0$ we can always find virtual coordinates (5.5) such that virtual space is empty. But proving this property requires the mathematical machinery of differential geometry that we set forth in Chapter 3.

Exercise 5.3
Consider a material where the refractive index n varies only in one direction, say the y direction, as in the mirage of Fig. 2.1 or in the refraction of light between two uniform media shown in Fig. 2.3. Find all profiles $n(y)$ that establish transformation media.
Solution
The curvature scalar (5.12) vanishes when $(\partial_y n)/n = (\partial_y^2 n)/(\partial_y n)$ or, equivalently, $\partial_y \ln n = \partial_y \ln(\partial_y n)$. Integration gives $\ln(\partial_y n) = \ln n + \text{const}$ and so leads to the differential equation $\partial_y n = an$ with the solution $n = n_0 \exp(ay)$ where both a and n_0 are real constants. Neither the interface in refraction (Fig. 1.1) nor the index profile of the mirage shown in Fig. 2.1 are of this type: such optical illusions are created by curved geometries.

Exercise 5.4
Find all transformation media in $2D$ with radially symmetric refractive–index profiles $n(r)$. Why is the Luneburg lens not a transformation medium?
Solution
Writing $r = \sqrt{zz^*}$ we express the condition $R = 0$ with curvature (5.13) as a differential equation in r. We get from the chain rule $2\partial_z n = (\partial_r n)\sqrt{z^*/z}$ and $2\partial_{z^*} n = (\partial_r n)\sqrt{z/z^*}$, and so $4(\partial_z n)(\partial_{z^*} n) = (\partial_r n)^2$ and $4\partial_z \partial_{z^*} n = \partial_r^2 n + (\partial_r n)/r$. The curvature (5.13) thus vanishes when $\partial_r \ln n = \partial_r \ln(\partial_r n) + 1/r = \partial_r \ln(r\partial_r n)$. We proceed as in Exercise 5.3 and obtain $n = ar^\nu$ with the real constants a and ν. The Hooke profile (3.14) of the Luneburg lens is radially symmetric but not of this type.

Transformations (5.6) change the measure of distance in space, but they do not alter angles between lines: they are *conformal* (Fig. 2.16). To see this, consider a curve starting from the point z_0 at the angle ϕ to the x–axis. In the vicinity of z_0 this curve is a straight line described by the formula

$$z = z_0 + \xi\, \mathrm{e}^{\mathrm{i}\phi} \tag{5.15}$$

for small parameters ξ. Substituting this expression into the transformation (5.6) we obtain in virtual space by Taylor expansion

$$w \sim w_0 + \xi \eta_0\, \mathrm{e}^{\mathrm{i}(\phi+\phi_0)}\,, \quad w_0 = w(z_0)\,, \quad \eta_0\, \mathrm{e}^{\mathrm{i}\phi_0} = \left.\frac{\mathrm{d}w}{\mathrm{d}z}\right|_{z_0}\,. \tag{5.16}$$

The line is rotated by ϕ_0 in virtual space, but this rotation is the same for all angles ϕ. Therefore, the angles between all curves crossing at any point z_0 are preserved when these curves are transformed to virtual space. This invariance of angles is the deeper reason why transformations (5.6) can be implemented by optically isotropic materials that, according to Fermat's principle, distort the measure of distance (1.1), but not the measure of angle.

The transformations (5.6) are well known in mathematics: they are the analytic functions of complex analysis, because the condition (5.6) is shorthand for the Cauchy–Riemann equations (Ablowitz and Fokas [1997])

$$\partial_x u = \partial_y v\,, \quad \partial_x v = -\partial_y u\,. \tag{5.17}$$

Exercise 5.5
Deduce the Cauchy–Riemann equations from the condition (5.6) that $\partial_{z^*} w$ vanishes.
Solution
Substitute Eqs. (5.2) and (5.5) into $\partial_{z^*} w = 0$ and write it in terms of real and imaginary parts.

The Cauchy–Riemann equations (5.17) define the differentiable functions of a complex variable (Ablowitz and Fokas [1997]); they establish the starting point of complex analysis, a subject that, like no other field of mathematics, combines analytical techniques with visual insight and furnishes a rich arsenal of elegant tools. (The reader is referred to the excellent texts Ablowitz and Fokas [1997], Needham [2002], Nehari [1952]). Here we just focus on some examples that illustrate the aspects of complex analysis that directly appear in the geometry of light.

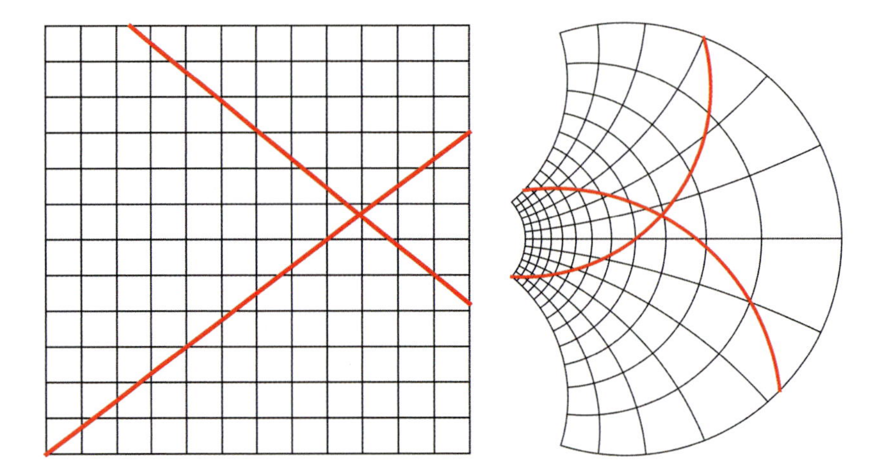

Figure 2.16: Möbius transformation. The figure illustrates the fact that conformal maps (5.6) preserve the angles between lines. Physical space (right) is transformed to virtual space (left) by a Möbius transformation (5.18). The angles between the red lines are preserved, and so are the right angles between the coordinate lines. The Möbius transformation turns straight lines into circles; the Cartesian grid of virtual space (left) appears in physical space (right) as an orthogonal grid of circles, and also the two red lines turn into circles intersecting each other at the same angle as in virtual space. For this figure, the specific Möbius transformation (9.34) with $\gamma = \pi/6$ has been applied.

Exercise 5.6

Traditionally, the Cauchy–Riemann equations are introduced as the conditions for uniquely defining the derivative of a complex function w in the plane. The derivative of w at point z_0 can be taken along various directions, but analytic functions are defined as the functions for which the derivative is independent of the direction in the complex plane. Show that from this requirement follow the Cauchy–Riemann equations (5.17).

Solution

Suppose that either x or y varies in $z = x + \mathrm{i}y$ and take the derivatives of $w = u + \mathrm{i}v$ in the two different directions—differentiate first with respect to x and then with respect to $\mathrm{i}y$. The two derivatives agree when the Cauchy–Riemann equations (5.17) are satisfied.

Consider some relatively simple conformal transformations as instructive examples, the *Möbius transformations*:

$$w = \frac{a\,z + b}{c\,z + d} \tag{5.18}$$

with complex constants a, b, c, d. Solving for z we obtain the inverse Möbius transformation:

$$z = \frac{d\,w - b}{-c\,w + a}\,. \tag{5.19}$$

Any Möbius transformation (5.18) can be expressed by the following chain of elementary transformations

$$w = z_4 = z_3 + \frac{a}{c}\,, \qquad z_3 = \frac{b\,c - a\,d}{c^2}\, z_2\,, \qquad z_2 = \frac{1}{z_1}\,, \qquad z_1 = z + \frac{d}{c}\,. \tag{5.20}$$

Exercise 5.7
Verify by direct calculation that $w = z_4(z_3(z_2(z_1(z))))$.

The elementary transformations (5.20) describe geometric operations: the Möbius transformation consists of a spatial shift, followed by an inversion, then a complex multiplication—geometrically, a rotation and re–scaling, finished by a further shift. The *inversion* $z \rightarrow 1/z$ inverts the distance $|z|$ from the origin and changes the sign of the angle of the point*, because

$$\text{for} \quad z_2 = \frac{1}{z} \quad \Rightarrow \quad |z_2| = \frac{1}{|z|}\,, \quad \arg z_2 = -\arg z\,. \tag{5.21}$$

Exercise 5.8
Sketch a drawing that shows the geometric effect of the inversion $1/z$.

Now, if $ad - bc$ vanishes so does z_3 in the chain (5.20) and the transformation (5.18) reduces to a constant—all z are projected into a single point. We shall exclude this trivial case. On the other hand, we can multiply all constants a, b, c, d by a common factor and not change the function (5.18). So we can normalize the constants of non–trivial Möbius transformations such that

$$a\,d - b\,c = 1\,. \tag{5.22}$$

Möbius transformations have a remarkable geometrical property we need later: they turn circles into circles. Let us prove this by considering the transformation of circles by the elementary operations in the Möbius chain (5.20). Clearly, the spatial shifts in the chain (5.20) preserve circles and the complex multiplication simply rotates and rescales them. The only non–trivial operation we really need to investigate in detail is the inversion. Imagine a circle with radius r_0 around the centre x_0 on the x axis.

Exercise 5.9
The circle is described by the equation $|z - x_0|^2 = r_0^2$. Show that the inversion $w = 1/z$ generates a new circle with centre at the x–axis and equation $|w - x_0'|^2 = {r_0'}^2$.
Solution
Expand the two equations of circles and insert the first one into the second:

$$\begin{aligned} r_0^2 &= x_0^2 - x_0(z + z^*) + |z|^2 \\ r_0'^2|z|^2 &= |1 - x_0' z|^2 = 1 - x_0'(z + z^*) + {x_0'}^2|z|^2 = 1 - \frac{x_0'}{x_0}\left(x_0^2 - r_0^2 + |z|^2\right) + {x_0'}^2|z|^2 \\ &\Rightarrow \; x_0' = \frac{x_0}{x_0^2 - r_0^2}\,, \quad r_0' = \frac{r_0}{|x_0^2 - r_0^2|}\,. \end{aligned} \tag{5.23}$$

With these parameters, the equation of the transformed circle is satisfied.

---

*The inversion $1/z$ is however not the so–called inversion in the unit circle; that turns out to be $1/z^*$ (Needham [2002]).

For $x_0^2 = r_0^2$ the circles are transformed into straight lines that, however, we may regard as circles with infinite radius. Conversely, straight lines in z space are mapped into circles in w space (Fig. 2.16), because we could run the same argument for $z = 1/w$. Having thus established the mapping of circles into circles by the inversion for real x_0, for circles centred at the x axis, we generalize it by rotation to any centre point. This proves that Möbius transformations map circles into circles.

The Möbius transformations map the entire plane into the entire plane, because the inverse transformation (5.19) exists and is unique, but they are the only analytic functions that do so. The *Riemann mapping theorem* (Nehari [1952]) says that all other analytic functions map the plane on to *Riemann sheets*. To understand the concept of Riemann sheets, consider an example, the Zhukovski transform (also spelled Joukowski):

$$w = z + \frac{1}{z}. \tag{5.24}$$

As we can replace z by $1/z$ and obtain the same w, the inverse of $w(z)$ must be at least double–valued.

Exercise 5.10
Solve Eq. (5.24) for z.

We obtain

$$z = \frac{1}{2}\left(w \pm \sqrt{w^2 - 4}\right). \tag{5.25}$$

The two signs of the square root indicate that we need two functions $z(w)$ to cover the entire z plane. Each function maps the whole w plane into a separate area, for example we may choose the two functions such that one projects onto the interior of the unit circle and the other onto the exterior; both areas combined cover the entire z plane. The inverse function $z(w)$ with two w planes thus tiles the z plane (Fig. 2.17).

Exercise 5.11
Show that the two solutions (5.25) are inverses of each other, one is $1/z$ of the other.

It is largely arbitrary where we draw the line between the tiles that represent the w sheets; the only fixtures are the points where the functions $z(w)$ coincide, the *branch points*. In our case, the two functions (5.25) differ by $\sqrt{w^2 - 4}$, so here the branch points are located at the zeros of this square root, at $w = \pm 2$. As we can freely move around between the tiles in the z plane, we ought to be able to do the same in w space. Hence the two w planes must be connected; their connections are called *branch cuts*. A branch cut simply represents on the w sheets the boundary between two tiles in the z plane, the *branches*. The w planes establish a *Riemann surface* made of Riemann sheets—stacks of planes pinned at branch points and connected at branch cuts (Fig. 2.17). In general, more than two branches can meet at a single branch point, even infinitely many, and the different branch points of a single function may connect different numbers of different branches. Riemann sheets can be incredibly complex and are perhaps best visualized by their tilings of

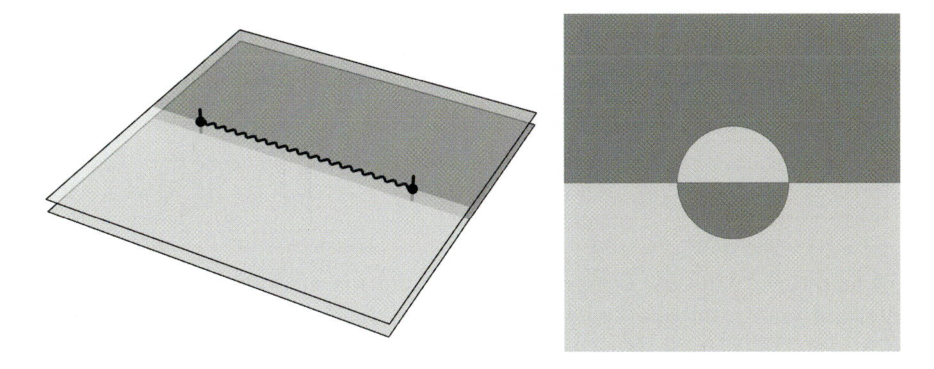

Figure 2.17: Riemann sheets. Apart from the Möbius transformations (5.18), all non-trivial conformal transformations in 2D map the plane (right) onto a stack of planes (left): Riemann sheets. The figure illustrates the Riemann sheets of the Zhukovski transform (5.24), the simplest example. Each sheet on the left corresponds to a region in the plane on the right, here to the inside and the outside of the unit circle. The different shades of grey indicate the upper and lower half sheets (left) and their representations in the plane (right); the Riemann sheets tile the plane. On the Riemann surface (left), the sheets are pinned at branch points. In the plane (right), these are the points where the upper and lower regions of the different Riemann sheets meet (the different shades of grey). The wavy line in the left picture indicates the branch cut.

the plane. Some functions known as automorphic functions (Erdélyi et al. [1981]) create truly Escheresque tilings (Fig. 2.18).

Figure 2.18: Tiles of the elliptic modular function known as the Klein invariant (Erdélyi et al. [1981] section 14.6). Such infinitely intricate patterns are made by a bewildering Riemann surface of infinitely many sheets. Elliptic modular functions are mysteriously connected to the most legendary part of Fermat's mathematical legacy, Fermat's Last Theorem (Wiles [1995], Singh [1997]).

Exercise 5.12
Sketch the Riemann sheets of $w = \exp(z) = \exp(x)[\cos y + \mathrm{i} \sin y]$ and the corresponding tiling of the plane.
Solution
As $w(z + 2\pi m\mathrm{i}) = w(z)$ with integer m, the z plane is tiled in stripes along the x–axis of height 2π in the y–direction (Fig. 2.25). The branch points are $z = 0$ and $z = \infty$. There are infinitely many Riemann sheets.

Let us return to optics. Consider a transformation medium (5.11) that performs a conformal map from physical space to virtual space, for example the Zhukovski transform (5.24). Virtual space—w space—is not a plane in general, but a Riemann surface, as we have learned. Suppose the outer branch, the top sheet agrees with physical space for $z \to \infty$. For example, in our specific case (5.24), the outer branch represents the exterior of the unit circle (Fig. 2.17). The optical deformation of the plane disappears for $z \to \infty$, because we see from formula (5.24) that $w \sim z$ for $z \to \infty$. The top sheet in virtual space merges with the physical plane for large distances from the origin. Imagine light rays coming from ∞ in the physical world. Consider their fate on the Riemann surface (Fig. 2.19). The rays are incident from

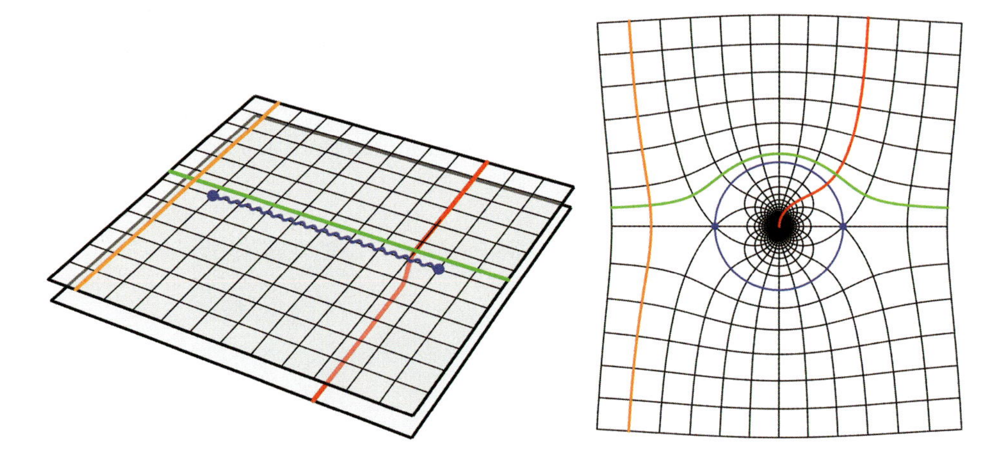

Figure 2.19: Optics on Riemann sheets. An isotropic transformation medium maps the physical plane (right) to Riemann sheets in virtual space (left). The figure illustrates the entire Zhukovski map (5.24), completing Fig. 2.15 where we had not considered Riemann sheets yet. The blue circle (right) represents the unit circle that corresponds to the branch cut, the wavy blue line (left). The picture illustrates the fate of various characteristic light rays. The rays are bent in physical space (right) but proceed along straight lines on the Riemann surface (left). Hence it is easier to discuss them in virtual space. The green and orange rays miss the branch cut (left). In physical space (right), they pass through the medium as if it were invisible. However, the red ray enters the branch cut (left). On the Riemann surface of virtual space, the ray continues towards ∞ on the lower sheet. In physical space (right) the ray is trapped at the singularity of the map (5.24).

the ∞ of the outer branch that corresponds to the ∞ of physical space. The rays may miss the branch cut, leaving as if nothing has happened. Some rays, however, strike the branch cut, entering the lower sheet, the "underworld". They continue to the ∞ of the lower branch, but this ∞ is no longer the ∞ of physical space, but one of the singularities of the conformal map, in our case $z = 0$ where $w \to \infty$ in the Zhukovski transform (5.24). Light entering the branch cut proceeds towards the singularity, if nothing else prevents it.

Yet one can shepherd the lost rays back to the outer branch by an additional refractive–index profile n' placed on the lower sheet (Leonhardt [2006a]). The profile should be designed such that light rays form closed loops. This is all that is needed. The light entering the "underworld" may be refracted at the border, but after a loop back to the outer branch it is refracted once more at the exit from the "underworld", leaving in the same direction it came from towards ∞ in physical space. If the light does not penetrate all areas of the "underworld"; these unexplored areas remain hidden. So if one has something to hide, one can hide it on Riemann sheets. The transformation medium combined with the additional profile n' makes an invisibility device. The only tangible trace of the detour through part of the "underworld" is a time delay. In §7 we show that the time delay is uniform though, for all rays entering the cloaking device. Only a phase slip at the edge of the cloaking device may indicate its presence. An example of a closed–loop medium is the Hooke profile (3.14) where light travels in ellipses. In §9 we determine a large class of refractive–index profiles with closed light loops. Such profiles are no longer ordinary transformation media; they do not map flat Euclidean space onto flat space, they rather establish a non–Euclidean geometry for light (see §9 for an illuminating example). Optical conformal mapping (Leonhardt [2006a]) forms the nucleus of non–Euclidean transformation optics (Leonhardt and Tyc [2009]).

In some cases of cloaking, complete optical invisibility may not be needed. Consider a simple example, the line $w = u + iv_0$ with constant v_0 on the outer branch of

Figure 2.20: From fugu to flatfish. Optical conformal mapping may turn round objects (right) to appear flat (left). (Credit: Maria Leonhardt.)

the Zhukovski map (5.24). This line corresponds to one of the horizontal coordinate lines on the left of Fig. 2.15. In physical space (right of Fig. 2.15), this line is curved; for $v_0 \to 0$ the line encloses a half circle (blue circle in Fig. 2.15). But in virtual space the line is straight. Curved objects thus may appear flat in optical conformal mapping (as if the fugu of Fig. 2.20 turns into a flatfish). Flat objects may be hidden by camouflage, assuming the colour and texture of the floor (as flatfish are masters of). Such partial cloaking is less demanding to implement* and may be equally effective (Li and Pendry [2008]).

§6. Transmutation

We have seen how optical conformal mapping transforms a refractive–index profile in physical space into a new profile in virtual space. In this Section, we discuss how the corresponding Newtonian potentials are transformed. We show how potentials can be transmuted into each other, for example the Hooke potential into Newton's. An application in optics is the Eaton lens.

Conformal mappings transform refractive–index profiles according to formula (5.9). In the Newtonian analogy of §3, index profiles appear as mechanical potentials (3.7). Writing the potential and energy associated with the transformed index profile n' as U' and E' we obtain from the transformation rule (5.9) how conformal transformations turn one potential into another:

$$U(z) - E = \left|\frac{\mathrm{d}w}{\mathrm{d}z}\right|^2 (U'(w) - E') \,. \tag{6.1}$$

In what follows we consider a special case of the transformation (6.1). Suppose each potential is the transformed energy of the other:

$$U(z) = -E'\left|\frac{\mathrm{d}w(z)}{\mathrm{d}z}\right|^2 , \quad U'(w) = -E\left|\frac{\mathrm{d}z(w)}{\mathrm{d}w}\right|^2 . \tag{6.2}$$

We call this transformation of potentials *transmutation*. As the trajectories are simply transformed by the functions $w(z)$ and $z(w)$, an analytic function establishes pairs of potentials with related trajectories. Moreover, seemingly unrelated potentials may turn out to be mere transmutations of each other (Arnold [1990], Needham [1993], Needham [2002]).

Consider the following instructive example, transmutation by the quadratic function (Fig. 2.21):

$$w = z^2 \,. \tag{6.3}$$

Assuming in transformed space an energy of

$$E' = -\frac{1}{2} \tag{6.4}$$

*One could apply quasiconformal transformations and approximately implement them with isotropic materials (Li and Pendry [2008]).

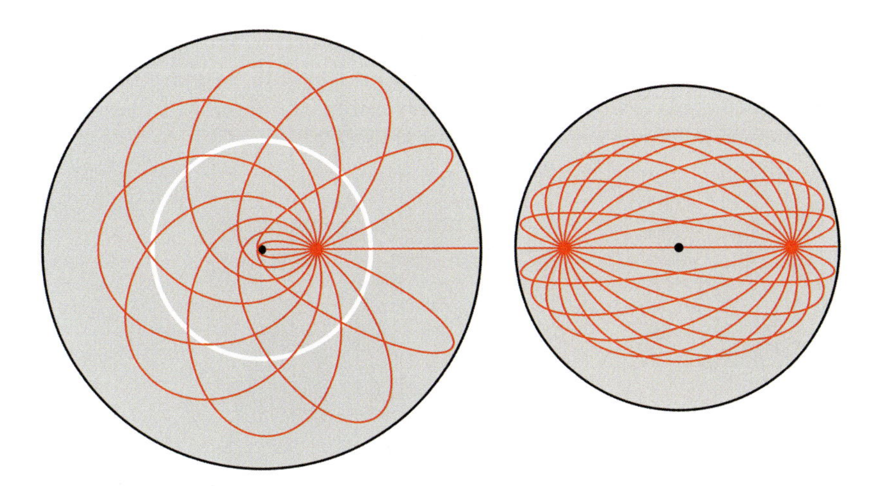

Figure 2.21: Transmutation. The right picture shows trajectories (red curves) in the Hooke potential of a harmonic oscillator. The trajectories are ellipses centred at the origin (dot). In the left picture Hooke's ellipses are transformed into Newton's by the conformal map (6.3). The trajectories (red curves) are still ellipses, but with one of their focal points at the centre (dot). The white circle indicates the unit circle (in our units). The left picture shows that each Newton ellipse crosses the unit circle twice in parallel directions. This property forms the basis of the Eaton lens (Fig. 2.23).

the recipe (6.2) produces the Hooke potential (3.15) of the harmonic oscillator:

$$U(z) = 2|z|^2 , \tag{6.5}$$

apart from an overall factor of 4 that is of no consequence to the trajectories if we multiply the energy (3.15) by 4 as well (see the discussion on p. 21):

$$E = 4 . \tag{6.6}$$

According to the recipe (6.2), the quadratic function (6.3) and the energy (6.6) produces the transmuted potential

$$U'(w) = -\frac{1}{|w|} , \tag{6.7}$$

the potential for Newton's inverse–square force of gravitational attraction. In optics, the potential (6.7) with the energy (6.4) constitutes the refractive–index profile (with n' denoted by n and $|w|$ by r):

$$n = \sqrt{2/r - 1} . \tag{6.8}$$

(Note that the profile (6.8) is only valid for $r \leq 2$ where n is real.)

Newton's and Hooke's potential are transmutations of each other. Ironically, Sir Isaac Newton and Robert Hooke were scientific rivals and bitter enemies, but, as we have seen, their rivaling laws of attraction turn out to be two sides of the same coin. Even more ironically, in 1987 the Bank of England issued a one–pound note in commemoration of Newton's 1687 *Principia*, but the family of planetary ellipses it shows are mostly Hooke's, not Newton's (Fig. 2.22).

Figure 2.22: One–pound note in honor of Sir Isaac Newton issued by the Bank of England. It shows planets orbiting the Sun in ellipses, but on the note the Sun appears in the centre of most ellipses, not in their focal points. In other words, the ellipses on the note are mostly those of Newton's archrival, Robert Hooke.

Yet the Bank's blunder is easily rectified by transmutation, as the following calculation proves. We know that trajectories in Hooke's profile (3.14) are ellipses (3.17) centred at the centre of attraction at $r = 0$. The transformation (6.3) of the equation (3.17) of Hooke's ellipses gives

$$w = \mathrm{e}^{2\mathrm{i}\alpha}\left(\frac{a'+b'}{2}\,\mathrm{e}^{2\mathrm{i}\xi} + \frac{a'-b'}{2}\,\mathrm{e}^{-2\mathrm{i}\xi} + f'\right) \tag{6.9}$$

with the new constants

$$a'+b' = \frac{(a+b)^2}{2}\,, \quad a'-b' = \frac{(a-b)^2}{2}\,, \quad f' = \frac{a^2-b^2}{2}\,. \tag{6.10}$$

Formula (6.9) is still the equation of an ellipse, but the new ellipse is shifted by f'. Let us work out how the constants (6.10) are connected and constrained. Expressing $a^2 - b^2$ as $(a+b)(a-b)$ we get

$$f'^2 = a'^2 - b'^2\,. \tag{6.11}$$

We see from Eq. (3.19) that f' is the focal length of the transformed ellipse. Furthermore, the condition (3.26) for a and b implies for the new constants (6.10):

$$\begin{aligned} a' &= \frac{(a+b)^2}{4} + \frac{(a-b)^2}{4} = \frac{a^2+b^2}{2} = 1\,, \\ b' &= \frac{(a+b)^2}{4} - \frac{(a-b)^2}{4} = ab = \sin(2\gamma) \end{aligned} \tag{6.12}$$

such that

$$w = e^{2i\alpha}\left[\cos(2\xi) + i\sin(2\gamma)\sin(2\xi) + \cos(2\gamma)\right] . \tag{6.13}$$

This equation describes displaced ellipses rotated by the angle 2α (Fig. 2.21). Both in Hooke's and in Newton's case the trajectories are ellipses, a fact Newton stated as remarkable in his 1687 *Principia*. But in Newton's case the ellipses are not centred at the origin; the centre of attraction lies at one of the focal points. Note that the recipe (6.2) for the various possible energies E' gives all the curves of planetary motion by relating them to trajectories in the harmonic–oscillator potential for negative E' and to trajectories in the inverted harmonic–oscillator potential for positive E': the circle, ellipses, parabolas and the hyperbolas of comets.

We discussed the transmutation of potentials where the spatial trajectories are related to each other by conformal transformations. Note that we also need to transform the parameter ξ if we wish the transmuted trajectory to obey Newton's equation of motion (3.6). The problem is that $dw/d\xi$ is not necessarily the correct velocity in Newton's law (3.6). We must re–parametrize the curve by introducing a new ξ' such that the modulus of the velocity, $|dw/d\xi'|$, agrees with the Newtonian speed (3.5) given by the transformed refractive index (5.9). The transformation

$$d\xi' = \left|\frac{dw}{dz}\right|^2 d\xi \tag{6.14}$$

does the trick, because

$$\left|\frac{dw}{d\xi'}\right| = \left|\frac{dz}{dw}\right|^2 \left|\frac{dw}{d\xi}\right| = \left|\frac{dz}{dw}\right|\left|\frac{dz}{d\xi}\right| = \left|\frac{dz}{dw}\right| n = n' , \tag{6.15}$$

where we substituted for the Newtonian speed (3.5) in z–space $|dz/d\xi| = n$ and used the transformation (5.9) of the refractive index in the last step. The new parameter ξ' is the integral of $d\xi'$ given by relation (6.14) along the trajectory.

Similar to the Luneburg lens introduced in §3, the Newton profile (6.8) has remarkable optical properties; it constitutes the *Eaton lens* (Fig. 2.23) (Eaton [1952]). Consider a spherically symmetric device with radius $r = 1$, having the Newtonian refractive–index profile (6.8) for $r \leq 1$. At the boundary of the lens, $r = 1$, the profile (6.8) agrees with the refractive index of empty space, $n = 1$, so light is not refracted at the boundary. Consider a family of light rays incident at the angle ϕ_0 to the x–axis. Like in our discussion of the Luneburg lens, we match the incident rays with sections of the Newton ellipses inside the device. Figure 2.21 suggests that the trajectory segments inside the unit circle are half ellipses. In the transmuted equation (6.13) for the Newton ellipses, 2ξ is the parameter of the trajectory, so half ellipses range from $-\pi/4$ to $+\pi/4$. We obtain at

$$\xi_\pm = \pm\frac{\pi}{4}, \quad \alpha = \frac{\phi_0}{2} \tag{6.16}$$

for the ellipses (6.13)

$$w(\xi_\pm) = w_\pm = e^{i(\phi_0 \pm 2\gamma)}, \quad \left.\frac{dw}{d\xi}\right|_\pm = \mp 2\, e^{i\phi_0} . \tag{6.17}$$

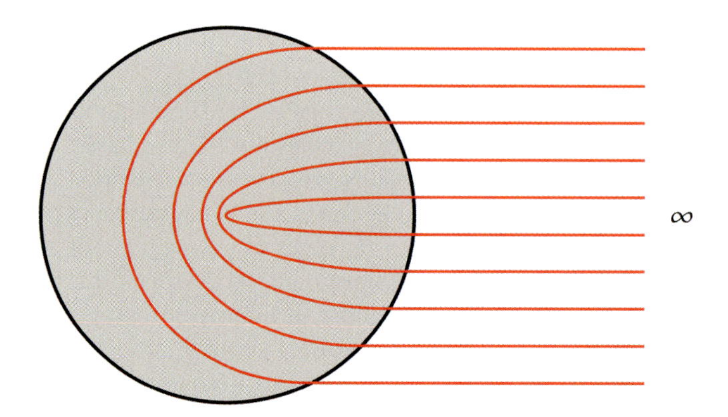

Figure 2.23: Eaton lens. Light rays coming from ∞ enter the refractive–index profile (6.8) for $r \leq 1$. The rays leave in the same direction they came from; they are retroreflected. The Eaton lens makes a perfect cat's eye.

As $|w_\pm|^2 = 1$, we see that the half ellipses indeed begin and end at the unit circle. Furthermore, (6.14) shows that $\mathrm{d}w/\mathrm{d}\xi' = |\mathrm{d}z/\mathrm{d}w|^2 \, \mathrm{d}w/\mathrm{d}\xi$, so the Newtonian velocity $\mathrm{d}w/\mathrm{d}\xi'$ (with transformed parameter) points in the direction of $\mathrm{d}w/\mathrm{d}\xi$ and hence in the ϕ_0 direction for ξ_-, the direction of the incident rays. The half ellipses thus match the incoming rays. After a half ellipse, when ξ reaches ξ_+, the angle of the trajectory is $\phi_0 + \pi$. Consequently, the incident light rays enter the device at w_-, perform a half ellipse with focal points at the centre of the Eaton lens, and finally leave at w_+ in precisely the opposite direction they came from. Light is retroreflected, regardless of direction: the Eaton lens makes an omnidirectional retroreflector (Fig. 2.23).

Exercise 6.1
Imagine an inside–out Eaton lens where you surround the space inside the unit sphere with a optical material of the Newton profile (6.8) for $1 \leq r \leq 2$ and $n = 1$ for $r < 1$. Figure 2.24 shows that this device is also a retroreflector. Why is that?

Exercise 6.2
Figure 2.24 also shows that all light rays emitted from one point z_0 are focused at the image $-z_0$ of this point. Argue why this has to be so.
Solution
Hint: use the mirror symmetry of the ray trajectories.

---

§7. Spherical symmetry

Luneburg and Eaton lenses are omnidirectional, because they are spherically symmetric. In this Section we study light rays in general spherically–symmetric index profiles or, equivalently, Newtonian particles in central potentials.

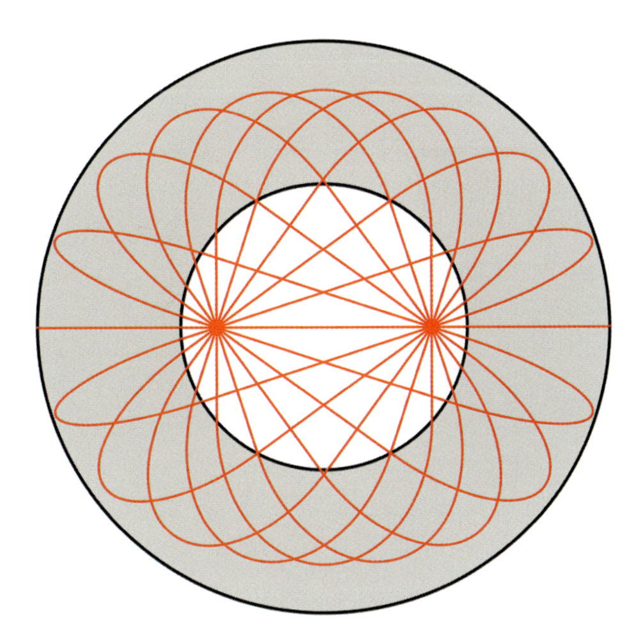

Figure 2.24: Inside–out Eaton lens. Light rays (red) in a device with Newton profile (6.8) (grey) for $1 \leq r \leq 2$ that encloses empty space inside the unit sphere.

Consider a spherically symmetric refractive–index profile $n(r)$. We know from the conservation of angular momentum (3.12) that each ray trajectory lies in a plane. Rays with the same direction of angular momentum (3.11) lie in the same plane. Imagine a family of such rays. As before, we orient the Cartesian coordinates such that x and y span the plane of trajectories, and we combine x and y in complex numbers (5.1). As the angular momentum vector (3.11) points orthogonal to this plane, the only non–vanishing component of $\boldsymbol{L}$ is

$$L_z = x\frac{\mathrm{d}y}{\mathrm{d}\xi} - y\frac{\mathrm{d}x}{\mathrm{d}\xi} = \frac{1}{2\mathrm{i}}\left(z^*\frac{\mathrm{d}z}{\mathrm{d}\xi} - z\frac{\mathrm{d}z^*}{\mathrm{d}\xi}\right). \tag{7.1}$$

Exercise 7.1
Verify the complex representation (7.1) of the angular momentum.

Let us apply ideas from optical conformal mapping to deduce some general features of light rays in spherically symmetric media. It is advantageous to transform the trajectories in physical space by the exponential map (Fig. 2.25)

$$z = \mathrm{e}^w\,, \quad w = u + \mathrm{i}v = \ln z\,. \tag{7.2}$$

Representing z in polar form $z = r\exp(\mathrm{i}\phi)$ we see that $u = \ln r$ and $v = \phi$:

$$w = \ln r + \mathrm{i}\phi\,. \tag{7.3}$$

The complex logarithm thus transforms the Cartesian coordinates to polar coordinates with logarithmic radius.

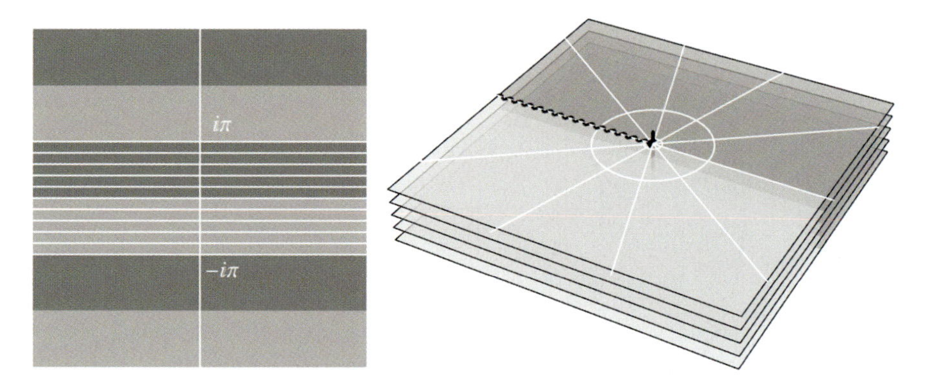

Figure 2.25: Exponential map. Physical space (right), the z plane, is represented in new coordinates (left), the w plane, by the exponential map (7.2). The new coordinates (white lines) are the polar coordinates (7.3) with logarithmic radius. As we can add any integer multiple of 2π to the angle without changing the physical coordinates, we can also add integer multiples of $2\pi\mathrm{i}$ to w without changing z: the stripes (left) in the w coordinates are indistinguishable in the z plane. In the language of conformal mapping, we represent z space by an infinite stack of Riemann sheets. Note that here physical space appears as a Riemann surface, not virtual space. The Riemann surface serves as a convenient mathematical construction for describing periodic motion and related phenomena.

We obtain from transformation rule (5.9) of optical conformal mapping

$$n' = \left|\frac{\mathrm{d}z}{\mathrm{d}w}\right| n = r\, n(r)\,. \tag{7.4}$$

Exercise 7.2
Verify expression (7.4) for the exponential map (7.2).

Hence in transformed space the potential (3.7) is independent of one of the Cartesian coordinates, the imaginary part v of w that describes the angle ϕ in physical space. Consider trajectories in transformed space. Here the right–hand side of Newton's law (3.6) vanishes for the ϕ component; no force is acting on the angular motion. The angle thus sweeps away at the constant rate

$$b = \frac{\mathrm{d}\phi}{\mathrm{d}\xi'}\,. \tag{7.5}$$

We assume b to be non–negative, because $-b$, where ϕ runs backwards, would simply correspond to the mirror image of the trajectory with positive b; we would not get anything new in this case. We have seen that the angle grows at a constant rate in

ξ'. However, ξ' is not the Newtonian evolution parameter in physical space, but is the transformed parameter (6.14) for the exponential map (7.2) with

$$d\xi' = \left|\frac{dw}{dz}\right|^2 d\xi = \frac{d\xi}{r^2} \tag{7.6}$$

so that from Eq. (7.5) follows

$$b = r^2 \frac{d\phi}{d\xi}. \tag{7.7}$$

This equation has a geometrical meaning that Johannes Kepler discovered for planetary motion in the 1610's. In polar coordinates, the area element dA is given by an infinitesimal square of sides dr and $r d\phi$,

$$dA = r\, dr\, d\phi. \tag{7.8}$$

Consider the area A swept by a line drawn from the origin to the trajectory (Fig. 2.26),

$$A = \int \int_0^{r(\phi)} r\, dr\, d\phi = \int \frac{r^2}{2}\, d\phi. \tag{7.9}$$

Hence from Eq. (7.7) we obtain *Kepler's second law* of motion,

$$A = \frac{b}{2}\xi, \tag{7.10}$$

equal areas are covered in equal times (equal evolution parameters ξ). Therefore, the rate of angular motion with respect to the transformed parameter (7.5) determines the increase in area (7.10) in Newtonian evolution.

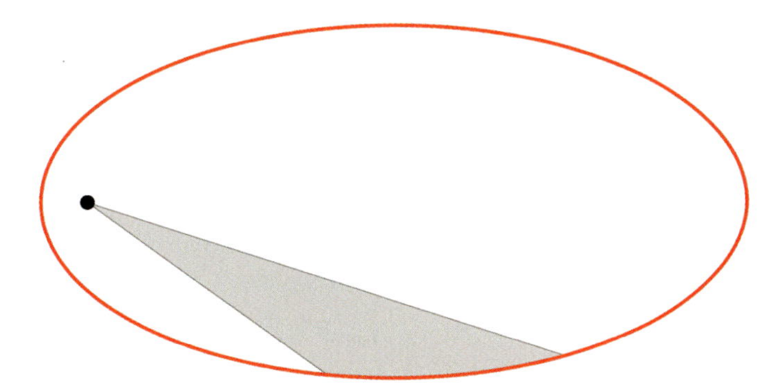

Figure 2.26: Kepler's area law. The area (grey) drawn from the moving object on the trajectory (red) to the centre of attraction (dot) grows proportionally to Newtonian time. Equal areas are covered in equal times.

But b has yet another, third, physical meaning: it is the angular–momentum component L_z. To see this, we transform the z–representation (7.1) of the angular momentum to w–space using the exponential map (7.2) and representation (7.3) and compare the result with relation (7.7):

$$L_z = \frac{1}{2\mathrm{i}}\left(z^* \frac{\mathrm{d}z}{\mathrm{d}w}\frac{\mathrm{d}w}{\mathrm{d}\xi} - z\frac{\mathrm{d}z^*}{\mathrm{d}w^*}\frac{\mathrm{d}w^*}{\mathrm{d}\xi}\right) = |z|^2 \frac{\mathrm{d}(w-w^*)}{2\mathrm{i}\,\mathrm{d}\xi} = |z|^2\frac{\mathrm{d}\phi}{\mathrm{d}\xi} = b\,. \tag{7.11}$$

Suppose the refractive–index profile $n(r)$ tends to 1 sufficiently fast for $r \to \infty$ such that asymptotic light rays are straight. Rays incident from ∞ along straight lines are bent in the optical medium and leave as straight lines in different directions, unless they are fatally attracted and absorbed in the material. In classical mechanics, the deflection of a beam is called *scattering*. In optics, scattering often implies multiple scattering, for example, Rayleigh scattering that creates the blue colour of the sky, and has connotations of light getting lost or becoming incoherent. Here we rather adopt the terminology of mechanics for scattering. The scattering angle depends on the speed and the offset of the beam to the centre of force, called the *impact parameter*. In optics, the speed (3.5) is given by the refractive index, and the impact parameter turns out to be b, as we now show. Assume that the incident ray initially travels on a straight line with direction ϕ_0 and offset b_0. We describe this straight line in the complex plane by the equation

$$z \sim \mathrm{e}^{\mathrm{i}\phi_0}(-\xi + \mathrm{i}b_0)\,. \tag{7.12}$$

Exercise 7.3
Why is (7.12) the equation for a ray incident with impact parameter b_0 at the angle ϕ_0?
Solution
For $\phi_0 = 0$, Eq. (7.12) describes a straight line parallel to the x–axis with offset b_0. As the Newtonian velocity $\mathrm{d}z/\mathrm{d}\xi$ is -1, the particle is incident from the right and moves to the left. For $\phi_0 \neq 0$, this trajectory is rotated by the angle ϕ_0.
Exercise 7.4
Calculate the angular momentum (7.1) for the straight trajectory (7.12).

Inserting expression (7.12) into formula (7.1) reveals that the impact parameter b_0 is the angular momentum in our units,* $b_0 = b$.

Having seen the various roles and disguises of the angular momentum, we turn to the ray trajectories. As the angle ϕ, the imaginary part v in w space, increases at the constant rate b in transformed time ξ' according to relation (7.5), we obtain from the Newtonian speed (3.5) in transformed space

$$n'^2 = \left(\frac{\mathrm{d}u}{\mathrm{d}\xi'}\right)^2 + \left(\frac{\mathrm{d}v}{\mathrm{d}\xi'}\right)^2 = \left(\frac{\mathrm{d}u}{\mathrm{d}\xi'}\right)^2 + b^2 = b^2\left(\frac{\mathrm{d}u}{\mathrm{d}v}\right)^2 + b^2\,, \tag{7.13}$$

where we applied $\mathrm{d}v = \mathrm{d}\phi = b\,\mathrm{d}\xi'$ in the last step, using relation (7.5) once more. We solve for $\mathrm{d}v$ and integrate

$$v = \pm b\int\frac{\mathrm{d}u}{\sqrt{n'^2 - b^2}}\,. \tag{7.14}$$

*In classical mechanics, the angular momentum is proportional to the impact parameter, but differs by a factor of mass and the speed at $r \to \infty$.

Equation (7.14) describes the trajectory in transformed space, because v appears as a function of u that is given by the integral on the right–hand side. From the exponential map (7.2) with the geometrical interpretation (7.3), and the relation (7.4), we immediately get the trajectory in physical space,

$$\phi = \pm \int \frac{(b/r)\,\mathrm{d}r}{\sqrt{\rho^2 - b^2}} \tag{7.15}$$

where ρ denotes

$$\rho = n' = nr\,. \tag{7.16}$$

Exercise 7.5
Verify Eqs. (7.15) and (7.16).

For example, for the Newton profile (6.8) we obtain

$$\phi = \int \frac{(b/r)\,\mathrm{d}r}{\sqrt{\rho^2 - b^2}} = \int \frac{(b/r)\,\mathrm{d}r}{\sqrt{2r - r^2 - b^2}} = \arccos\left(\frac{b^2/r - 1}{\sqrt{1-b^2}}\right) \tag{7.17}$$

and hence

$$\cos(\phi - \phi_0) = \frac{b^2/r - 1}{\sqrt{1-b^2}}\,. \tag{7.18}$$

The angular momentum is restricted to the range $b^2 \leq 1$, because otherwise the square root in Eqs. (7.17) and (7.18) would be imaginary.

Exercise 7.6
Check the result (7.18) by differentiating the right–hand side of integral (7.17) and transforming it to the expression for the integrand of ϕ.
Exercise 7.7
Show that formula (7.18) describes the Newton ellipses (6.13) with $b = b' = \sin(2\gamma)$ and $\phi_0 = 2\alpha$. The angle ϕ_0 simply rotates the curve drawn by Eq. (7.18). You may thus identify ϕ_0 with the rotation angle 2α. It is sufficient to show that, for $\phi_0 = 0$, formula (7.18) agrees with the equation of an ellipse with one focal point (3.19) at the origin:

$$(x+f)^2 + \frac{y^2}{b^2} = 1\,, \quad f = \sqrt{1-b^2}\,. \tag{7.19}$$

Solution
Equation (7.18) and $x = r\cos\phi$ gives $r = b^2 - x\sqrt{1-b^2}$. Consequently,

$$x^2 + y^2 = r^2 = b^4 - 2b^2\sqrt{1-b^2}\,x + (1-b^2)x^2\,, \tag{7.20}$$

from which follows the equation (7.19) of the ellipse.

Note that the radial motion is limited in general. The argument of the square root in integral (7.15) for the angle must not be negative, for ϕ to remain real. Radii r_a where ρ equals b define *radial turning points*,

$$\rho(r_a) = b \quad \text{or, equivalently} \quad n(r_a)r_a = b\,. \tag{7.21}$$

The notion of radial turning points also adds physical meaning to ρ, the refractive index (7.16) in transformed space: $\rho(r)$ describes the angular momentum or impact parameter b for which r is a radial turning point. We call ρ the *turning parameter.*

Maximally two radii can limit a given trajectory, the inner and the outer radius r_- and r_+ with $r_- \leq r_+$. In this case, the trajectory lies in a ring (Fig. 2.27). If the $r_\pm$ coincide, the trajectory is reduced to a circle. In general the ray draws a rosetta (Fig. 2.27), covering the entire ring over time, unless the trajectories are closed. To determine the condition for closed trajectories, consider the angle between two turning points:

$$\Delta = \phi(r_+) - \phi(r_-) \,. \tag{7.22}$$

The trajectory is a closed curve if the angle Δ is exactly a rational fraction p/q of π with integer p and q:

$$\text{closed loop} \quad \Longleftrightarrow \quad \Delta = \frac{p}{q}\pi \,, \tag{7.23}$$

for the following reason. Suppose the condition (7.23) holds. After q loops from r_- to r_+ and back to r_- the angle ϕ reaches $2p\pi$, an integer multiple of 2π: the trajectory is closed. For example, the Hooke and Newton ellipses are closed, with $\Delta = \pi/2$ in Hooke's case and $\Delta = \pi$ in Newton's. Figure 2.24 shows another example, the inside–out Eaton lens.

In optics, closed ray trajectories have a remarkable property: to perform a closed loop in a given profile $n(r)$ always takes the same physical time t (Leonhardt [2006b]).

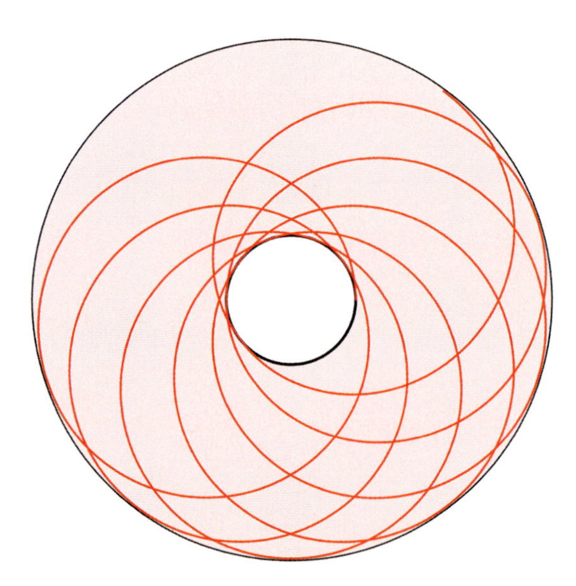

Figure 2.27: Rosetta between inner and outer radial turning point. The trajectory (red curve) lies in a ring between the radial turning points. (The figure shows a ray trajectory in the transmuted Maxwell fish–eye (9.20) with $\nu = 1.1$.)

To be precise, the time t in Hamilton's equations (4.10) between two turning points turns out to be independent of the angular momentum b. (Note that the delay with respect to the Newtonian evolution parameter ξ is not uniform in general.) To prove this property of closed loops, we use the integral (7.15) to express the angle Δ between two turning points

$$\Delta = \int_{r_-}^{r_+} \frac{(b/r)\,\mathrm{d}r}{\sqrt{\rho^2 - b^2}} . \tag{7.24}$$

We took the positive sign in the integral (7.15), because for positive angular momentum b the angle ϕ increases from the inner to the outer turning point, from r_- to r_+. Then we represent the Hamiltonian time increment $\mathrm{d}t$ by the Newtonian increment (4.13) and the latter by the angle (7.7):

$$c\,\mathrm{d}t = n^2 \mathrm{d}\xi = \frac{n^2 r^2}{b}\,\mathrm{d}\phi\,. \tag{7.25}$$

For $\mathrm{d}\phi$ we use the integrand of formula (7.15) with positive sign and replace ρ by nr according to relation (7.16). We integrate between the two turning points and obtain:

$$\begin{aligned} ct &= \int_{r_-}^{r_+} \frac{n^2 r\,\mathrm{d}r}{\sqrt{n^2r^2 - b^2}} = \int_{r_-}^{r_+} \frac{n^2r^2 - b^2 + b^2}{\sqrt{n^2r^2 - b^2}} \frac{\mathrm{d}r}{r} \\ &= \int_{r_-(b)}^{r_+(b)} \sqrt{n^2r^2 - b^2}\,\frac{\mathrm{d}r}{r} + b\,\Delta\,, \end{aligned} \tag{7.26}$$

where we used the integral (7.24) for Δ in the last step. We differentiate ct with respect to b and get from Eqs. (7.21) and (7.24)

$$\begin{aligned} c\,\frac{\partial t}{\partial b} &= \frac{\sqrt{n^2r^2 - b^2}}{r} \frac{\partial r}{\partial b}\bigg|_{r_-}^{r_+} - \int_{r_-}^{r_+} \frac{b}{\sqrt{\rho^2 - b^2}} \frac{\mathrm{d}r}{r} + \Delta + b\frac{\partial \Delta}{\partial b} \\ &= b\,\frac{\partial \Delta}{\partial b}\,. \end{aligned} \tag{7.27}$$

Yet for closed loops (7.23) the angle difference Δ between two turning points is constant. Consequently,

$$\frac{\partial t}{\partial b} = 0\,, \tag{7.28}$$

the time delay is uniform. If we employ profiles with closed loops in optical conformal mapping the time delay in virtual space is the same as the time delay in physical space, because only space is transformed, not physical time. (The Newtonian parameter (6.14) is transformed however, as we have seen in the transmutations of potentials.) As we discussed in §5, such closed loops in virtual space are needed in invisibility devices. Cloaking causes uniform time delays.

Consider now light rays incident from ∞ with positive impact parameter b. As they approach, the angle ϕ increases according to Eq. (7.7), but r decreases. Therefore we must first take the negative branch of integral (7.15) until r reaches the radial turning point r_0 at the closest distance to the centre of force. In empty space, the closest distance of a straight line is equal to the impact parameter—light rays in empty space thus have radial turning points as well. Having reached the closest distance to the origin, the ray leaves, taking then the positive branch of integral (7.15). The incoming and the outgoing halves of the trajectory are mirror–symmetric with respect to the axis of the closest distance (Fig. 2.28). This description holds for both attractive and repulsive profiles; Fig. 2.28 illustrates a case of attraction.

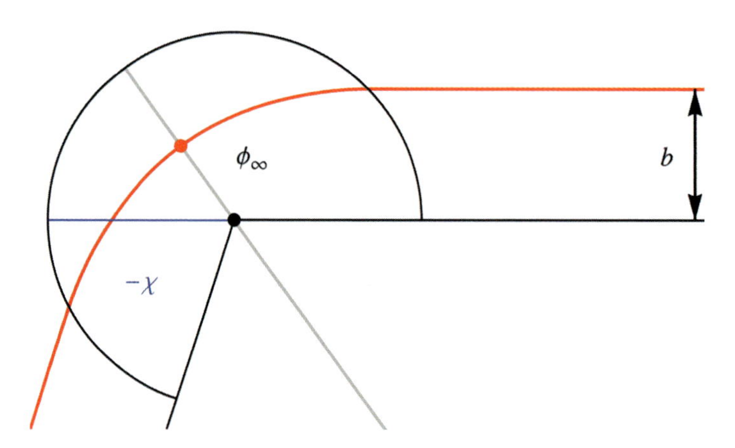

Figure 2.28: Scattering. A trajectory (red curve) is incident from ∞ with impact parameter b. The curve approaches its closest distance to the origin at the radial turning point (red dot) and leaves to ∞. The curve is symmetric around the axis of the radial turning point (grey line), because the potential (or refractive–index profile) is radially symmetric. At the turning point the trajectory has covered the angle ϕ_∞; so the final angle is $2\phi_\infty$. The scattering angle χ measures the deviation of the trajectory from the direction of incidence (blue line); χ describes the deviation (8.1) of the total angle $2\phi_\infty$ from π.

In scattering, the ray trajectory has only one radial turning point. In some cases, however, the rays may not have any radial turning points at all. In such a case, nothing stops the light ray from falling into the centre of attraction (Fig. 2.29). For this to happen, ρ must approach a non–zero constant or diverge for $r \to 0$ such that the condition (7.21) for a radial turning point is never satisfied; this implies according to definition (7.16) that the refractive index must diverge at least like r^{-1} and the potential (3.7) like r^{-2} for $r \to 0$. Note that such cases of fatal attraction do not make optical analogues of black holes. Although the incident light disappears in the singularity at the origin, we could always send light back in opposite direction from any point on the incident trajectory. Such media do not establish horizons; they are naked singularities (forbidden by Cosmic Censorship in general relativity (Penrose [2004])). The situation is fundamentally different in moving media. If a medium moves faster than the speed of light in the material, light is trapped behind

a horizon. The horizon may exist without a singularity (something forbidden in general relativity as well (Penrose [2004])). Optical black holes, their horizons and Hawking radiation are discussed in the final Section of this book.

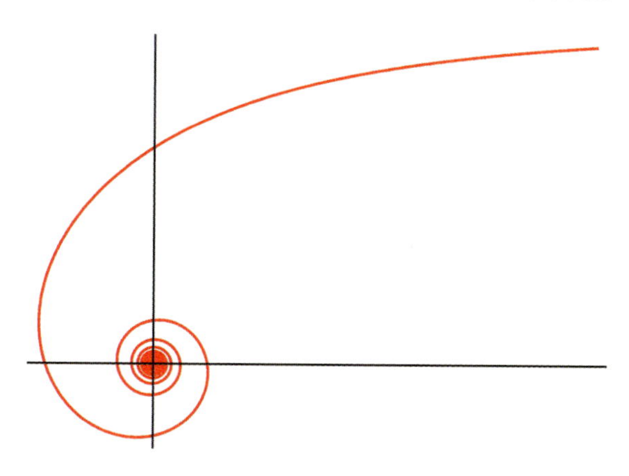

Figure 2.29: Spiral of fatal attraction. Trajectories (red) may fall into the centre of attraction when the potential grows at least like r^{-2} (and the refractive index like r^{-1}).

Exercise 7.8

Consider the potential of fatal attraction, $U = -1/(2r^2)$. In this case the Newtonian equation of motion reads, in complex notation,

$$\frac{\mathrm{d}^2 z}{\mathrm{d}\xi^2} = -\frac{z}{|z|^4}\,. \tag{7.29}$$

Show that the solution of the equation of motion is

$$z = z_+ z_- \quad \text{with} \quad z_\pm = w_\pm^{(1\pm\mu)/2}, \quad w_\pm = -\xi \pm \mathrm{i}\frac{b}{\mu}, \quad \frac{1}{\mu^2} = 1 - \frac{1}{b^2} \tag{7.30}$$

for real b. Note that μ is real for $b^2 > 1$ and purely imaginary for $b^2 < 1$.

Solution

We calculate $|z|^4$ and $\mathrm{d}^2 z/\mathrm{d}\xi^2$ that occur in the differential equation (7.29). It is wise to express $|z|^4$ in terms of $w_\pm$, with the result

$$|z|^4 = |w_+ w_-|^2 = (\xi^2 + b^2\mu^{-2})^2 = w_+^2 w_-^2 \tag{7.31}$$

for both real and imaginary μ. We calculate the second derivative of z according to the Leibnitz rule, $\mathrm{d}^2(z_+ z_-) = (\mathrm{d}^2 z_+) z_- + 2(\mathrm{d}z_+)(\mathrm{d}z_-) + z_+(\mathrm{d}^2 z_-)$, use

$$\frac{\mathrm{d}z_\pm}{\mathrm{d}\xi} = -\frac{1\pm\mu}{2}\,\frac{z_\pm}{w_\pm}, \quad \frac{\mathrm{d}^2 z_\pm}{\mathrm{d}\xi^2} = -\frac{(1\pm\mu)(-1\pm\mu)}{4}\,\frac{z_\pm}{w_\pm^2} \tag{7.32}$$

for the expressions (7.30) and thus obtain:

$$\begin{aligned} \frac{\mathrm{d}^2 z}{\mathrm{d}\xi^2} &= \frac{(\mu+1)(\mu-1)\,z}{4w_+^2} + \frac{(\mu+1)(1-\mu)\,z}{2w_+ w_-} + \frac{(1-\mu)(-\mu-1)\,z}{4w_-^2} \\ &= \frac{\mu^2-1}{4}\,\frac{(w_+ - w_-)^2 z}{w_+^2 w_-^2} = -(\mu^2-1)\frac{b^2}{\mu^2}\,\frac{z}{|z|^4} = -\frac{z}{|z|^4} \end{aligned} \tag{7.33}$$

where the explicit expressions (7.30) for $w_\pm$ and μ are applied in the last line. This proves that formula (7.30) describes the solution of the differential equation (7.29) (Leonhardt and Piwnicki [2001]).

Exercise 7.9

The notation of the parameter b in the solution (7.30) suggests that b is the impact parameter. Show that this is indeed the case by comparing the solution in the limit $\xi \to -\infty$ with Eq. (7.12) that describes the straight line of the incoming trajectory. Argue why the trajectory of exercise 7.8 falls into the centre of attraction for $|b| < 1$ (when it aims too close). Show that, in the limiting case $b \to 1$, the solution for the incoming trajectory with $\xi < 0$ approaches the formula of a logarithmic spiral in complex representation

$$z = -\xi \exp(-\mathrm{i}/\xi)\,. \tag{7.34}$$

Solution

Combining the terms of the solution (7.30) in one formula we obtain

$$z = \sqrt{\xi^2 + b^2\mu^{-2}} \left(\frac{1 - \mathrm{i}b/(\mu\xi)}{1 + \mathrm{i}b/(\mu\xi)}\right)^{\mu/2} \sim |\xi| \left(1 - \mathrm{i}\frac{b}{\xi}\right), \tag{7.35}$$

replacing the square root by $|\xi|$ and using first–order Taylor expansion in b/ξ for the second term. The asymptotics (7.35) of the solution (7.30) for $\xi \to -\infty$ thus agrees with the equation (7.12) of a straight line, which reveals that b is the impact parameter. For $|b| < 1$ the parameter μ is purely imaginary and so $|z|^2 = \xi^2 + b^2\mu^{-2}$ reaches zero at $\xi_0 = -\sqrt{1 - b^2}$, which proves that the trajectory falls into the origin. For $b \to 1$ and $\xi < 0$ one gets from expressions (7.30) $w_+ w_- = -\xi$, and so

$$z = -\xi \lim_{\mu\to\infty} \frac{w_+^{\mu/2}}{w_-^{\mu/2}} = -\xi\, \frac{\lim\limits_{\mu\to\infty} [1 - \mathrm{i}/(\mu\xi)]^{\mu/2}}{\lim\limits_{\mu\to\infty} [1 + \mathrm{i}/(\mu\xi)]^{\mu/2}} = -\xi\, \frac{\exp[-\mathrm{i}/(2\xi)]}{\exp[+\mathrm{i}/(2\xi)]} \tag{7.36}$$

according to Gauss' limit, which gives the desired result (7.34).

---

§8. Tomography

In this Section we consider spherically symmetric media and develop a visual, tomographic way of reconstructing the refractive–index profile or, equivalently, the potential from scattering data or the angle between the radial turning points.

A spherically symmetric optical medium deflects incident light. Suppose we know the deflection angle for all impact parameters—can we use this information to infer the refractive–index profile $n(r)$? In other words, suppose we know what the device does, can we find out how it does it? An answer to this question would solve a classic design problem: if we know what the device should do, how can we make it happen? In 1944, Rudolf Luneburg delivered a series of lectures on the mathematical theory of optics at Brown University where, among other original ideas, he solved the design problem for spherically symmetric optical media. The lecture notes were later published in his posthumous book (Luneburg [1964]). In 1953, Firsov independently developed the same procedure for the equivalent problem in classical mechanics (Firsov [1953]). Here we cast Luneburg's formula in a visual form (Hendi, Henn and Leonhardt [2006]). We also apply this visualization to another inverse problem: how to deduce $n(r)$ from the angle (7.22) between turning points? The answer to this question will give us a large class of spherically symmetric media with closed loops of light.

Consider ray propagation in a spherically symmetric refractive–index profile $n(r)$. We assume that $n(r) \to 1$ sufficiently fast for $r \to \infty$ such that rays follow asymptotically straight lines at ∞. Ray trajectories with negative impact parameter b are mirror–symmetric to the ones with positive b and so we only need to discuss rays with non–negative b. Consider a light ray incident on $n(r)$ with non–negative impact parameter b. The ray comes in from ∞, reaches the radial turning point r_0 and leaves to ∞ with the deflection angle χ (Fig. 2.28):

$$\chi = \pi - 2\phi_\infty = \pi - 2\int_{r_0}^{\infty} \frac{(b/r)\,\mathrm{d}r}{\sqrt{\rho^2 - b^2}} \tag{8.1}$$

according to formula (7.15) with positive sign, because the angle ϕ_∞ is positive. Sufficiently far away from the medium, for $b \to \infty$, the deflection angle χ tends to zero, because $n \to 1$ there. We introduce two new functions that turn out to be related to each other, the integrated scattering angle

$$\Phi = \int_{\infty}^{b} \chi\,\mathrm{d}b \tag{8.2}$$

and the logarithm of the refractive index n regarded as a function of the turning parameter (7.16),

$$W = \ln n(\rho) = \ln(\rho/r) = \ln\rho - \ln r(\rho)\,. \tag{8.3}$$

The turning parameter ρ deviates from the radius r by a factor of the refractive index n, but for large r the index tends to 1 and so ρ approaches r at ∞. Furthermore, as $n \to 1$ sufficiently fast, both W and $\mathrm{d}W/\mathrm{d}\rho$ vanish for $\rho \to \infty$.

In order to see how Φ and W are connected, we use the following trick: in formula (8.1) we express π as a definite integral similar in structure to the integral for ϕ_∞ that occurs there:

$$\begin{aligned}\pi &= 2\lim_{\rho\to b}\arctan\left(\frac{b}{\sqrt{\rho^2-b^2}}\right) = -2\arctan\left(\frac{b}{\sqrt{\rho^2-b^2}}\right)\Bigg|_{\rho\to b}^{\rho\to\infty} \\ &= 2b\int_b^{\infty}\frac{\mathrm{d}\rho}{\rho\sqrt{\rho^2-b^2}}\end{aligned} \tag{8.4}$$

and obtain from Eqs. (8.1) and (8.3)

$$\chi = 2b\int_b^{\infty}\left(\frac{1}{\rho} - \frac{1}{r}\frac{\mathrm{d}r}{\mathrm{d}\rho}\right)\frac{\mathrm{d}\rho}{\sqrt{\rho^2-b^2}} = 2\int_b^{\infty}\frac{\mathrm{d}W}{\mathrm{d}\rho}\frac{b}{a}\,\mathrm{d}\rho \tag{8.5}$$

with the abbreviation

$$a = \sqrt{\rho^2 - b^2}\,, \tag{8.6}$$

or, equivalently,

$$\rho = \sqrt{a^2 + b^2}\,. \tag{8.7}$$

We see that the factor b/a in expression (8.5) can be written as the derivative $-\mathrm{d}a/\mathrm{d}b$. Furthermore, since $a = 0$ for $\rho = b$, we can supplement expression (8.5) by $2a(\mathrm{d}W/\mathrm{d}\rho)$ at $\rho = b$ to get a total derivative:

$$\chi = -2\int_b^{\infty} \frac{\mathrm{d}W}{\mathrm{d}\rho}\frac{\mathrm{d}a}{\mathrm{d}b}\,\mathrm{d}\rho + 2a\left.\frac{\mathrm{d}W}{\mathrm{d}\rho}\right|_{\rho=b} = -2\frac{\mathrm{d}}{\mathrm{d}b}\int_b^{\infty}\frac{\mathrm{d}W}{\mathrm{d}\rho}a\,\mathrm{d}\rho\,. \tag{8.8}$$

By definition (8.2) we obtain Φ by integrating expression (8.8) with respect to b from $b = \infty$ to $b = b$:

$$\Phi = -2\int_b^{\infty}\frac{\mathrm{d}W}{\mathrm{d}\rho}a\,\mathrm{d}\rho\,. \tag{8.9}$$

We then obtain by partial integration:

$$\Phi = 2\int_b^{\infty} W\frac{\mathrm{d}a}{\mathrm{d}\rho}\,\mathrm{d}\rho\,. \tag{8.10}$$

Finally we represent this expression as an integral with respect to a, where a ranges from 0 to ∞ when ρ runs from b to ∞ according to definition (8.6). It is advantageous, however, to admit also negative values of a (according to relation (8.7) ρ depends on $+a$ and $-a$ in the same way) for including the prefactor 2 in the elegant integral

$$\Phi = \int_{-\infty}^{+\infty} W\,\mathrm{d}a\,. \tag{8.11}$$

We give this formula a geometrical meaning if we imagine the turning parameter ρ as the radius (8.7) in a fictitious, two–dimensional space of impact parameters a and b. Instead of only one impact parameter we have two; b is the actual impact parameter defined with respect to the particular direction of incidence and a is a fictitious impact parameter. If we consider a different direction of incidence we can simply rotate the actual impact parameter in the (a, b) plane. Introducing the plane of impact parameters thus seems to be a convenient and natural way to include the rotational symmetry of our scattering problem. In an actual ray trajectory, only one direction in impact–parameter space becomes apparent, the other one is hidden. Formula (8.11) says that the integrated deflection angle (8.2) appears as the projection of the logarithmic profile (8.3) along the hidden parameter, similar to the image of a semi–transparent object with opacity W projected to a screen: Φ is the "shadow" of W (Fig. 2.30).

Tomography, from the Greek word τομος = slice, is a method to infer the shape and structure of a hidden object from its shadows (projections) under various angles. Our case is called *scattering tomography* where we reconstruct W from scattering data. For asymmetric objects, we need projections from all directions for complete tomographic reconstruction. For rotationally symmetric objects, all projections are identical, and the reconstruction formula turns out to be rather simple (Luneburg [1964]):

$$W = -\frac{1}{\pi}\int_{\rho}^{\infty}\frac{\chi\,\mathrm{d}b}{\sqrt{b^2-\rho^2}}\,. \tag{8.12}$$

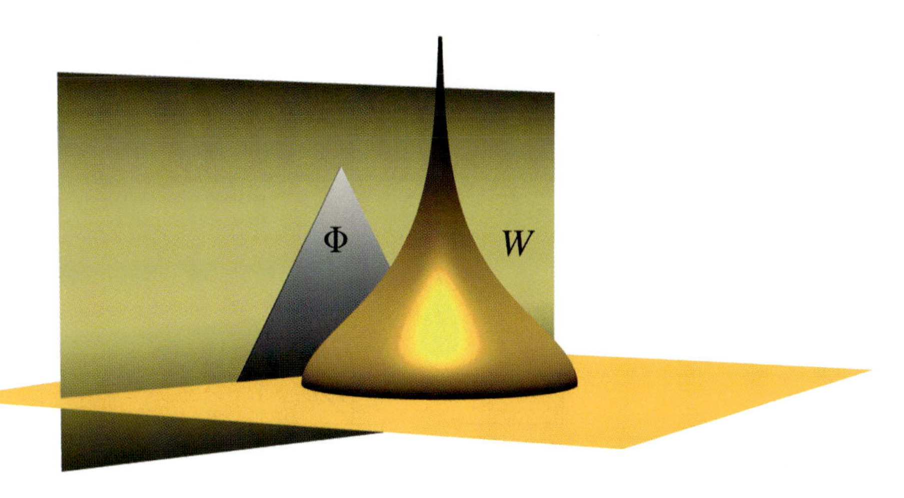

Figure 2.30: Tomography. The picture illustrates how an object with opacity W in the (a, b)–plane casts a shadow with profile Φ. In tomography, the density profile of the object is reconstructed from its shadows (projections) cast from different directions. We consider a case when the object is rotationally symmetric, where the shadow does not depend on direction.

Relations (8.5) and (8.12) are known as *Abel transformations*. To prove Luneburg's reconstruction formula (8.12) we substitute expression (8.5) with definition (8.6) for the deflection angle χ with ρ replaced by the variable σ and exchange the order of integration:

$$\begin{aligned} \int_\rho^\infty \frac{\chi\,\mathrm{d}b}{\sqrt{b^2-\rho^2}} &= \int_\rho^\infty \frac{2b}{\sqrt{b^2-\rho^2}} \int_b^\infty \frac{\mathrm{d}W}{\mathrm{d}\sigma}\frac{\mathrm{d}\sigma}{\sqrt{\sigma^2-b^2}}\,\mathrm{d}b \\ &= \int_\rho^\infty \frac{\mathrm{d}W}{\mathrm{d}\sigma} \int_\rho^\sigma \frac{2b\,\mathrm{d}b}{\sqrt{(b^2-\rho^2)(\sigma^2-b^2)}}\,\mathrm{d}\sigma\,. \end{aligned} \tag{8.13}$$

Figure 2.31 explains why the integration limits are replaced the way they are. Finally, from the integral

$$\int_{x_1}^{x_2} \frac{\mathrm{d}x}{\sqrt{(x_2-x)(x-x_1)}} = 2\arctan\sqrt{\frac{x-x_1}{x_2-x}}\Bigg|_{x_1}^{x_2} = \pi \tag{8.14}$$

follows the reconstruction formula (8.12): in the inner integral of expression (8.13) we substitute x for b^2, use integral (8.14), and the remaining integral over $\pi\mathrm{d}W/\mathrm{d}\sigma$ from ρ to ∞ gives $-\pi W(\rho)$, because $W(\infty)$ vanishes. Luneburg's formula is proven.

Given the deflection angles χ for all impact parameters b, we obtain the refractive–index profile from Eqs. (8.3) and (8.12)

$$n = \exp\left(-\frac{1}{\pi}\int_\rho^\infty \frac{\chi\,\mathrm{d}b}{\sqrt{b^2-\rho^2}}\right). \tag{8.15}$$

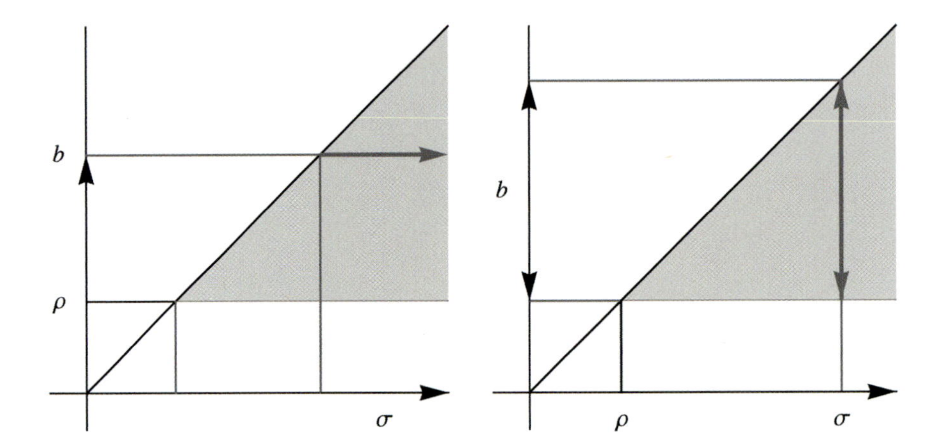

Figure 2.31: Two equivalent representations of the integration area (grey) used in the calculation (8.13). In the left picture b is the outer integration variable that runs from ρ to ∞ and σ is the inner variable that runs from b to ∞. In the right picture the area is covered by letting σ run as outer integration variable with ρ as inner variable. We see that σ runs from ρ to ∞ and b from ρ to σ.

Note, however, that $n(r)$ is given in an indirect, implicit form, as a function of ρ that itself is the function nr of n and r. Alternatively, we express r as a function of ρ,

$$r = \rho \exp\left(\frac{1}{\pi}\int_\rho^\infty \frac{\chi\,\mathrm{d}b}{\sqrt{b^2-\rho^2}}\right) . \tag{8.16}$$

We plot the curve $r(\rho)$ given by this equation in the (r,ρ) plane and read off ρ from r (Fig. 2.32). Numerically, one could store a sufficiently large number of data points for this curve and get $\rho(r)$ by interpolation; $n(r)$ is then given by ρ/r. For some artificial $\chi(b)$, however, $\rho(r)$ is multivalued (Fig. 2.32), which is of course impossible in reality. No real medium could implement such a scattering behavior.

Another type of multivaluedness adds another twist to scattering tomography. The radius surely must be singlevalued, but $r(\rho)$ may be multivalued, as $\rho(r)$ may have the same values for different r (Fig. 2.33). Imagine we plot a three–dimensional representation of $W(\rho)$ over the plane of impact parameters a and b (the right part of Fig. 2.33 shows a cut through W for a fixed angle in the (a,b) plane). The turning parameter ρ is the radius (8.7) and so we may represent a and b as $\rho\cos\phi$ and $\rho\sin\phi$ with polar angle ϕ. According to definition (8.3) we regard W as the function $\ln\rho(r) - \ln r$ and parameterize the surface $\{\rho(r)\cos\phi, \rho(r)\sin\phi, W(r)\}$ by r and ϕ. Suppose $\rho(r)$ has identical values for different intervals of r. There the $\rho(r)$ values are the same, but the $W(r)$ are different; the surface $\{\rho(r)\cos\phi, \rho(r)\sin\phi, W(r)\}$ covers some regions of the impact–parameter plane multiple times (Fig. 2.33). In such a case we have to be careful in understanding the correct meaning of the projection integral (8.11): the integration does not go straight from $-\infty$ to $+\infty$

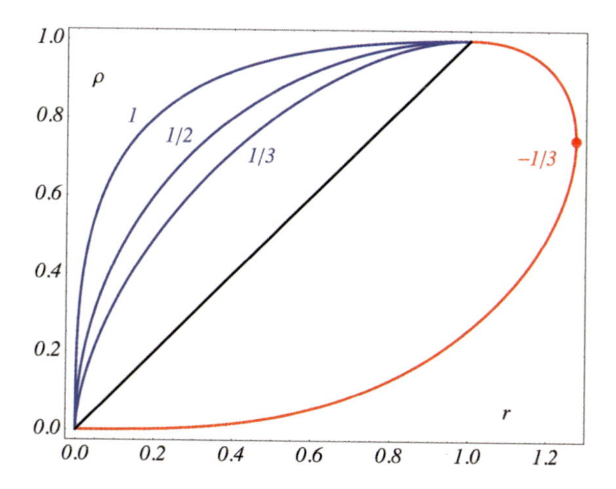

Figure 2.32: Turning parameter versus radius. Graph of the turning parameter ρ as a function of the radius r or, equivalently, r as a function of ρ, for the case (9.3) with different values of $\nu \in \{-1/3,\ 1/3,\ 1/2,\ 1\}$. The function $\rho(r)$ reconstructed from scattering angles is not always singlevalued and so the refractive–index profile that produces the desired scattering data is not always physically possible. For example, we see that $\rho(r)$ is multivalued for negative ν around the maximal value (red dot).

but in a multiple–valued region the integral (8.11) goes back and forth to cover all the branches of ρ. Now imagine we add some function $W_1(\rho)$ to both the top sheet and the sheet below (Fig. 2.33). On one sheet the integral of this function is added, but on the other sheet it is subtracted, because there we integrate in the opposite direction. Hence the extra function $W_1(\rho)$ does not make a net contribution to the integral (8.11); it does not appear in the scattering data. This means that several refractive–index distributions may lead to exactly the same scattering data, the same $\chi(b)$. Scattering is thus ambiguous: different optical media may produce the same image in ray optics.* Luneburg's formula (8.12) gives only one of the possible index profiles.

We can also apply tomography for finding the refractive–index profiles for bound light rays (Fig. 2.34). Suppose we know the angle (7.22) between the inner and the outer radial turning points r_- and r_+ for all relevant angular momenta b. The turning points are given as solutions of Eq. (7.21) for given b. For zero angular momentum the trajectory is a straight line segment through the origin, so $r_- = 0$ in this case. With increasing angular momentum the gap between the inner and the outer tuning points r_- and r_+ decreases until the $r_\pm$ coincide at the maximal angular momentum b_c. As the trajectory is confined between r_- and r_+ (Fig. 2.27) it must form a circle of radius r_c in the limiting case of $b = b_c$. We thus conclude that b ranges from 0 to b_c. Suppose we know the turning angle (7.22) in this range, can we reconstruct the refractive–index profile? We follow a similar procedure as

*However, the inverse scattering problem is uniquely solvable for isotropic optical media in wave optics (Nachman [1988]).

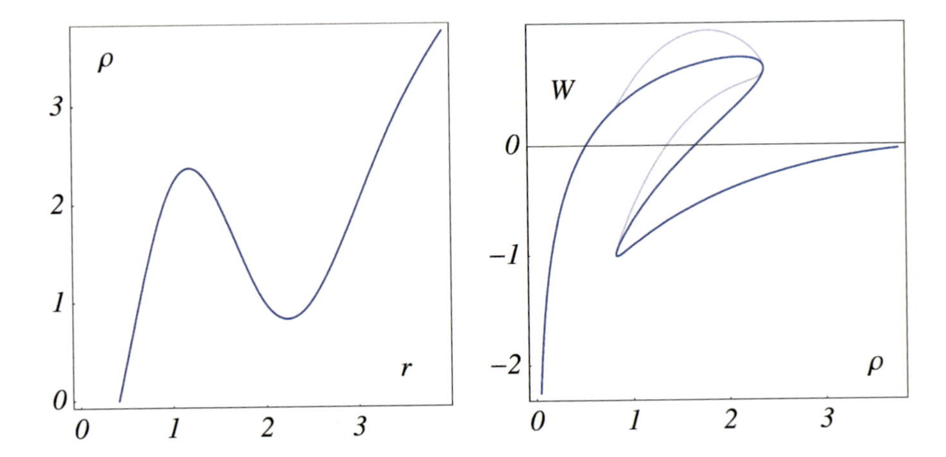

Figure 2.33: Multivalued turning parameter and scattering ambiguity. The left figure shows the curve of $\rho(r)$ with turning parameter ρ according to definition (7.16) in a case where several r give the same ρ. If we read this picture as r depending on ρ the function $r(\rho)$ is multivalued. The right picture shows the scattering function $W(\rho)$ defined by formula (8.3) that (blue curve) corresponds to the turning–parameter profile of the left picture. We could deform $W(\rho)$ by adding a common function $W_1(\rho)$ to two branches of W (light–blue curve) without changing the projection integral (8.11) of W and hence the integrated scattering angle Φ: scattering is ambiguous.

in the derivation of scattering tomography, except that b is restricted by b_c and the integration boundaries of the deflection angle (8.1) are replaced by r_- and r_+. In this way we obtain

$$\ln(\rho/r_+) - \ln(\rho/r_-) = -\frac{1}{\pi}\int_\rho^{b_c} \frac{\chi\,\mathrm{d}b}{\sqrt{b^2-\rho^2}} \tag{8.17}$$

and from Eq. (7.22) and the modified Eq. (8.1) the formula (Demkov and Osherov [1968]):

$$\frac{r_+}{r_-} = \exp\left(-\frac{2}{\pi}\int_\rho^{b_c} \frac{\Delta\,\mathrm{d}b}{\sqrt{b^2-\rho^2}}\right) . \tag{8.18}$$

Exercise 8.1
Derive Demkov's and Osherov's reconstruction formula (8.18). Write down the explicit calculations that we only sketched in words. Follow a similar procedure as in the derivation of relation (8.5) and then use the inverse Abel transformation (8.12).

Suppose we represent r_+ as an arbitrary function $f(r_-)$, requiring only that $f(r_-) \geq r_-$ and $f(r_c) = r_c$. We get from formula (8.18):

$$\frac{f(r)}{r} = \exp\left(-\frac{2}{\pi}\int_\rho^{b_c} \frac{\Delta\,\mathrm{d}b}{\sqrt{b^2-\rho^2}}\right) . \tag{8.19}$$

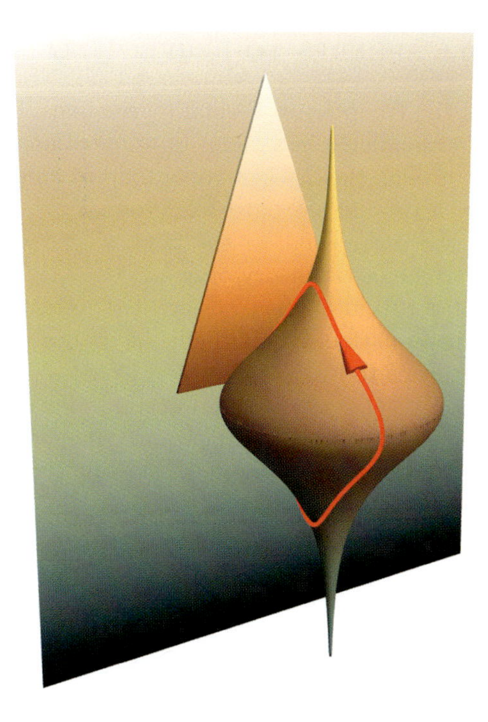

Figure 2.34: Tomography of the turning angle. Consider a bound trajectory (Fig. 2.27). The turning angle (7.22) is the difference between the angles of the two radial turning points. Like in the case of scattering tomography (Fig. 2.30), the integrated angles are the projections of W, one for the inner turning points (lower part of the object) and one for the outer turning points (upper part). As we take the difference between the angles, we subtract the two projections; the difference of the projections is given by the closed contour in the figure (red, with arrow). We could deform the object by adding values to both the lower and the upper part, without changing the difference of the projections: different index profiles may have the same turning angles (7.22) as a function of angular momentum.

This is an implicit equation for $\rho(r)$. If we know $\Delta(b)$ we can calculate the integral on the right–hand side and solve the equation for ρ (we discuss several examples in §9). Now, $\rho(r)$ gives via the definition (7.16) of the turning parameter the refractive–index profile $n(r)$, provided $\rho(r)$ is singlevalued. We can thus determine the media that produce a given distribution of turning angles (7.22) as a function of the angular momentum, if that distribution is physically possible. The arbitrary function $f(r)$ represents the ambiguity of the optical medium, the infinitely many profiles $n(r)$ with identical angles Δ between turning points. For example, formula (8.19) establishes a wide range of optical media with closed loops of light. Such media are useful for invisibility, but also for perfect imaging, as we show next.

§9. From invisible spheres to perfect lenses

This Section concludes the Chapter on Fermat's principle. The ideas accumulated so far make their final appearance in a series of applications before the curtain to the next act, where we develop the mathematical tools of differential geometry. The Section shows how much ground we have already covered. It also gives a taste of what is to come in the second half of the book where we combine differential geometry with the electromagnetic theory of light.

Luneburg's formula (8.12) allows us to design the refractive–index profile for a spherically symmetric medium that deflects light the way we wish. Suppose we want all rays that strike the device to be deflected by a fixed angle; rays that miss it are not deflected of course. Optical media with refractive index larger than 1 bend light inwards, according to Snell's law (1.3), giving a negative angle χ. It will be proved below that the uniform deflection of rays we are seeking can only be performed for $\chi \leq 0$. We thus require:

$$\chi = \begin{cases} -2\pi\nu & \text{for} \quad b \leq 1 \\ 0 & \text{for} \quad b > 1 \end{cases} \quad \text{with} \quad \nu > 0\,, \tag{9.1}$$

where we take the radius of the device as spatial unit—rays with impact parameters larger than the radius 1 miss it. Substituting this $\chi(b)$ in the right–hand side of Luneburg's expression (8.16) and using the integral

$$\int_\rho^1 \frac{\mathrm{d}b}{\sqrt{b^2 - \rho^2}} = \ln\left(\frac{1 + \sqrt{1 - \rho^2}}{\rho}\right) \tag{9.2}$$

we find the radius as a function of the turning parameter,

$$r = \frac{\rho}{\left(\rho^{-1} + \sqrt{\rho^{-2} - 1}\right)^{2\nu}} \quad \text{for} \quad r \leq 1 \tag{9.3}$$

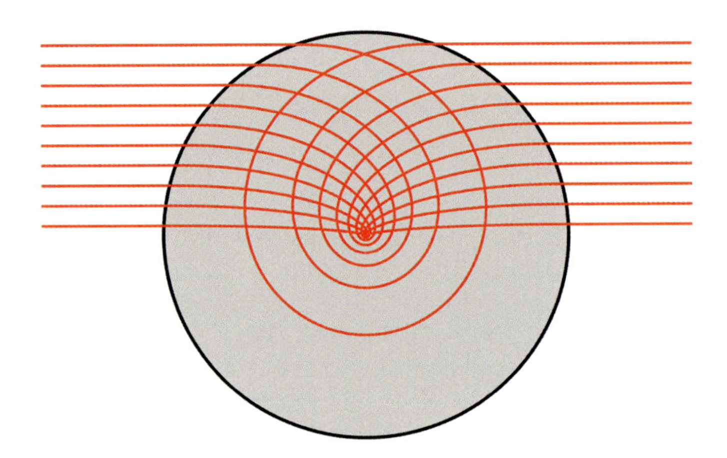

Figure 2.35: Invisible sphere.

and $r = \rho$ for $r > 1$. Figure 2.32 shows that $\rho(r)$ is singlevalued for $\nu \geq 0$, whereas $\rho(r)$ is multivalued and hence unphysical for $\nu < 0$ (positive deflection angles χ). A positive deflection angle χ results from devices with $n < 1$, as can be seen from Snell's law, but we have now shown that a *uniform* deflection angle for all rays that strike the device cannot be achieved for positive χ. In order to obtain the refractive–index profile $n(r)$, we substitute ρ/n for r in the result (9.3), solve for ρ and divide by r, with the result

$$r = \frac{2\,n^{\mu-1}}{1+n^{2\mu}}\,, \quad \mu = \frac{1}{2\,\nu}\,. \tag{9.4}$$

Exercise 9.1
Deduce expression (9.4) for $r(n)$ from Eqs. (7.16) and (9.3).
Solution
From $n = (\rho^{-1} + \sqrt{\rho^{-2}-1})^{2\nu}$ follows $n^\mu = \rho^{-1} + \sqrt{\rho^{-2}-1}$ with μ defined in Eq. (9.4), and hence $\sqrt{\rho^{-2}-1} = n^\mu - \rho^{-1}$. Squaring this relationship produces ρ^{-2} terms on both sides of the equation. The rest is linear in ρ^{-1} such that one can easily solve for ρ. Division by r gives n.

Let us discuss some specific examples. First, consider $\nu = 1/2$ that according to Eq. (9.1) corresponds to the deflection of a ray by π. In this case each ray that strikes the device leaves in precisely the opposite direction it came from. Deflection by π is thus the same as retroreflection; so for $\nu = 1/2$ we obtain an omnidirectional retroreflector, the Eaton lens.

Exercise 9.2
Verify that formula (9.4) with $\nu = 1/2$ indeed gives the Newton profile (6.8) of the Eaton lens.

Consider another case, $\nu = 1$, that corresponds to the deflection by 2π. The light rays perform one loop inside the device and leave in the same direction they came from, as if nothing had happened (Fig. 2.35). Deflection by 2π is equivalent to no deflection at all; the device appears as an invisible sphere. For $\nu = 1$, formula (9.4) produces a cubic equation in n, $(1+n)^2\,n\,r = 4$, that turns out to have the real solution (Miñano [2006])

$$n = \left(Q - \frac{1}{3Q}\right)^2\,, \quad Q = \sqrt[3]{-\frac{1}{r} + \sqrt[2]{\frac{1}{r^2} + \frac{1}{3^3}}}\,. \tag{9.5}$$

We see that the refractive index of the invisible sphere diverges at the centre, developing a singularity for $r \to 0$, and so does the profile (6.8) of the Eaton lens, which turns out to be a general feature of uniformly deflecting devices (9.1). For uniform deflection, we always need media with singularities, as we verify as follows. For small radii the turning parameter (7.16) tends to zero, unless n diverges stronger than r^{-1}. Let us assume that ρ does indeed tend to zero for small r and consider the solution (9.3) for small ρ. Equation (9.3) shows that for $\rho \to 0$ the radius $r \sim \rho^{1+2\nu}\,2^{-2\nu} = (nr)^{1+2\nu}\,2^{-2\nu}$. We solve for n and obtain

$$n \sim (2/r)^{1/(1+\mu)} \quad \text{for} \quad r \to 0\,,\ \mu = \frac{1}{2\nu}\,. \tag{9.6}$$

As n does not diverge stronger than r^{-1} since $\nu > 0$, our procedure for calculating the asymptotic behavior of n for $r \to 0$ is justified *a posteriori*. Uniform deflection thus requires singularities of type (9.6). Implementing invisible spheres, omnidirectional retroreflectors and other uniform deflectors in practice takes tricks from transformation optics we discuss later. In Exercise 33.2 we show that the singularity (9.6) can be transmuted into a harmless topological defect in an anisotropic material.

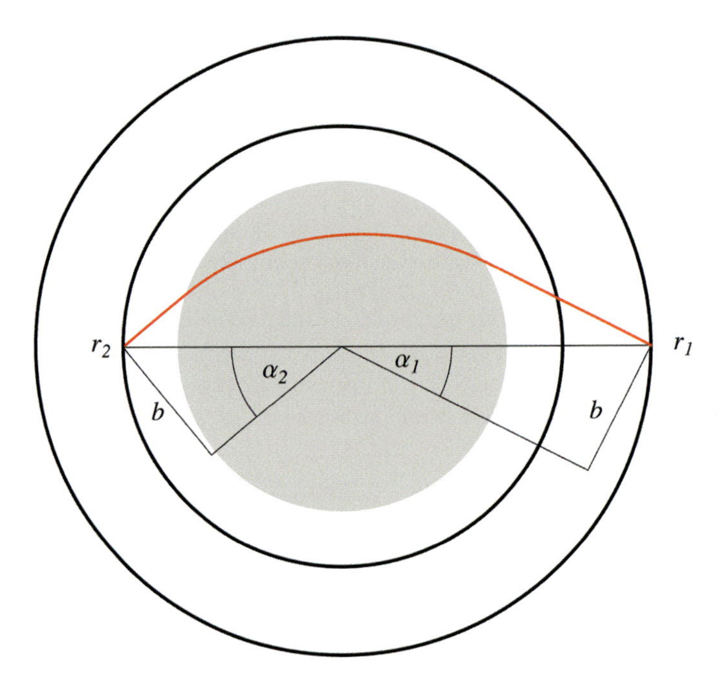

Figure 2.36: Luneburg's design problem. Consider light rays (red) that may pass through a spherically symmetric medium (grey). We require that all rays that do pass through the medium from radius r_1 arrive at radius r_2 exactly opposite to their starting point (meaning the starting point and end point lie on a straight line through $r = 0$.). We characterize the rays by their impact parameters b. The initial and final segments of the trajectories are inclined by the angles α_1 and α_2. The drawing shows that $b = r_1 \sin\alpha_1 = r_2 \sin\alpha_2$, which is all we need to know for solving Luneburg's design problem by tomography.

Consider a case where Luneburg's formula (8.12) becomes immediately useful in practice, *Luneburg's design problem* illustrated in Fig. 2.36. Imagine a spherically symmetric device of radius 1 (in our units) with the following property: all light rays from points at radius r_1 arrive at radius r_2 such that the starting and end points are on a straight line through $r = 0$ (Fig. 2.36), provided they pass through the medium. Figure 2.36 shows that the deflection angle χ (Fig. 2.28) in Luneburg's problem is given by

$$\chi = -\alpha_1 - \alpha_2\,, \quad \sin\alpha_i = \frac{b}{r_i}\,. \tag{9.7}$$

Let us discuss a few prominent examples. Suppose that

$$r_1 = \infty\,, \quad r_2 = 1\,. \tag{9.8}$$

In this case, all incident light rays from ∞ are focused on points on the surface of the device in the direction they came from: the device makes a Luneburg lens. One verifies that Luneburg's formula (8.15) indeed gives the profile (3.14) of the Luneburg lens, using the integral

$$\int_\rho^1 \frac{\arcsin b}{\sqrt{b^2-\rho^2}}\,\mathrm{d}b = \frac{\pi}{2}\ln\left(1+\sqrt{1-\rho^2}\right)\,. \tag{9.9}$$

Exercise 9.3
Deduce the profile of the Luneburg lens from Eqs. (8.15), (9.7) and (9.8) with $\rho = nr$.
Solution
We obtain from the requirements (9.8) for the deflection angles (9.7) $\alpha_1 = 0$ and $\alpha_2 = \arcsin b$. We insert these angles in Luneburg's formula (8.15), use integral (9.9) and get

$$n = \sqrt{1+\sqrt{1-\rho^2}}\,, \tag{9.10}$$

which indeed produces Hooke's profile (3.14) of the Luneburg lens, if we substitute nr for ρ and solve for n.

Consider another prominent case (Fig. 2.37) that will become important for imaging,

$$r_1 = r_2 = 1\,. \tag{9.11}$$

Imagine we surround this device by a reflecting surface—silver it with a spherical mirror. Figure 2.37 shows that rays going up or down between two opposite points

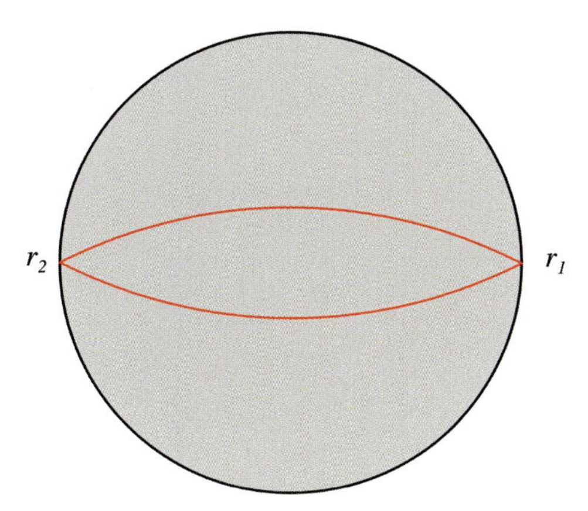

Figure 2.37: Maxwell's fish–eye from Luneburg's design. The initial and final radii r_1 and r_2 of the light rays (red) are identical and the medium (grey) fills the entire device. This device is known as Maxwell's fish–eye (Maxwell [1854]).

on the sphere are mirror images of each other. They are the light trajectories in the silvered sphere; the ray that follows the upper path, is reflected towards the lower path and after another reflection is on the upper path again, etc. The light is confined on closed loops. Consider all the light rays emitted at one point $\boldsymbol{r}_0$ inside the device. The mirror symmetry of the reflection implies that the rays must also go through $-\boldsymbol{r}_0$: all light rays are focused there (Fig. 2.38). Furthermore, we proved in

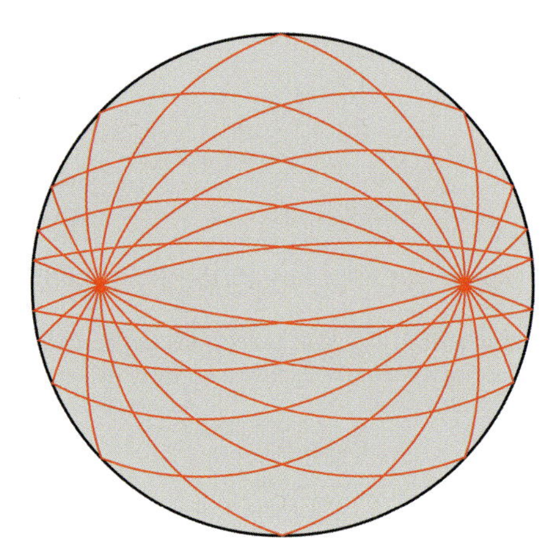

Figure 2.38: Perfect imaging with fish–eye mirror. Suppose Maxwell's fish–eye (Fig. 2.37) is surrounded by a reflecting surface. All rays (red) emitted from an arbitrary point inside the medium are focused on the opposite point.

§7 that the traveling times of light rays on closed loops are always the same (for a given spherically symmetric device). Consequently, all the focused light rays arrive at the same time; they carry the same phase, forming a coherent, perfect image (Leonhardt [2009]). Let us calculate the refractive–index profile that does the trick. For case (9.11) we obtain the angles (9.7) $\alpha_1 = \alpha_2 = \arcsin b$, so the deflection angle χ is $2 \arcsin b$, and we get from Luneburg's formula (8.15) and integral (9.9):

$$n = 1 + \sqrt{1 - \rho^2}\,. \tag{9.12}$$

As before, we substitute nr for ρ and solve for n. In this way we obtain the refractive–index profile of *Maxwell's fish–eye*

$$n = \frac{2}{1 + r^2}\,. \tag{9.13}$$

James Clerk Maxwell imagined such an optical medium in 1854, but without the spherical mirror. As Maxwell wrote 'the possibility of the existence of a medium of this kind possessing remarkable optical properties, was suggested by the contemplation of the crystalline lens in fish'—hence fish–eye—'and the method of searching for these properties was deduced by analogy from Newton's *Principia*, Lib. I Prop. VII'. The mirror adds the practical advantage that the refractive–index profile is finite; the ratio between the highest value at the centre of the device and the lowest at the surface $r = 1$ is 2.*

The fish–eye mirror makes a perfect lens, but it is a rather peculiar lens that contains both the object and the image inside the optical medium. This lens does not magnify, but one could reach an arbitrary optical resolution, independent of the wavelength of light. The fish–eye mirror could transfer embedded images with details significantly smaller than the wavelength of light over distances much larger than the wavelength, a useful feature for nanolithography. Provided of course, the lens is not only perfect for rays, but also for waves. We return to this point in §36 after we have honed the mathematical tools of differential geometry and gained experience with Maxwell's electromagnetism. Solving Maxwell's equations, we show that Maxwell's fish–eye perfectly images electromagnetic waves.

Consider another application of tomographic design for spherically symmetric optical media. Suppose we wish to implement a device that confines light rays on trajectories with uniform turning angles as follows: we require that the angles (7.22) between the outer and inner turning points are identical,

$$\Delta = -2\pi\nu\,, \tag{9.14}$$

for all bound trajectories with angular momenta b, until b reaches b_c where the curve approaches a circle of radius r_c and the two radial turning points become the same. For simplicity, we set our spatial units such that

$$b_c = 1\,. \tag{9.15}$$

According to condition (7.23), when ν is a rational fraction $p/(2q)$ with integers p and q the rays trajectories are closed: light propagates along closed loops. Formula (8.19) with integral (9.2) gives

$$\frac{f(r)}{r} = \left(\rho^{-1} + \sqrt{\rho^{-2} - 1}\right)^{4\nu}\,. \tag{9.16}$$

where $f(r)$ is an arbitrary real function with $f(r) \geq r$ and $f(r_c) = r_c$. We solve for ρ and obtain

$$\rho = \frac{2}{\sqrt[4\nu]{f(r)/r} + \sqrt[4\nu]{r/f(r)}}\,. \tag{9.17}$$

Exercise 9.4
Derive Eq. (9.17) from Eq. (9.16).

Relationship (9.17) directly gives the desired refractive–index profiles $n = \rho/r$, in particular for a large class of optical media with closed loops of light (when $\nu = p/(2q)$), depending on the function $f(r)$. For example, we may put

$$f(r) = \frac{\rho^2}{r}\,. \tag{9.18}$$

*The required refractive–index range is lower for a medium that solves the Luneburg design problem with $r_1 = r_2 > 1$, so that, in contrast to the fish–eye mirror in Fig. 2.38, there is a gap between the surrounding mirror and the device of radius 1.

Comparing Eqs. (9.3) and (9.16) we notice that case (9.18) corresponds to the uniform–deflection devices we considered before, except that now the refractive–index profile is not cut at the unit sphere. Uncut uniform deflectors thus confine light on closed loops when the deflection angle is a rational fraction of π. Consider another example of media (9.17) with closed loops; put

$$f(r) = \frac{1}{r}\,. \tag{9.19}$$

We obtain from Eqs. (7.16), (9.17) and (9.19) the refractive–index profile

$$n = \frac{2\,r^{\mu-1}}{1+r^{2\mu}}\,, \quad \mu = \frac{1}{2\,\nu}\,, \tag{9.20}$$

which, in the propagation plane with complex numbers (5.1) as coordinates, turns out to be a mere transmutation (5.9) of Maxwell's fish–eye (9.13)

$$w = z^{\mu}\,, \quad n' = \frac{2/\mu}{1+|w|^2}\,. \tag{9.21}$$

Exercise 9.5
Show that the transformation (5.9) with the functions (9.21) gives the profile (9.20).

Furthermore, comparing Eqs. (9.4) and (9.20) we notice that these two relationships are identical, except that r and n are exchanged. In Exercise 4.4 we found that light rays in spherically symmetric media have dual trajectories in position and momentum space. By momentum we mean the wavevector $\boldsymbol{k}$; as a solution of Hamilton's equations (4.10), a light ray draws a trajectory in position and momentum space. We have seen that in k space r plays the role of the refractive index, while n gives the radius of $\boldsymbol{k}$. Consequently, the physical curves of light rays in the transmuted fish–eye (9.20) are the same as the momentum trajectories in uniform deflectors, and vice versa (Miñano [2006]). The transmuted ray trajectories are simply the transformed curves of rays in the original fish–eye (9.13), according to transformation (9.21). But what are the light trajectories in Maxwell's fish–eye? Geometry gives the answer.

In his 1944 lectures, Luneburg also established a beautiful geometrical interpretation for Maxwell's fish–eye (9.13): he discovered that the fish–eye performs the *stereographic projection* of the surface of a sphere to the plane. The stereographic projection, invented by Ptolemy, lies at the heart of the *Mercator projection* used in cartography. Figure 2.39 shows how the stereographic projection works. The points on the surface of the sphere are projected to the plane cut through the Equator. Through each point a line is drawn from the North Pole that intersects the plane at one point, the projected point. To describe the stereographic projection in analytical geometry, we employ the following coordinates. For the plane we use complex numbers (5.1) as coordinates; for the surface of the unit sphere we take the Cartesian coordinates X, Y, Z with

$$X^2 + Y^2 + Z^2 = 1\,. \tag{9.22}$$

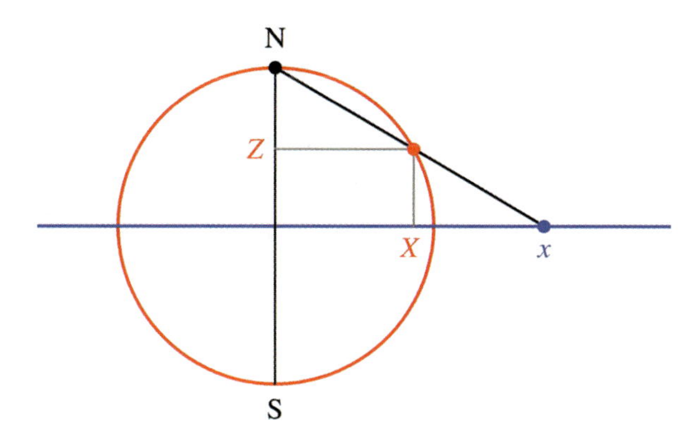

Figure 2.39: Stereographic projection. The sphere (red) is projected to the plane (blue). The figure shows a cut through the sphere. A point (red dot) on the sphere with coordinates X, Y, Z is projected from the North Pole N (black dot) to the plane (blue dot) with coordinates $z = x + \mathrm{i}y$ (here $Y = 0$ and $y = 0$). The South Pole S is projected to 0 and the North Pole to ∞.

Let us work out the formulae for the stereographic projection. Imagine first $Y = 0$ and hence $y = 0$. Figure 2.39 shows that the triangle between the North Pole N, the point on the sphere and Z is geometrically similar to the larger triangle between N, the point on the plane and the origin. The similarity of the triangles implies that the ratio of the horizontal side to the vertical side is the same for both, i.e. $X/(1 - Z) = x/1$. Now suppose we rotate the points with $Y = 0$ on the sphere and $y = 0$ on the plane to points of an arbitrary fixed angle. As the triangles are simply rotated, we follow the same recipe as before with x replaced by $x + \mathrm{i}y$ and X by $X + \mathrm{i}Y$, and get the same relationship. In this way we arrive at the complex representation of the stereographic projection

$$z = \frac{X + \mathrm{i}Y}{1 - Z} . \tag{9.23}$$

The Mercator projection of cartography is related to the stereographic projection: it is given by the complex logarithm of the stereographic projection multiplied by i:

$$z_{\mathcal{M}} = \mathrm{i} \ln \left(\frac{X + \mathrm{i}Y}{1 - Z} \right) . \tag{9.24}$$

The complex logarithm maps the complex plane to a stripe with imaginary parts between $\pm\pi$ (see the left picture of Fig. 2.25); multiplication by i rotates this stripe by 90°. Typically, the projection is then cut in the polar regions where the imaginary part of $z_{\mathcal{M}}$ approaches $\pm\infty$, to keep the cartographer's map finite.

Formula (9.23) describes how the stereographic projection maps the points on the sphere to points on the plane. In turn, the coordinates on the sphere are given

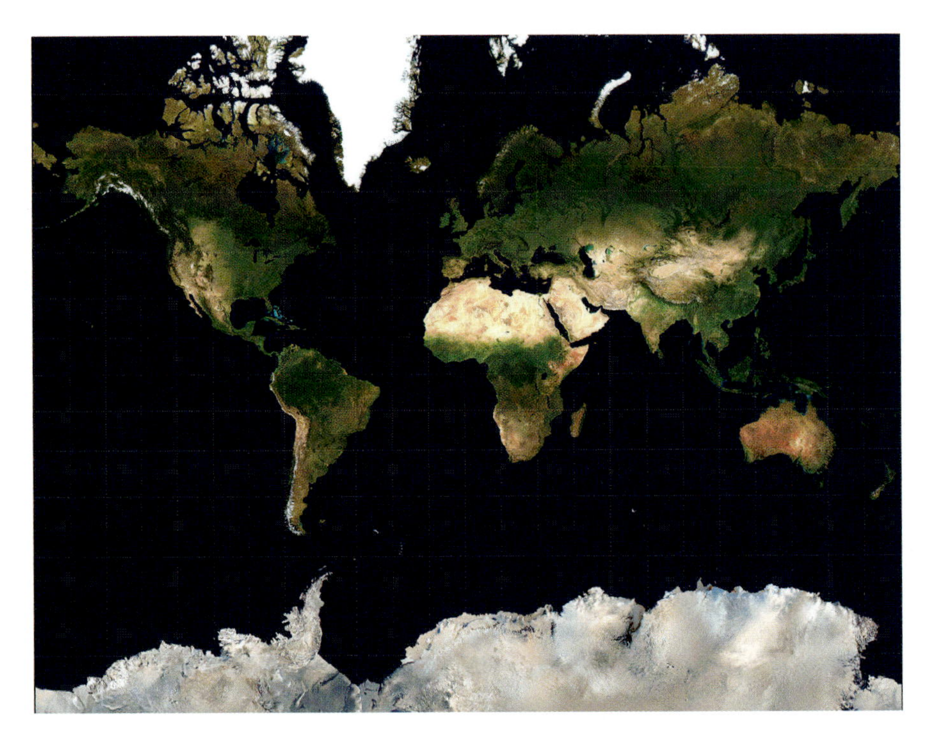

Figure 2.40: Mercator projection. In the Mercator projection, the surface of Earth is mapped by the complex logarithm (9.24) of the stereographic projection in appropriate units (Fig. 2.39) cut in the polar regions. As the stereographic projection and the complex logarithm are both conformal transformations, the Mercator projection preserves angles between lines, for example the right angles between the lines of longitude and latitude. The measure of distance and area is distorted however. According to Fermat's principle, an optically isotropic material has a similar effect; it changes the measure of distance, but not the measure of angle. In fact, the Mercator projection can be optically implemented by the refractive–index profile (9.31). (Credit: NASA's Earth Observatory.)

by the *inverse stereographic projection*

$$X + \mathrm{i}Y = \frac{2z}{1+|z|^2}\,, \quad Z = \frac{|z|^2 - 1}{|z|^2 + 1}\,. \tag{9.25}$$

Exercise 9.6
Prove that Eq. (9.25) inverts the stereographic projection (9.23). For example, substitute the expressions (9.25) for X, Y, Z in formula (9.23) and show that the result is z.

Exercise 9.7
Show that $X^2 + Y^2 + Z^2 = 1$ in the inverse stereographic projection (9.25).

Let us work out how distances on the sphere appear in stereographic coordinates $\{z\}$ in the complex plane given by the projection formula (9.23). The distances on the sphere are measured with respect to the usual infinitesimal line element $\mathrm{d}s$ in three–dimensional space (1.2) with Cartesian coordinates $\{X, Y, Z\}$ that, by condition (9.22), lie on the unit sphere:

$$\mathrm{d}s^2 = \mathrm{d}X^2 + \mathrm{d}Y^2 + \mathrm{d}Z^2 = \mathrm{d}(X + \mathrm{i}Y)\,\mathrm{d}(X - \mathrm{i}Y) + \mathrm{d}Z^2 . \tag{9.26}$$

We express this line element in the stereographic coordinates (9.23). For this we need $\mathrm{d}(X+\mathrm{i}Y)$ and $\mathrm{d}Z$. We obtain from formula (9.25) by the chain rule of differentiation

$$\mathrm{d}(X + \mathrm{i}Y) = \frac{2\,(\mathrm{d}z - z^2\mathrm{d}z^*)}{(1 + |z|^2)^2}\,, \quad \mathrm{d}Z = \frac{2\,(z^*\mathrm{d}z + z\mathrm{d}z^*)}{(1 + |z|^2)^2}\,. \tag{9.27}$$

Exercise 9.8
Verify expressions (9.27) using $\mathrm{d}X = (\partial_z X)\mathrm{d}z + (\partial_{z^*} X)\mathrm{d}z^*$ etc.

We thus obtain from the line element (9.26) and the refractive–index profile (9.13)

$$\mathrm{d}s^2 = n^2 \mathrm{d}z\,\mathrm{d}z^* = n^2(\mathrm{d}x^2 + \mathrm{d}y^2)\,, \quad n = \frac{2}{1 + |z|^2}\,. \tag{9.28}$$

Exercise 9.9
Prove formula (9.28).

Equation (9.28) says that the line element (9.26) on the sphere in stereographic projection is precisely the same as the optical path increment (1.1) in Maxwell's fish–eye. Consequently, according to Fermat's principle, Maxwell's fish–eye implements the stereographic projection of the sphere (Luneburg [1964]). Note that the stereographic projection preserves angles, in particular the right angles between longitude and latitude, because the refractive index n does not change the measure of angle, although it changes the measure of distance: the stereographic projection is a conformal map. As the Mercator projection (9.24) is an analytic function of the stereographic projection, it also is conformal, which is the reason why the lines of longitude and latitude cross at right angles in the Mercator map of Earth (Fig. 2.40).

Exercise 9.10
Determine the refractive–index profile that implements the Mercator projection (9.24) of the geometry of the sphere.
Solution
The Mercator projection is the complex logarithm (9.24) of the stereographic projection, and the latter is implemented by Maxwell's fish–eye (9.13). In terms of optical conformal mapping (§5), for the Mercator projection the stereographic coordinates (9.23) are the coordinates w of virtual space and the Mercator coordinates $z_{\mathcal{M}}$ the coordinates of physical space $z = x + \mathrm{i}y$, with

$$w = \exp(-\mathrm{i}z) = \exp(y - \mathrm{i}x) \tag{9.29}$$

according to definition (9.24). The refractive index profile in virtual space is the profile (9.13) of Maxwell's fish–eye:

$$n' = \frac{2}{1+|w|^2}\,. \tag{9.30}$$

We thus obtain from the transformation rule (5.9) of the refractive–index profile in optical conformal mapping

$$n = \left|\frac{\mathrm{d}w}{\mathrm{d}z}\right| n' = \frac{2}{|w|^{-1}+|w|} = \frac{2}{\exp(y)+\exp(-y)} = \operatorname{sech} y\,. \tag{9.31}$$

Maxwell's fish–eye creates the illusion that light propagates on the surface of the virtual sphere (9.22), a curved space. The curvature (5.13) for the fish–eye geometry (9.13) is

$$R = 2\,. \tag{9.32}$$

Exercise 9.11
Show that the curvature scalar (5.13) of Maxwell's fish–eye (9.13) with $r = \sqrt{z^*z}$ is exactly 2.

The surface of the sphere is a space of constant, positive curvature. Maxwell's fish–eye thus represents a beautiful, simple example of a non–Euclidean medium where the virtual space of light is curved. As we have seen, non–Euclidean transformation optics may find applications in cloaking and perfect imaging. Curved space could also create fascinating optical illusions. In Fig. 2.41 Hebe, the Greek goddess of

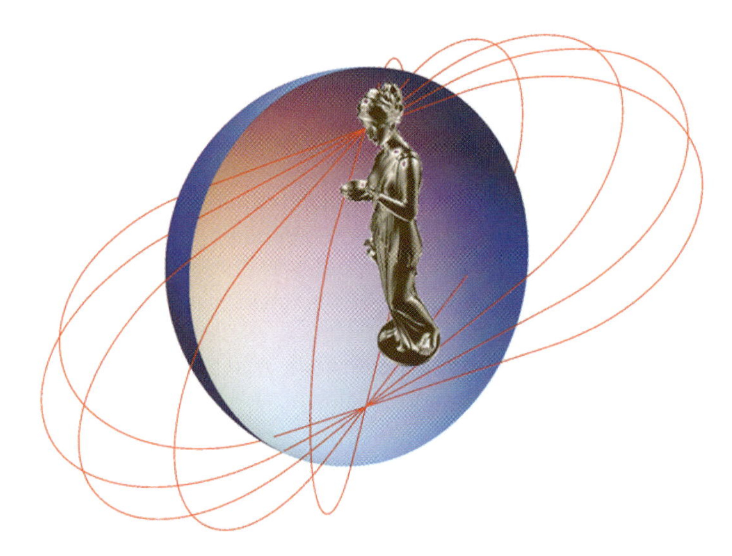

Figure 2.41: Hebe, Greek goddess of youth, looking at herself in curved space. Here the curved–space geometry is created by the optical material of an inside–out Eaton lens (Fig. 2.24). Inside the sphere, light rays (red) propagate along straight lines, but as soon as the rays enter the non–Euclidean mirror they turn along ellipse segments and are retroreflected. (Credit: Aaron Danner.)

youth, is looking at herself in a curved space where light goes around in loops. Figure 2.42 shows what she sees: the cup in her hand, but also the back of her head upside down. Her picture always remains the same, regardless where she stands and looks into the non–Euclidean mirror.

Figure 2.42: Hebe's image. Hebe (Fig. 2.41) sees the back of her head upside down in the non–Euclidean mirror. (Credit: Aaron Danner.)

Consider light rays in the curved virtual space of Maxwell's fish–eye where they are confined to the surface of the sphere. Clearly, the light rays on the sphere are no longer straight, but light still propagates along the shortest geometrical paths. On the sphere, the shortest paths are the great circles (Fig. 2.43), the typical flight paths of commercial aircraft on the globe (Fig. 3.18). The great circles are the circles with the greatest radius, 1, on the unit sphere. Now imagine several light rays are emitted from one point: the rays disperse along their great circles, but they all meet again at the antipodal point. As Maxwell's fish–eye faithfully maps the surface of the sphere to the plane by stereographic projection (9.23), the rays must also meet in physical space. Light emitted from point z_0 is focused at

$$z_0' = -\frac{1}{z_0^*}\,. \tag{9.33}$$

Exercise 9.12
The antipodal point to (X, Y, Z) is given by $(-X, -Y, -Z)$. Show that z_0' given by Eq. (9.33) is the stereographic projection of the antipodal point.

Even without the mirror in Fig. 2.38, Maxwell's fish–eye perfectly images light rays. But what are the ray trajectories? The stereographic projection has the property that circles on the sphere are mapped to circles in the plane, and vice versa, so in Maxwell's fish–eye light goes around in circles. Moreover, as we have seen, all the light circles from one point z_0 also intersect at the image (9.33). These are the two optical properties of the fish–eye Maxwell found so remarkable (Maxwell [1854]).

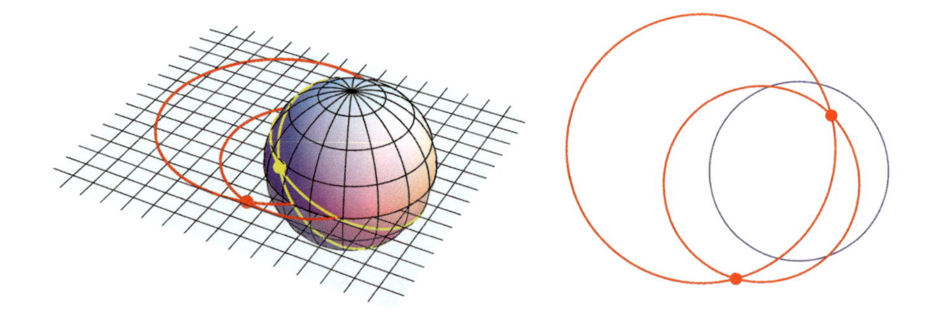

Figure 2.43: Circles of light. In Maxwell's fish–eye in 2D, virtual space (left) is the surface of a sphere and physical space (right) a plane. Circles on the sphere appear as circles in the plane by stereographic projection (Fig. 2.39). Light rays, shown in yellow on the sphere and in red on the plane, are the geodesics, the great circles on the sphere. All the rays emitted from one point meet again at the antipodal point, forming a perfect image there. The right picture illustrates how the light circles meet in the stereographic projection to physical space with the Equator shown in blue.

Let us prove that circles are mapped to circles in the stereographic projection. First, imagine a circle around the top of the sphere, not necessarily a great circle, centred at the North Pole (or, equivalently, at the South Pole). Clearly, as the stereographic projection is rotationally symmetric, the image of this circle is a circle in the plane. Now imagine we rotate the circle on the sphere. We can reach any other circle of the same size. It turns out that the rotation on the sphere corresponds to a particular Möbius transformation (5.18) of the projected points (9.23) in the complex plane:

$$z' = \mathrm{e}^{\mathrm{i}\beta}\frac{z\cos\gamma - \mathrm{e}^{\mathrm{i}\alpha}\sin\gamma}{z\,\mathrm{e}^{-\mathrm{i}\alpha}\sin\gamma + \cos\gamma}\,. \tag{9.34}$$

Let us first analyze this formula and then justify it. We write formula (9.34) as

$$z' = \mathrm{e}^{\mathrm{i}\alpha+\mathrm{i}\beta} z'_0\left(z\mathrm{e}^{-\mathrm{i}\alpha}\right) \quad \text{with} \quad z'_0 = \frac{z\cos\gamma - \sin\gamma}{z\sin\gamma + \cos\gamma}\,. \tag{9.35}$$

Here z'_0 denotes the Möbius transformation (9.34) with $\alpha = \beta = 0$. We see that the angle α rotates the points on the z plane, then the transformation z'_0 is performed, followed by a rotation by $-\alpha-\beta$. Inspecting the stereographic projection (9.23), we also see that rotations on the complex plane correspond to rotations of the sphere around the Z–axis. The only non–trivial case is the transformation (9.34) for zero α and β:

$$z' = \frac{z\cos\gamma - \sin\gamma}{z\sin\gamma + \cos\gamma}\,. \tag{9.36}$$

In this case the inverse stereographic projection (9.25) gives a rotation around the Y axis of the sphere,

$$
\begin{aligned}
X' &= X\cos 2\gamma - Z\sin 2\gamma\,, \\
Y' &= Y\,, \\
Z' &= X\sin 2\gamma + Z\cos 2\gamma\,.
\end{aligned}
\tag{9.37}
$$

Exercise 9.13

Derive the result (9.37) from the inverse stereographic projection (9.25) of the Möbius transformation (9.34).

The Möbius transformation (9.34) thus has the following geometrical meaning on the sphere: it describes a rotation around the Z–axis by the angle α, followed by a rotation around the Y–axis by γ and finally a rotation around the Z–axis by $-\alpha-\beta$. The three angles α, γ and $-\alpha-\beta$ correspond to the *Euler angles* that describe any rotation on the sphere (Landau and Lifshitz [1982]). Our result proves that the Möbius transformation (9.34) describes any rotation on the sphere. Finally, we showed in §5 that Möbius transformations always map circles to circles. Consequently, the light rays in Maxwell's fish–eye are circles.

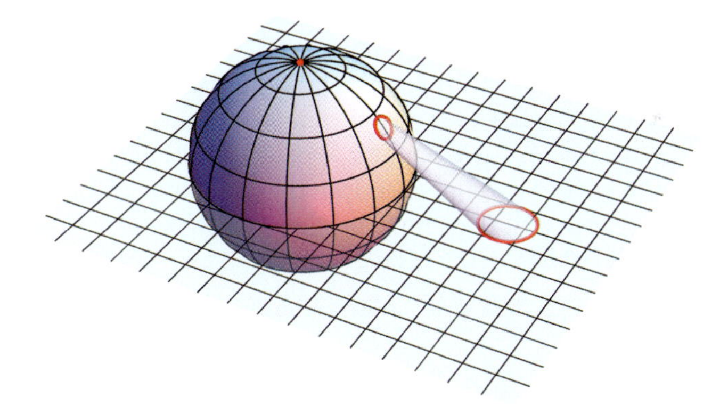

Figure 2.44: Geometrical meaning of the refractive index. The ratio between infinitesimal distances in virtual space and real space gives the refractive index n. One can find this as follows. Virtual space (here the surface of the sphere) is mapped on to real space (here the plane). When light propagates a distance $\mathrm{d}l$ in real space during the time $\mathrm{d}t$, it passes the distance $\mathrm{d}l'$ in virtual space during the same time. In virtual space, the speed of light is c. Consequently, the speed of light in physical space, $\mathrm{d}l/\mathrm{d}t$, is given by $(\mathrm{d}l/\mathrm{d}l')c$, or, expressed in terms of the refractive index, $\mathrm{d}l/\mathrm{d}t = c/n$ where the refractive index n is the ratio between $\mathrm{d}l'$ and $\mathrm{d}l$. In general, n depends on the direction of the line element. But, in this picture, circles in virtual space appear as circles in physical space: no direction of the line element is distinguished. The optical material that creates this geometry has a scalar refractive index n independent of propagation direction; it is optically isotropic.

Any refractive–index profile establishes a geometry for light rays, because we may regard the optical–path element ds in Fermat's principle (1.1) as the line element dl' in a virtual space. The refractive index is the ratio between the infinitesimal length in virtual space ds and the infinitesimal line increment dl in physical space (Fig. 2.44). In some cases, virtual space is a mere transformation of Euclidean, flat space. As people are primed to perceive light as propagating along straight lines—in flat space—they would have the impression that such Euclidean transformation media simply shift the points in space to new locations. Figure 2.6 shows that a region in physical space may be squashed into a single, invisible point in virtual space, although not with a conformal transformation that preserves the angles between the coordinate lines (as seen in Fig. 2.6). Implementing this transformation therefore takes an anisotropic medium. We may also imagine transformations that require *negative refraction*, (Veselago [1967], Pendry [2000]) a concept that goes beyond the optical media with simple refractive indices considered in this Chapter. Also, in order to prove that non–Euclidean devices, such as Maxwell's fish–eye, perfectly image light, we need to focus on the full complexity of electromagnetic waves and not only consider the trajectories of light rays, because the resolution of optical instruments is governed by the wave nature of light. To give another example of how our ideas may develop further, imagine we transform not only space, but space and time, something we have not considered yet. We could alter the flux of time for light and create horizons where light comes to a standstill.

In most cases, the geometry of light is not flat, but curved. For example, the 3D version of Maxwell's fish–eye represents the 3D surface of the 4D hypersphere. Optical materials may appear as hyperspace to light. But to achieve the full potential of the geometry of light, both for practical applications and theoretical insights, we need to study differential geometry first. For, the entrance to advanced electromagnetics appears to carry the same inscription as the door of Plato's Academy at Athens: 'Let no one enter who does not know geometry' —
ΑΓΕΩΜΕΤΡΗΤΟΣ ΜΗΔΕΙΣ ΕΙΣΙΤΩ.

FURTHER READING

The *Principles of Optics* by Born and Wolf [1999] is the definitive monograph in optics that also contains sections on Fermat's principle and perfect imaging. Luneburg's *Mathematical Theory of Optics* was far ahead of its time and developed many original ideas that enjoy a renaissance today. The book is out of print, but may be available in libraries or second–hand.

On mechanics, we recommend Goldstein, Poole and Safko [2001] and in particular Landau and Lifshitz [1982] that is probably the best textbook on theoretical physics ever written.

Path integrals are explained in Feynman and Hibbs [1965] and have found applications in a wide range of subjects, see e.g. *Path Integrals in Quantum Mechanics, Statistics, Polymer Physics, and Financial Markets* (Kleinert [2009]).

Needham's *Visual Complex Analysis* [2002] explains beautifully and pedagogically the visual aspects of complex analysis that, for example, appear in optical conformal mapping. Ablowitz and Fokas [1997] introduce complex analysis in a way that appeals to physicists and other practitioners of applied mathematics. Nehari [1952] is a superb and deep book on conformal mapping.

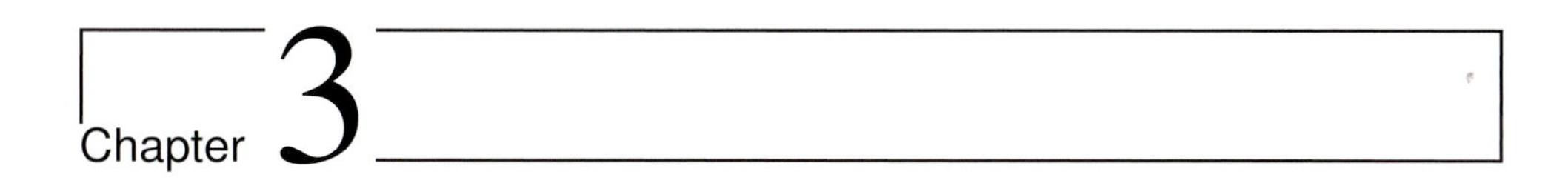

Differential geometry

The theory of media that act as effective geometries for light requires consideration of Maxwell's equations in arbitrary coordinates and in non–Euclidean spaces. This means that the natural mathematical language for describing such media is *differential geometry*, the mathematics that also underlies Einstein's general relativity. Here we introduce the reader to the mathematics of arbitrary coordinates and curvature, to the extent necessary to explore the geometry of light. Applied scientists may be tempted to take the connection to general relativity as an assurance that this mathematics is esoteric and of no relevance to them, but if they take a little time to peruse this Chapter they will find that the opposite is the case. They will see that differential geometry provides the most transparent way of describing something as mundane as non–Cartesian coordinate systems. Basic material such as Maxwell's equations in spherical polar coordinates is much simpler in the language of differential geometry than in the standard treatment found in the electromagnetism textbooks. As well as elucidating the apparently familiar, this Chapter provides the foundation for understanding the innovations in optics described in the remainder of this book.

§10. Coordinate transformations

We seek to describe quantities such as electric fields in a manner that is not tied to any specific coordinate system and to develop a set of rules for manipulating these quantities that hold for arbitrary coordinates. The familiar vector calculus is extremely limited in the quantities it can describe; for example it can handle the gradient or curl of a vector, but not the general idea of the derivative of a vector. Nor does it provide an efficient computational formalism once a coordinate system is chosen, unless the system is Cartesian. Substantial supplementation to standard vector calculus is required to understand fully the differentiation of vectors and to achieve general rules that are indifferent to the coordinates used in the calculation. This general coordinate–indifferent calculus is called *tensor analysis* and we shall discover it by carefully considering the effect of coordinate transformations.

We deal first with spatial coordinates; the extension to space–time coordinates

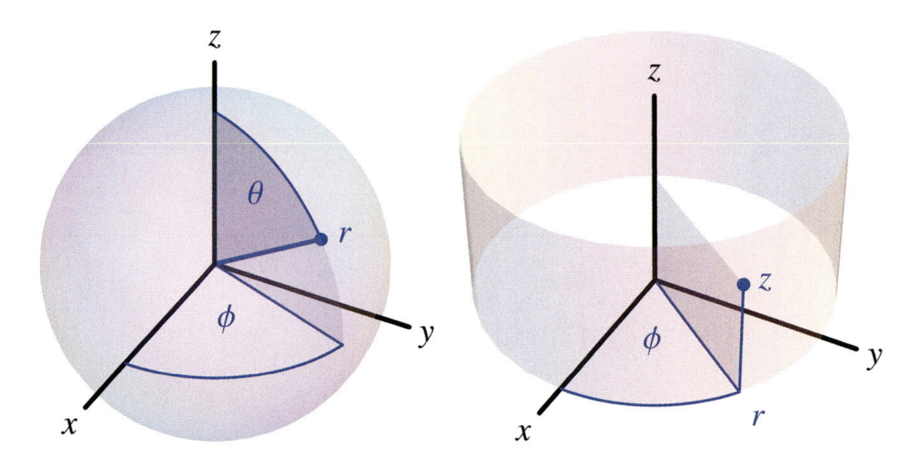

Figure 3.1: Spherical polar and cylindrical polar coordinates. The left picture shows how the spherical polar system $\{r, \theta, \phi\}$ is related to right–handed Cartesian coordinates $\{x, y, z\}$, illustrating Eqs. (10.1). The right picture shows the same for the cylindrical polar system $\{r, \phi, z\}$, illustrating Eqs. (10.2).

will then be straightforward. Our interest is in writing equations that are valid in an arbitrary spatial coordinate system $\{x^i,\ i = 1, 2, 3\}$ and in performing an arbitrary transformation to another set of coordinates that we distinguish from the original by a prime on the index: $\{x^{i'},\ i' = 1, 2, 3\}$. It is essential to the enterprise that the unprimed and primed coordinates are completely arbitrary, but we shall be illustrating interesting quantities that emerge from our considerations by computing them in the case of some familiar coordinate systems (Fig. 3.1). Thus, we take a special interest in seeing how things look when the unprimed coordinates are Cartesian $\{x, y, z\}$ and the primed coordinates are spherical polar $\{r, \theta, \phi\}$, so that we have

$$\begin{aligned} &\{x^i\} = \{x, y, z\}\,, && \{x^{i'}\} = \{r, \theta, \phi\}\,, \\ &x = r\sin\theta\cos\phi\,, && r = \sqrt{x^2 + y^2 + z^2}\,, \\ &y = r\sin\theta\sin\phi\,, && \theta = \arctan(\sqrt{x^2 + y^2}/z)\,, \\ &z = r\cos\theta && \phi = \arctan(y/x)\,. \end{aligned} \tag{10.1}$$

The case where the primed coordinates in Eqs. (10.1) are replaced by cylindrical polar $\{r, \phi, z\}$ is also of interest; here we have

$$\begin{aligned} &\{x^i\} = \{x, y, z\}\,, && \{x^{i'}\} = \{r, \phi, z\}\,, \\ &x = r\cos\phi\,, && r = \sqrt{x^2 + y^2}\,, \\ &y = r\sin\phi\,, && \phi = \arctan(y/x)\,. \end{aligned} \tag{10.2}$$

At this point we introduce the *Einstein summation convention* in which a summation is implied over repeated indices; for example

$$A^i B_i \equiv \sum_{i=1}^{3} A^i B_i = A^1 B_1 + A^2 B_2 + A^3 B_3 \,. \tag{10.3}$$

This convention allows us to dispense with writing summation signs, which are completely unnecessary. For reasons that will become clear later on, our summations will generally, but not always, be over a pair of indices in which one index is a subscript and one is a superscript, as in Eq. (10.3). We also introduce the *Einstein range convention* by which a free index (i.e. an index that is not summed over) is understood to range over all possible values of the index, for example

$$A^i \equiv \{A^i \,,\ {\scriptstyle i} = 1, 2, 3\} \,, \tag{10.4}$$

Together, the summation and range conventions allow an economy of notation such as the following:

$$R^i{}_{jik} \equiv \left\{ \sum_{i=1}^{3} R^i{}_{jik} \,,\ {\scriptstyle j,k} = 1, 2, 3 \right\} \,. \tag{10.5}$$

The differentials of our two arbitrary sets of coordinates, x^i and $x^{i'}$, are related by the chain rule (remember the summation and range conventions!):

$$\mathrm{d}x^i = \frac{\partial x^i}{\partial x^{i'}} \mathrm{d}x^{i'} \,, \quad \mathrm{d}x^{i'} = \frac{\partial x^{i'}}{\partial x^i} \mathrm{d}x^i \,, \tag{10.6}$$

with similar relations holding for the differential operators:

$$\frac{\partial}{\partial x^{i'}} = \frac{\partial x^i}{\partial x^{i'}} \frac{\partial}{\partial x^i} \,, \quad \frac{\partial}{\partial x^i} = \frac{\partial x^{i'}}{\partial x^i} \frac{\partial}{\partial x^{i'}} \,. \tag{10.7}$$

We denote the transformation matrices in Eqs. (10.6) and (10.7) by

$$\Lambda^i{}_{i'} = \frac{\partial x^i}{\partial x^{i'}} \,, \quad \Lambda^{i'}{}_i = \frac{\partial x^{i'}}{\partial x^i} \,. \tag{10.8}$$

Note carefully that $\Lambda^i{}_{i'}$ and $\Lambda^{i'}{}_i$ are different matrices, despite sharing the same "kernal" Λ: the upper and lower indices refer to different coordinate systems and the coordinate represented by the upper index is differentiated with respect to the coordinate represented by the lower index. When writing and viewing equations involving the matrices (10.8) the reader should always "see" the partial derivatives for which they are a shorthand.

Exercise 10.1

Verify that for the example (10.1) the transformation matrices (10.8) are

$$\Lambda^i{}_{i'} = \begin{pmatrix} \sin\theta\cos\phi & r\cos\theta\cos\phi & -r\sin\theta\sin\phi \\ \sin\theta\sin\phi & r\cos\theta\sin\phi & r\sin\theta\cos\phi \\ \cos\theta & -r\sin\theta & 0 \end{pmatrix} \tag{10.9}$$

$$= \begin{pmatrix} \frac{x}{r} & \frac{xz}{\sqrt{x^2+y^2}} & -y \\ \frac{y}{r} & \frac{yz}{\sqrt{x^2+y^2}} & x \\ \frac{z}{r} & -\sqrt{x^2+y^2} & 0 \end{pmatrix}, \tag{10.10}$$

$$\Lambda^{i'}{}_{i} = \begin{pmatrix} \frac{x}{r} & \frac{y}{r} & \frac{z}{r} \\ \frac{xz}{r^2\sqrt{x^2+y^2}} & \frac{yz}{r^2\sqrt{x^2+y^2}} & -\frac{\sqrt{x^2+y^2}}{r^2} \\ -\frac{y}{x^2+y^2} & \frac{x}{x^2+y^2} & 0 \end{pmatrix} \tag{10.11}$$

$$= \begin{pmatrix} \sin\theta\cos\phi & \sin\theta\sin\phi & \cos\theta \\ \frac{\cos\theta\cos\phi}{r} & \frac{\cos\theta\sin\phi}{r} & -\frac{\sin\theta}{r} \\ -\frac{\csc\theta\sin\phi}{r} & \frac{\csc\theta\cos\phi}{r} & 0 \end{pmatrix}. \tag{10.12}$$

Exercise 10.2
Compute the transformation matrices (10.8) for the example (10.2).
Solution
As an alternative to direct partial differentiation this can be done by using Exercise 10.1, since cylindrical coordinates are obtained from spherical polars by first putting $\theta = \pi/2$, whereupon the system collapses to a two–dimensional one, and then supplementing this two–dimensional system by the Cartesian coordinate z. The result is

$$\Lambda^{i}{}_{i'} = \begin{pmatrix} \cos\phi & -r\sin\phi & 0 \\ \sin\phi & r\cos\phi & 0 \\ 0 & 0 & 1 \end{pmatrix} \tag{10.13}$$

$$= \begin{pmatrix} \frac{x}{r} & -y & 0 \\ \frac{y}{r} & x & 0 \\ 0 & 0 & 1 \end{pmatrix}, \tag{10.14}$$

$$\Lambda^{i'}{}_{i} = \begin{pmatrix} \frac{x}{r} & \frac{y}{r} & 0 \\ -\frac{y}{r^2} & \frac{x}{r^2} & 0 \\ 0 & 0 & 1 \end{pmatrix} \tag{10.15}$$

$$= \begin{pmatrix} \cos\phi & \sin\phi & 0 \\ -\frac{\sin\phi}{r} & \frac{\cos\phi}{r} & 0 \\ 0 & 0 & 1 \end{pmatrix}. \tag{10.16}$$

From Eqs. (10.6) and (10.8) we find $\mathrm{d}x^i = \Lambda^i{}_{i'}\,\mathrm{d}x^{i'} = \Lambda^i{}_{i'}\Lambda^{i'}{}_j\,\mathrm{d}x^j$ and $\mathrm{d}x^{i'} = \Lambda^{i'}{}_i\,\mathrm{d}x^i = \Lambda^{i'}{}_i\Lambda^i{}_{j'}\,\mathrm{d}x^{j'}$, which imply

$$\Lambda^i{}_{i'}\Lambda^{i'}{}_j = \delta^i{}_j\,, \quad \Lambda^{i'}{}_i\Lambda^i{}_{j'} = \delta^{i'}{}_{j'}\,, \tag{10.17}$$

where $\delta^i{}_j$ and $\delta^{i'}{}_{j'}$ are the Kronecker delta, the identity matrix ($\delta^i{}_j = 1$ for $i = j$ and $\delta^i{}_j = 0$ for $i \neq j$). We position the indices on the Kronecker delta in (10.17) so that they line up with the positions of those indices on the left–hand side of the equations. Equations (10.17) state that the matrices $\Lambda^i{}_{i'}$ and $\Lambda^{i'}{}_i$ are the inverses of each other.

Exercise 10.3
Deduce the property (10.17) directly from the definitions (10.8) and the chain rule for partial derivatives.

Exercise 10.4
Verify the inverse relationship (10.17) between the matrices $\Lambda^i{}_{i'}$ and $\Lambda^{i'}{}_i$ when they are given by (*i*) the result of Exercise 10.1, (*ii*) the result of Exercise 10.2.

---

§11. THE METRIC TENSOR

Once a coordinate system is chosen, distances between points can be expressed in terms of the coordinate values assigned to those points. Although we have awarded ourselves the freedom of covering space with any coordinate system we wish, the distances between points are inherent properties of space that are independent of any idea of a coordinate system. This means the distance between any two points is coordinate *invariant*—it is the same number no matter which coordinates are used to express it. We require a general method of calculating distances in space using arbitrary coordinates, bearing in mind that the result is coordinate invariant.

The basic quantity is the square of the infinitesimal distance $\mathrm{d}s$ between the points x^i and $x^i + \mathrm{d}x^i$ (Fig. 3.2). For Cartesian coordinates $x^i = \{x, y, z\}$ this is given by the 3–dimensional Pythagoras theorem:

$$\mathrm{d}s^2 = \mathrm{d}x^2 + \mathrm{d}y^2 + \mathrm{d}z^2 = \delta_{ij}\,\mathrm{d}x^i\mathrm{d}x^j \qquad (x^i \text{ Cartesian}), \tag{11.1}$$

where δ_{ij} is again the Kronecker delta and we have arranged the indices so that each summation is over a subscript/superscript pair. The quantity $\mathrm{d}s^2$ is known as the *line element*; it is the square of a coordinate–invariant distance and we need to re–express it in a form valid for arbitrary coordinates. Using the general relations (10.6) and (10.8) we write the Cartesian coordinate differentials in Eq. (11.1) in terms of those for an arbitrary system $x^{i'}$:

$$\mathrm{d}s^2 = \delta_{ij}\,\Lambda^i{}_{i'}\Lambda^j{}_{j'}\,\mathrm{d}x^{i'}\mathrm{d}x^{j'} \qquad (x^i \text{ Cartesian}), \tag{11.2}$$

where we must remember that this is only valid when x^i are Cartesian. Comparing Eq. (11.2) with Eq. (11.1) we see that the line element expressed in arbitrary coordinates differs from the Cartesian form in that the Kronecker delta δ_{ij} gets replaced

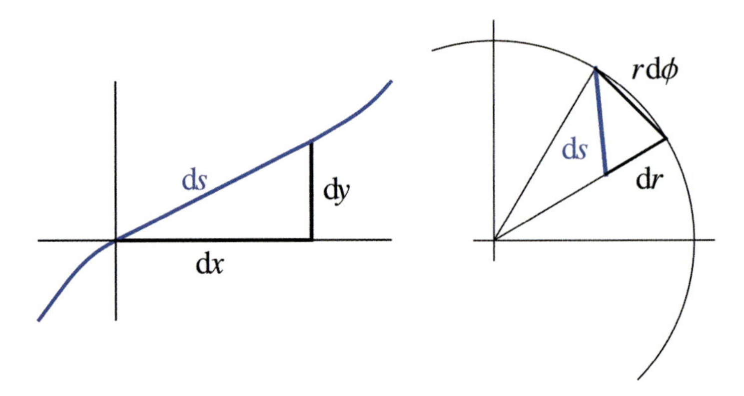

Figure 3.2: The line element in two dimensions. On the left the length $\mathrm{d}s$ of the straight blue line is related to the difference in the Cartesian coordinate values of its endpoints; the picture shows that $\mathrm{d}s^2 = \mathrm{d}x^2 + \mathrm{d}y^2$ according to Pythagoras' theorem (compare Eq. (11.1)). On the right the length of the blue line is related to the difference in the polar coordinate values of its endpoints; the picture shows that $\mathrm{d}r$ and $r\mathrm{d}\phi$ are the legs of the triangle with hypotenuse $\mathrm{d}s$, so $\mathrm{d}s^2 = \mathrm{d}r^2 + r^2\mathrm{d}\phi^2$ (compare Eq. (11.22)). The line element is obtained by taking the distance $\mathrm{d}s$ to be infinitesimal.

by the more complicated quantity $\delta_{ij}\Lambda^i{}_{i'}\Lambda^j{}_{j'}$. The Λ matrices in Eq. (11.2) can be written in terms of the coordinates $x^{i'}$, thus removing any appearance of the Cartesian x^i (cf. Exercises 10.1 and 10.2); denoting the resulting $x^{i'}$–dependent quantity $\delta_{ij}\Lambda^i{}_{i'}\Lambda^j{}_{j'}$ by $g_{i'j'}$ we have the desired general form of the line element:

$$\mathrm{d}s^2 = g_{i'j'}\,\mathrm{d}x^{i'}\mathrm{d}x^{j'}. \tag{11.3}$$

The two–index object $g_{i'j'}$ is the *metric tensor*, the quantity that allows us to calculate distances in space. In Cartesian coordinates (11.1) the metric tensor takes the simple form δ_{ij}, known as the *Euclidean metric.*

We can write the relation (11.3) in another arbitrary coordinate system x^i, denoting the metric tensor in this system by g_{ij}; from the coordinate invariance of $\mathrm{d}s$ we have

$$\begin{aligned}\mathrm{d}s^2 = g_{i'j'}\,\mathrm{d}x^{i'}\mathrm{d}x^{j'} &= g_{ij}\,\mathrm{d}x^i\mathrm{d}x^j \\ &= g_{ij}\,\Lambda^i{}_{i'}\Lambda^j{}_{j'}\,\mathrm{d}x^{i'}\mathrm{d}x^{j'}\,,\end{aligned} \tag{11.4}$$

where we have used Eqs. (10.6) and (10.8) in the second line. Equation (11.4) reveals how the metric tensor transforms upon changing from one arbitrary coordinate system to another:

$$g_{i'j'} = \Lambda^i{}_{i'}\Lambda^j{}_{j'}\,g_{ij}\,. \tag{11.5}$$

The metric tensor is always symmetric in its indices, i.e.

$$g_{ij} = g_{ji}\,, \tag{11.6}$$

a property that follows from its form in Cartesian coordinates.

Exercise 11.1
Show that Eq. (11.6) holds in any coordinate system.
Solution
As noted above, the metric tensor in any coordinate system $x^{i'}$ is given in terms of a Cartesian system x^i by

$$g_{i'j'} = \Lambda^i{}_{i'}\Lambda^j{}_{j'}\,\delta_{ij}\,. \tag{11.7}$$

The required relation $g_{i'j'} = g_{j'i'}$ now follows by a simple but powerful trick of "index gymnastics", a relabeling of the dummy summation indices i and j in Eq. (11.7):

$$g_{i'j'} = \Lambda^i{}_{i'}\Lambda^j{}_{j'}\,\delta_{ij} = \Lambda^j{}_{i'}\Lambda^i{}_{j'}\,\delta_{ji} = \Lambda^i{}_{j'}\Lambda^j{}_{i'}\,\delta_{ji} = \Lambda^i{}_{j'}\Lambda^j{}_{i'}\,\delta_{ij} = g_{j'i'}\,. \tag{11.8}$$

The relabeling of dummy indices occurs after the second equality in Eq. (11.8) (the dummy i is replaced by j while the dummy j is replaced by i; the third equality is obtained by interchanging the Λ matrices; the fourth equality uses the obvious symmetry $\delta_{ji} = \delta_{ij}$ of the Euclidean metric; and the last equality follows from (11.7). It is easy to see that this proof shows that the metric tensor is symmetric in any coordinate system because it is symmetric in Cartesian coordinates.

Although the metric tensor g_{ij} has $3 \times 3 = 9$ components, the symmetry (11.6) in its indices means that in any coordinate system at most 6 of these components are independent of each other.

Writing the metric tensors $g_{i'j'}$ and g_{ij} as matrices $\mathbf{g}'$ and $\mathbf{g}$, we can display the transformation (11.5) in the matrix form

$$\mathbf{g}' = \boldsymbol{\Lambda}^T\mathbf{g}'\boldsymbol{\Lambda} \qquad \text{(matrix equation)}, \tag{11.9}$$

where $\boldsymbol{\Lambda}$ denotes the transformation matrix $\Lambda^i{}_{i'}$ (so that the matrix $\Lambda^{i'}{}_i$, the inverse of $\Lambda^i{}_{i'}$ would be denoted $\boldsymbol{\Lambda}^{-1}$). Consider a transformation from a Cartesian coordinate system, in which $g_{ij} = \delta_{ij}$, to another Cartesian system, by means of a rotation. As the new coordinates are Cartesian, the transformed metric must also be Euclidean and (11.5) shows that this is the case:

$$g_{i'j'} = \Lambda^i{}_{i'}\Lambda^j{}_{j'}\,\delta_{ij} = \Lambda^i{}_{i'}\Lambda^i{}_{j'} = \delta_{i'j'} \quad \text{(Cartesian} \longrightarrow \text{Cartesian)}, \tag{11.10}$$

where the final equality follows from the fact that rotations are performed by *orthogonal matrices* (Goldstein, Poole and Safko [2001]), which are defined by the property $\Lambda^i{}_{i'}\Lambda^i{}_{j'} = \delta_{i'j'}$, i.e. $\boldsymbol{\Lambda}^T = \boldsymbol{\Lambda}^{-1}$ in the notation of Eq. (11.9). Rotations thus preserve the Euclidean metric. In contrast, any transformation to a non–Cartesian system will result in a non–trivial expression for the metric tensor.

The length of a curve in space (not necessarily a straight line) is computed in arbitrary coordinates using the metric tensor, as follows. Points on the curve are given by $x^i(\xi)$ where ξ is a parameter. The length $s(P,Q)$ of the curve between a point P with coordinates $x^i(\xi_1)$ and a point Q with coordinates $x^i(\xi_2)$ is given by integrating the infinitesimal distances $\mathrm{d}s$ along the curve, so that from (11.4) we have

$$s(P,Q) = \int_{\xi_1}^{\xi_2} \mathrm{d}s = \int_{\xi_1}^{\xi_2} \sqrt{g_{ij}\,\mathrm{d}x^i(\xi)\,\mathrm{d}x^j(\xi)} = \int_{\xi_1}^{\xi_2} \sqrt{g_{ij}\frac{\mathrm{d}x^i(\xi)}{\mathrm{d}\xi}\frac{\mathrm{d}x^j(\xi)}{\mathrm{d}\xi}}\,\mathrm{d}\xi\,. \tag{11.11}$$

This length is of course coordinate invariant.

To compute volumes in arbitrary coordinates we require an expression for the volume element $\mathrm{d}V$. Volumes are coordinate invariant and are built from lengths so, like the length $\mathrm{d}s$, the volume element $\mathrm{d}V$ must be determined by the metric tensor. We again use the device of choosing x^i to be a Cartesian system and then transforming to an arbitrary system $x^{i'}$. In Cartesian coordinates $x^i = \{x, y, z\}$, $\mathrm{d}V$ is of course $\mathrm{d}x\,\mathrm{d}y\,\mathrm{d}z = \mathrm{d}^3x$. Multivariable calculus tells us that the volume element in arbitrary coordinates $x^{i'} = \{x^{1'}, x^{2'}, x^{3'}\}$ is given by the Jacobian of the variable transformation from Cartesian coordinates to $x^{i'}$:

$$\begin{aligned} \mathrm{d}V = \mathrm{d}x\,\mathrm{d}y\,\mathrm{d}z &= \left|\frac{\partial(x,y,z)}{\partial(x^{1'},x^{2'},x^{3'})}\right| \mathrm{d}x^{1'}\,\mathrm{d}x^{2'}\,\mathrm{d}x^{3'} \\ &= \left|\frac{\partial(x,y,z)}{\partial(x^{1'},x^{2'},x^{3'})}\right| \mathrm{d}^3x'\,. \end{aligned} \tag{11.12}$$

The Jacobian

$$\left|\frac{\partial(x,y,z)}{\partial(x^{1'},x^{2'},x^{3'})}\right| \tag{11.13}$$

is the determinant of the matrix $\partial x^i/\partial x^{i'}$, which is the now–familiar $\Lambda^i{}_{i'}$ of Eq. (10.8), in the case where x^i are Cartesian. Denoting the matrix $\Lambda^i{}_{i'}$ by $\mathbf{\Lambda}$, as in Eq. (11.9), we can re–write Eq. (11.12) as

$$\mathrm{d}V = \mathrm{d}x\,\mathrm{d}y\,\mathrm{d}z = \left|\det\mathbf{\Lambda}\right| \mathrm{d}^3x' \qquad (x^i \text{ Cartesian}), \tag{11.14}$$

where we must remember that x^i in $\Lambda^i{}_{i'}$ are Cartesian. In the matrix equation (11.9) we have $\mathbf{g} = \mathbb{1}$ when x^i are Cartesian, since the Euclidean metric δ_{ij} is the identity matrix $\mathbb{1}$; taking the determinant of both sides of (11.9) in this case gives

$$g' = (\det\mathbf{\Lambda})^2 \qquad (x^i \text{ Cartesian}), \tag{11.15}$$

where g' denotes the determinant of the matrix formed by the metric tensor $g_{i'j'}$. Note from Eq. (11.15) that g' is always positive, for arbitrary $x^{i'}$. Inserting Eq. (11.15) in Eq. (11.14) we obtain the volume element in arbitrary coordinates in terms of the metric, without any reference to Cartesian coordinates:

$$\mathrm{d}V = \sqrt{g}\,\mathrm{d}^3x\,. \tag{11.16}$$

We dropped the prime, because expression (11.16) is valid in general coordinates, including of course the Cartesian ones we started from.

Exercise 11.2
Use the results of this Chapter to find the metric tensor, the line element and the volume element in spherical polar coordinates.
Solution
The required metric tensor is obtained from Eq. (11.5) by taking x^i to be Cartesian, so that $g_{ij} = \delta_{ij}$, and $x^{i'}$ to be spherical polars. The transformation matrix $\Lambda^i{}_{i'}$ in Eq. (11.5) is then given by Eq. (10.9). It is computationally simplest to take Eq. (11.5) in the matrix form (11.9), where $\mathbf{g} = \mathbb{1}$ in this case. The result is

$$g_{i'j'} = \begin{pmatrix} 1 & 0 & 0 \\ 0 & r^2 & 0 \\ 0 & 0 & r^2\sin^2\theta \end{pmatrix} = \operatorname{diag}\left(1,\ r^2,\ r^2\sin^2\theta\right), \tag{11.17}$$

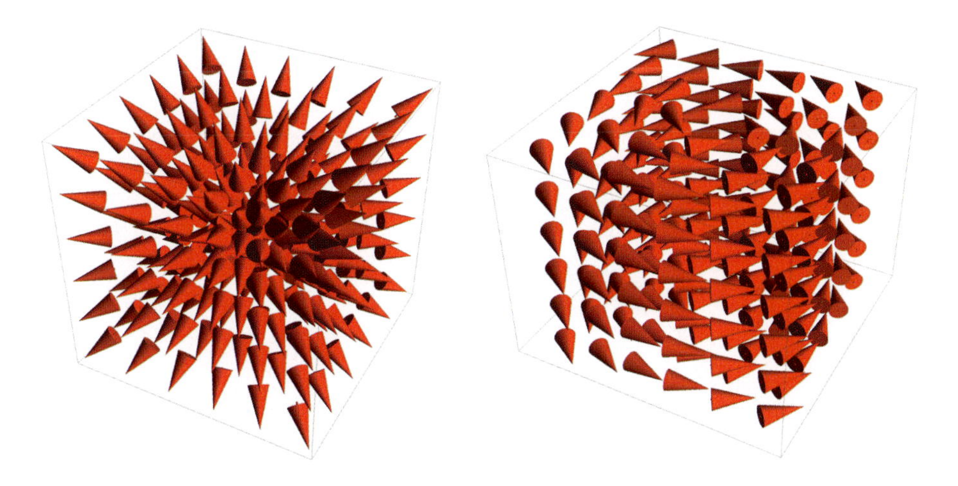

Figure 3.3: Vector fields. The arrowheads represent the direction of the vector field, but not its length, at the point where the arrowhead is located. The left picture represents the basis vector field $\boldsymbol{e}_r$ of the spherical polar system (see Eqs. (12.7)), while the right picture represents the basis vector field $\boldsymbol{e}_\phi$ of the cylindrical polar system (see Eqs. (12.8)).

giving the line element

$$\mathrm{d}s^2 = g_{i'j'}\,\mathrm{d}x^{i'}\mathrm{d}x^{j'} = \mathrm{d}r^2 + r^2(\mathrm{d}\theta^2 + \sin^2\theta\,\mathrm{d}\phi^2)\,. \tag{11.18}$$

The volume element (11.16) contains the square root of the determinant of the matrix (11.17); this determinant is easily seen to be

$$g' = r^4 \sin^2\theta \tag{11.19}$$

and we obtain the familiar spherical–polar volume element

$$\mathrm{d}V = r^2 \sin\theta\,\mathrm{d}r\,\mathrm{d}\theta\,\mathrm{d}\phi\,. \tag{11.20}$$

Exercise 11.3
Find the metric tensor, the line element and the volume element in cylindrical polar coordinates.
Solution
Proceeding as Exercise 11.2, or using the shortcut described in Exercise 10.2, one finds

$$g_{i'j'} = \mathrm{diag}\,(1,\ r^2,\ 1)\,, \tag{11.21}$$
$$\mathrm{d}s^2 = g_{i'j'}\,\mathrm{d}x^{i'}\mathrm{d}x^{j'} = \mathrm{d}r^2 + r^2\mathrm{d}\theta^2 + \mathrm{d}z^2\,, \tag{11.22}$$
$$g' = r^2\,, \tag{11.23}$$
$$\mathrm{d}V = r\,\mathrm{d}r\,\mathrm{d}\theta\,\mathrm{d}z\,. \tag{11.24}$$

The reader should note that the concept of the metric tensor and its properties were all derived from the simple relations (10.6)–(10.7), which are basic results of multivariable calculus. Indeed, all of the material in this Chapter is constructed on the modest foundations of Eqs. (10.6) and (10.7). We now proceed to develop the general concept of objects like the metric tensor, and their calculus. In the first

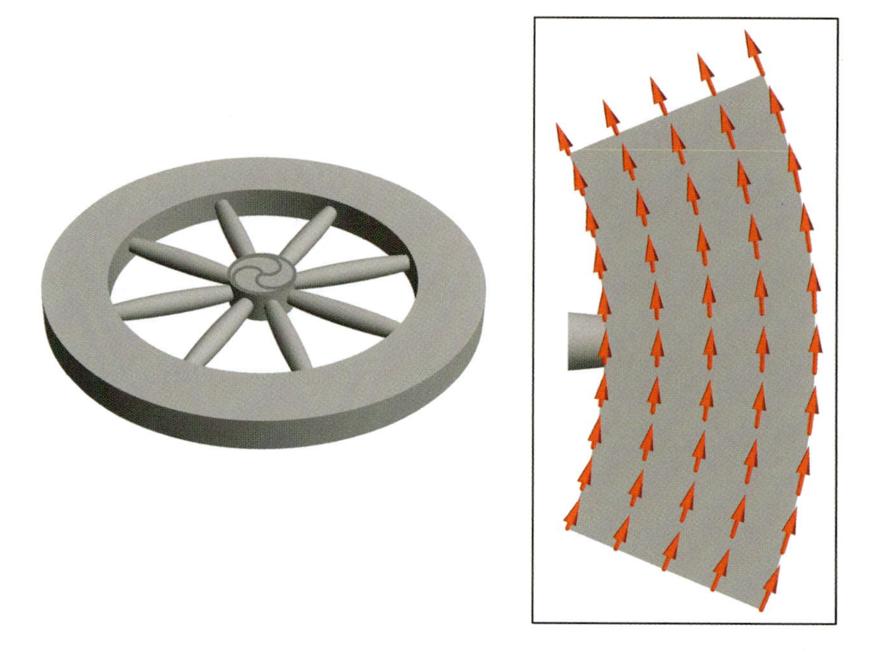

Figure 3.4: The velocity of a moving medium is a vector field. The picture shows a wheel that is rotating counter–clockwise at a constant angular velocity. The arrows in the inset show the velocity profile of the rim; the direction of the arrows is that of the instantaneous velocity and the length is the speed. Points on the rim have a velocity tangent to the spokes with a magnitude (speed) proportional to their distance from the centre.

instance, we take a step back to reconsider the most familiar example of this kind of object, a *vector field* (Fig. 3.3).

§12. Vectors and bases

The coordinate displacements $\mathrm{d}x^i$ are the components of a vector field in space: at each point with coordinates x^i they define a vector that points from x^i to $x^i + \mathrm{d}x^i$. Equations (10.6) reveal how the components of this particular vector field transform under a change of coordinate system. Now any vector field in space, such as the velocity profile of a moving medium (Fig. 3.4), will have the same transformation properties as every other vector field, so the components V^i of a general vector in the x^i coordinate system are related to the components $V^{i'}$ of this vector in the $x^{i'}$ coordinate system by the same equations (10.6), with the shorthand (10.8):

$$V^{i'} = \Lambda^{i'}_{\ i} V^i \,, \quad V^i = \Lambda^{i}_{\ i'} V^{i'} \,. \tag{12.1}$$

The components of a vector field are meaningless unless we specify the triad of basis vector fields at each point to which the components refer. In Cartesian coordinates

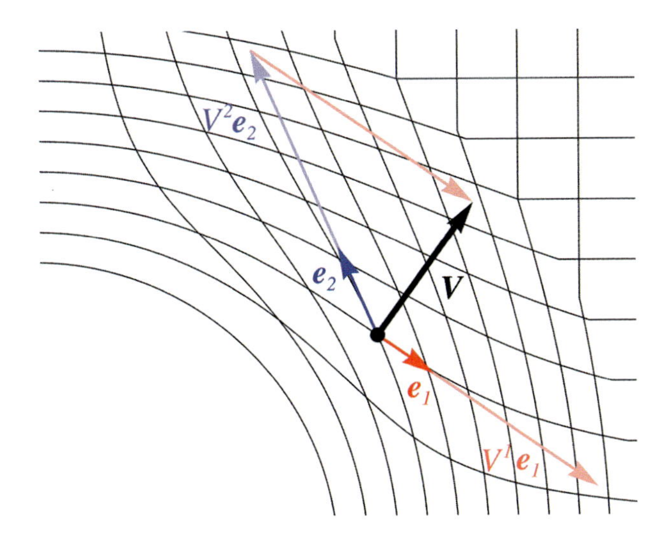

Figure 3.5: Vector and basis. A visualization of the representation (12.3) of a vector in terms of a basis in two dimensions. The value of the vector field $\boldsymbol{V}$ at every point is given by the vector sum of $V^1\boldsymbol{e}_1$ and $V^2\boldsymbol{e}_2$, where $\{\boldsymbol{e}_1, \boldsymbol{e}_2\}$ are the basis vector fields at the point and $\{V^1, V^2\}$ are numbers, the components of $\boldsymbol{V}$ in the basis $\{\boldsymbol{e}_1, \boldsymbol{e}_2\}$ at the point. See also Figs. 3.6 and 3.7.

$x^i = \{x, y, z\}$ the basis vectors are the familiar unit vectors $\{\boldsymbol{i}, \boldsymbol{j}, \boldsymbol{k}\}$ in the x–, y– and z–directions:

$$\{\boldsymbol{i}, \boldsymbol{j}, \boldsymbol{k}\} = \{\boldsymbol{e}_x, \boldsymbol{e}_y, \boldsymbol{e}_z\} = \boldsymbol{e}_i \qquad (x^i \text{ Cartesian}), \tag{12.2}$$

where we introduce the notation $\boldsymbol{e}_i$ to denote the basis vector fields in x^i coordinates. The vector field $\boldsymbol{V}$ can be expanded (Fig. 3.5) in terms of its components V^i in the basis $\boldsymbol{e}_i$ associated with any coordinate system x^i, as

$$\boldsymbol{V} = V^i \boldsymbol{e}_i \,. \tag{12.3}$$

Since the same vector field $\boldsymbol{V}$ can be written in a different coordinate system $x^{i'}$ we have

$$\begin{aligned} \boldsymbol{V} &= V^i \boldsymbol{e}_i = V^{i'} \boldsymbol{e}_{i'} \\ &= \Lambda^i{}_{i'} V^{i'} \boldsymbol{e}_i \,. \end{aligned} \tag{12.4}$$

The second line in Eq. (12.4) was obtained by use of Eq. (12.1) and comparison with the first line reveals how the basis vectors transform when we change coordinate system:

$$\boldsymbol{e}_{i'} = \Lambda^i{}_{i'} \boldsymbol{e}_i \,. \tag{12.5}$$

Note the appearance in Eq. (12.5) of the matrix $\Lambda^i{}_{i'}$ that is the inverse of the matrix $\Lambda^{i'}{}_i$ appearing in the first of Eqs. (12.1); this ensures that $V^{i'}\boldsymbol{e}_{i'} = V^i\boldsymbol{e}_i$ so the vector

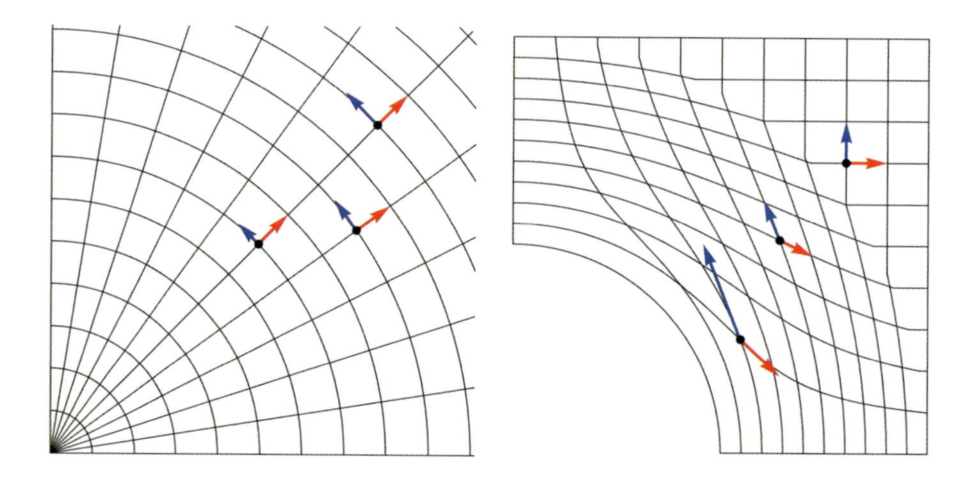

Figure 3.6: Coordinate basis vectors change with position, following the coordinate lines. Polar coordinates in two dimensions (spherical polar coordinates with $\theta = \pi/2$) are depicted on the left; the red arrows are the vector field $\boldsymbol{e}_r$ and the blue $\boldsymbol{e}_\phi$ (see Eqs. (12.7)). The basis vectors are everywhere orthogonal but they rotate as one moves through the plane. While $\boldsymbol{e}_r$ remains a unit vector, the length of $\boldsymbol{e}_\phi$, equal to r, varies from zero to infinity (see Eq. (12.14)). A coordinate system associated with cloaking is shown on the right (Fig. 2.6). Here the orientations and lengths of both basis vectors vary, as does the angle between them. Since the dot product $\boldsymbol{e}_1 \cdot \boldsymbol{e}_2$ of the basis vectors does not vanish everywhere, the metric tensor is not diagonal in these coordinates: $g_{12} \neq 0$ (see Eq. (12.12)).

$\boldsymbol{V}$ is the same whatever coordinate system we choose. Definition (10.8) of $\Lambda^i{}_{i'}$ also shows that the components of the new basis system (12.5) in terms of the old one $\boldsymbol{e}_i$ are

$$e^i_{i'} = \frac{\partial x^i}{\partial x^{i'}} \,. \tag{12.6}$$

The basis vectors thus point in the direction in which the coordinates change; they follow the coordinate lines (Fig. 3.6) and, in orthogonal systems, are orthogonal to the coordinate surfaces (Fig. 3.7).

Exercise 12.1
Compute the basis vectors of spherical polar coordinates in term of the Cartesian basis (12.2).
Solution
From Eqs. (12.5), (12.2) and (10.9) we obtain for $\boldsymbol{e}_{i'} = \{\boldsymbol{e}_r, \boldsymbol{e}_\theta, \boldsymbol{e}_\phi\}$

$$\begin{aligned} \boldsymbol{e}_r &= \sin\theta\cos\phi\,\boldsymbol{i} + \sin\theta\sin\phi\,\boldsymbol{j} + \cos\theta\,\boldsymbol{k}\,, \\ \boldsymbol{e}_\theta &= r\cos\theta\cos\phi\,\boldsymbol{i} + r\cos\theta\sin\phi\,\boldsymbol{j} - r\sin\theta\,\boldsymbol{k}\,, \\ \boldsymbol{e}_\phi &= -r\sin\theta\sin\phi\,\boldsymbol{i} + r\sin\theta\cos\phi\,\boldsymbol{j}\,. \end{aligned} \tag{12.7}$$

Exercise 12.2
Compute the basis vectors of cylindrical polar coordinates in term of the Cartesian basis (12.2).

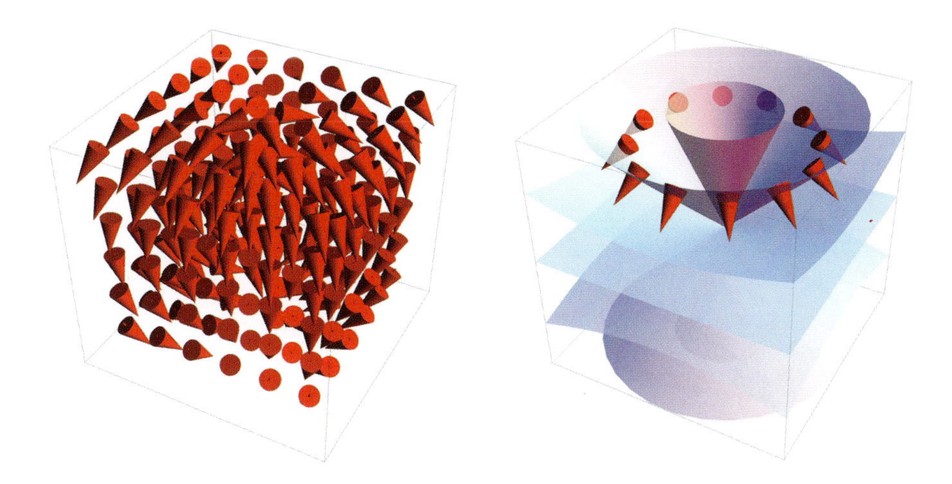

Figure 3.7: The basis vector field $\boldsymbol{e}_\theta$ of the spherical polar system (see Eqs. (12.7)). On the left, the arrowheads show the direction, but not the length, of $\boldsymbol{e}_\theta$ at various points. The surfaces in the right picture are surfaces of constant θ, to which $\boldsymbol{e}_\theta$ is orthogonal. In an orthonormal coordinate basis, like the spherical polar basis (12.7), each basis vector $\boldsymbol{e}_i$ is orthogonal to surfaces of constant x^i.

Solution
From Eqs. (12.5), (12.2) and (10.13) we obtain for $\boldsymbol{e}_{i'} = \{\boldsymbol{e}_r, \boldsymbol{e}_\phi, \boldsymbol{e}_z\}$

$$\begin{aligned} \boldsymbol{e}_r &= \cos\phi\, \boldsymbol{i} + \sin\phi\, \boldsymbol{j}\,, \\ \boldsymbol{e}_\phi &= -r\sin\phi\, \boldsymbol{i} + r\cos\phi\, \boldsymbol{j}\,, \\ \boldsymbol{e}_z &= \boldsymbol{k}\,. \end{aligned} \tag{12.8}$$

The expression (11.3) is the squared length of the vector dx^i, and the metric tensor similarly gives the squared length of a general vector:

$$\left|\boldsymbol{V}\right|^2 = \boldsymbol{V}\cdot\boldsymbol{V} = g_{ij}V^iV^j = g_{i'j'}V^{i'}V^{j'}\,. \tag{12.9}$$

In the final equality in Eq. (12.9) we have used the fact that the length of a vector is an invariant quantity, a number independent of the coordinates used to compute it. The equality of the expressions $g_{ij}V^iV^j$ and $g_{i'j'}V^{i'}V^{j'}$ evaluated in the coordinate systems x^i and $x^{i'}$, respectively, can be explicitly shown from the transformation rules of the metric tensor and vector components.

Exercise 12.3
Prove that

$$g_{ij}V^iV^j = g_{i'j'}V^{i'}V^{j'} \tag{12.10}$$

using the transformation properties (11.5) and (12.1) together with the inverse relations (10.17).

The reader will see very clearly upon performing Exercise 12.3 that the invariance of the length (12.9) under coordinate transformations works because the vector V^i and

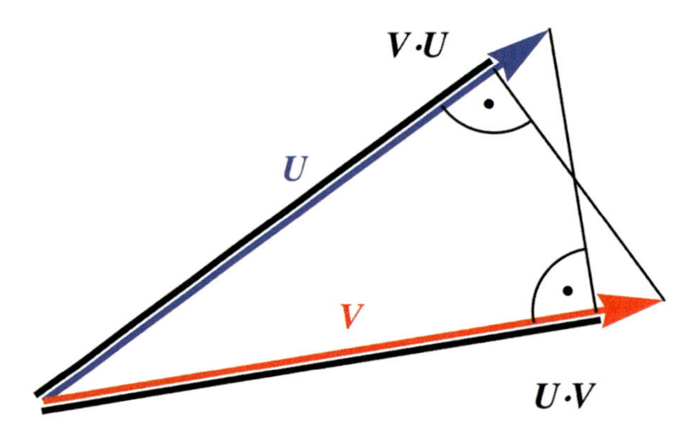

Figure 3.8: The scalar product of two vectors $\boldsymbol{V}$ and $\boldsymbol{U}$ of unit length, $|\boldsymbol{V}|^2 = 1$ and $|\boldsymbol{U}|^2 = 1$. The two black lines show the result of orthogonally projecting one of the vectors on to the other; because the vectors are unit vectors these black lines have length equal to the scalar product $\boldsymbol{V} \cdot \boldsymbol{U}$. It is a standard exercise of vector algebra to show that in Cartesian coordinates the expression $\delta_{ij} V^i U^j$ agrees with this geometrical construction of the scalar product; Eq. (12.11) generalizes the expression to arbitrary coordinates.

metric g_{ij} are transformed by matrices that are inverse to each other. This inverse relationship of their transformations is to be associated with the fact that V^i is an upper–index object whereas g_{ij} is a lower–index object. Each summation in Eq. (12.9) over an upper and lower index, called a *contraction*, is a coordinate–invariant operation because of this inverse property. The expression (12.9) is a special case of the scalar product of two vectors $\boldsymbol{V}$ and $\boldsymbol{U}$ (Fig. 3.8),

$$\boldsymbol{V} \cdot \boldsymbol{U} = g_{ij} V^i U^j \,, \tag{12.11}$$

the coordinate invariance of which is again a consequence of the contraction of the indices, just as in Eq. (12.9).

The scalar product of two basis vectors is seen from the definition (12.11) to be

$$\boldsymbol{e}_i \cdot \boldsymbol{e}_j = g_{ij} \,, \tag{12.12}$$

since the components $(\boldsymbol{e}_i)^j$ of a basis vector $\boldsymbol{e}_i$ are δ_i^j, i.e. $\boldsymbol{e}_i = \delta_i^j \boldsymbol{e}_j$. In the case of Cartesian coordinates the metric is δ_{ij} and Eq. (12.12) gives

$$\boldsymbol{e}_i \cdot \boldsymbol{e}_j = \delta_{ij} \qquad (x^i \text{ Cartesian}). \tag{12.13}$$

Note that Eq. (12.13) simply says that the Cartesian basis (12.2) is orthonormal. By contrast, Eq. (12.12) shows that in a non–Cartesian system, where the metric g_{ij} is non–trivial, the basis will not be orthonormal. Consider for example the case of spherical polar coordinates. For the spherical polar basis we can compute the

dot products of the basis vectors using the right–hand sides of Eqs. (12.7) or, much more simply, by using (12.12) and the metric (11.17) in spherical polar coordinates. We see that the basis vectors are orthogonal to each other, but they are not all unit vectors:

$$|\boldsymbol{e}_r|^2 = 1\,, \quad |\boldsymbol{e}_\theta|^2 = r^2\,, \quad |\boldsymbol{e}_\phi|^2 = r^2 \sin^2\theta\,. \tag{12.14}$$

One can of course easily construct an orthonormal basis $\hat{\boldsymbol{e}}_i$ by rescaling $\boldsymbol{e}_\theta$ and $\boldsymbol{e}_\phi$:

$$\hat{\boldsymbol{e}}_r = \boldsymbol{e}_r\,, \quad \hat{\boldsymbol{e}}_\theta = \frac{1}{r}\,\boldsymbol{e}_\theta\,, \quad \hat{\boldsymbol{e}}_\phi = \frac{1}{r\sin\theta}\,\boldsymbol{e}_\phi\,. \tag{12.15}$$

The reader unfamiliar with the material of this Section will only have encountered spherical polar coordinates in combination with the orthonormal basis (12.15). How did we end up with the non–orthonormal basis $\{\boldsymbol{e}_r, \boldsymbol{e}_\theta, \boldsymbol{e}_\phi\}$? The answer is that we let the coordinates induce our basis through their differentiable structure. Recall that the components of a vector were introduced by analogy with the coordinate differentials $\mathrm{d}x^i$. The transformation rule for vector components was induced by that of $\mathrm{d}x^i$ and this discussion of vector components determined the basis as in Eqs. (12.1)–(12.5). Such a basis, induced naturally by the coordinates, is called a *coordinate basis*. The fact that the differentiable properties of the coordinates completely determine the coordinate basis is the reason why the coordinate bases (12.5) transform exactly like the partial derivative operators (10.7).* The orthonormal basis (12.15), by contrast, is not induced in a similar manner by *any* coordinate system—it is a *non–coordinate basis* (see Schutz [2009] and Misner, Thorne and Wheeler [1973] for more details). As noted above, the only coordinate system that induces an orthonormal coordinate basis is the Cartesian system, so the use of coordinate bases in general means dealing with non–orthonormal basis vectors. It might be suspected that it is always simpler to work in an orthonormal basis; in fact, for most purposes a coordinate basis is much simpler, in particular for the manipulations in curvilinear coordinates performed in electromagnetism textbooks. The reason these texts use the more complicated non–coordinate bases is that to exploit the simplicity of coordinate bases requires a little knowledge of tensor analysis.

Exercise 12.4
Show explicitly that the coordinate basis for cylindrical polar coordinates is not orthonormal. Construct an orthonormal basis.
Solution
From Eq. (12.12) and the metric (11.21) in cylindrical polar coordinates follows

$$|\boldsymbol{e}_r|^2 = 1\,, \quad |\boldsymbol{e}_\theta|^2 = r^2\,, \quad |\boldsymbol{e}_z|^2 = 1\,. \tag{12.16}$$

The basis is not orthonormal because of the variation of $|\boldsymbol{e}_\theta|^2$ with r; an orthonormal basis is clearly

$$\hat{\boldsymbol{e}}_r = \boldsymbol{e}_r\,, \quad \hat{\boldsymbol{e}}_\theta = \frac{1}{r}\,\boldsymbol{e}_\theta\,, \quad \hat{\boldsymbol{e}}_z = \boldsymbol{e}_z\,. \tag{12.17}$$

---

*This is not just a pleasant correspondence; in rigorous differential geometry the partial derivative operators $\partial/\partial x^i$ *are* the coordinate basis vectors.

Exercise 12.5

One of the most widely used vectors in flat space is the position vector, a vector that points from the origin to any point of interest (it would, for example, describe Hooke's force that grows linearly with distance). This is a somewhat special example of a vector field and some care is required in its use. The first thing to note is that there is no unique concept of the position vector; it is dependent on the origin of the particular coordinate system used. Nevertheless the position vector associated with any origin can be raised to the status of a proper vector field. Thus, consider an arbitrary Cartesian coordinate system $x^i = \{x, y, z\}$; we can uniquely define a vector field $\boldsymbol{r}$ by specifying that its components r^i at every point x^i are given by $\{x, y, z\}$, i.e.

$$\boldsymbol{r} = x\,\boldsymbol{e}_x + y\,\boldsymbol{e}_y + z\,\boldsymbol{e}_z\,. \tag{12.18}$$

This completely determines the vector $\boldsymbol{r}$ because its components in any other coordinate system are found from the transformation law (12.1).

(a) Let $x^{i'} = \{x', y', z'\}$ be a Cartesian system related to $x^i = \{x, y, z\}$ by a rotation; show that the components $r^{i'}$ of $\boldsymbol{r}$ in the coordinates $x^{i'}$ are given by $\{x', y', z'\}$, i.e.

$$\boldsymbol{r} = x'\boldsymbol{e}_{x'} + y'\boldsymbol{e}_{y'} + z'\boldsymbol{e}_{z'}\,. \tag{12.19}$$

(b) Verify that in the spherical polar system defined by Eqs. (10.1), which has the same origin as $x^i = \{x, y, z\}$, the components of $\boldsymbol{r}$ are given by $\{r, 0, 0\}$, i.e.

$$\boldsymbol{r} = r\,\boldsymbol{e}_r\,. \tag{12.20}$$

(b) Show that in the cylindrical polar system defined by Eqs. (10.2), which also has the same origin as $x^i = \{x, y, z\}$, the components of $\boldsymbol{r}$ are given by $\{r, 0, z\}$, i.e.

$$\boldsymbol{r} = r\,\boldsymbol{e}_r + z\,\boldsymbol{e}_r\,. \tag{12.21}$$

These results show that the position vector associated with each point must be carefully distinguished from the coordinates of that point. In Cartesian systems (that share an origin) the components of $\boldsymbol{r}$ at each point are indeed given by the coordinates x^i of that point, but this is not true in other coordinate systems: in spherical polars the position vector at the point $\{r, \theta, \phi\}$ has components $\{r, 0, 0\}$, while in cylindrical polars the position vector at the point $\{r, \phi, z\}$ has components $\{r, 0, z\}$. The notation $\boldsymbol{x}$ that is sometimes used for the position vector derives from its components in Cartesian coordinates and has the drawback that it suggests that the components of this vector are always x^i, which is not the case. The other common notation $\boldsymbol{r}$, which we adopt here, derives from the component values of the position vector in spherical polars and is also not entirely logical in view of the fact that r is commonly used to denote the radial coordinate in cylindrical polars (see Eqs. (12.20) and (12.21)).

Solution

(a) It is a well–known property of a rotation from one Cartesian system $x^i = \{x, y, z\}$ to another $x^{i'} = \{x', y', z'\}$ that the transformation equations $x^{i'} = x^{i'}(x, y, z)$ are linear in the coordinates x^i (Goldstein, Poole and Safko [2001]). Hence the matrices (10.8) are independent of the coordinates; they contain only the (three independent) parameters specifying the rotation. This means that, in this case, Eq. (10.6) can be immediately integrated to give

$$x^{i'} = \Lambda^{i'}{}_i x^i \qquad \text{(rotation from Cartesian to Cartesian)}, \tag{12.22}$$

where there is no constant of integration because in a rotation the origin is the same in both systems. Now since we have $r^i = x^i$ from Eq. (12.18), the right–hand side of Eq. (12.22) is $\Lambda^{i'}{}_i r^i$, which we recall from (12.1) is precisely $r^{i'}$. Equation (12.22) thus shows that $r^{i'} = x^{i'}$, as in Eq. (12.19).

(b) This follows from the transformation rule (12.1) for the vector $\boldsymbol{r}$, namely $r^{i'} = \Lambda^{i'}{}_i r^i$. In the case at hand, $r^i = x^i = \{x, y, z\} = \{r\sin\theta\cos\phi, r\sin\theta\sin\phi, r\cos\theta\}$ and $\Lambda^{i'}{}_i$ is given by Eq. (10.9). A straightforward calculation gives $r^{i'} = \{r, 0, 0\}$, which is Eq. (12.20).

(c) We proceed as in (b), but here $r^i = x^i = \{x, y, z\} = \{r\cos\phi, r\sin\phi, z\}$ and $\Lambda^{i'}{}_i$ is given by Eq. (10.13). The transformation $r^{i'} = \Lambda^{i'}{}_i r^i$ gives $r^{i'} = \{r, 0, z\}$, which is Eq. (12.21).

---

§13. ONE–FORMS AND GENERAL TENSORS

Vector fields and the metric tensor are examples of geometrical objects whose components have simple transformation rules involving the matrices (10.8). We now develop the notion of such objects, called *tensors*, in full generality. In the expression $g_{ij}U^iV^j$ for the scalar product (12.11), we isolate the quantity $g_{ij}V^j$ which is contracted with U^i to give $\boldsymbol{U}\cdot\boldsymbol{V}$. There is one free index on $g_{ij}V^j$, in the lower position, and we denote this object by V_i, as follows

$$V_i = g_{ij}V^j\,. \tag{13.1}$$

Note carefully that although we have used the same "kernal" V in the definition (13.1) of V_i as appears in the vector V^i, these are two different objects, albeit related by Eq. (13.1). The transformation rule for V_i is easily established from those of V^i and g_{ij}; one easily verifies from Eqs. (12.1) and (11.5) and the inverse relation (10.17) that

$$V_{i'} = g_{i'j'}V^{j'} = \Lambda^i{}_{i'}g_{ij}V^j = \Lambda^i{}_{i'}V_i\,. \tag{13.2}$$

We see again that a contraction of indices, such as that of j in (13.1), is invariant, so that the transformation of V_i results from the transformation of the first index on the metric in (13.1). The quantities V_i are the components of a *covariant vector* or *one–form*; the ordinary vector V^i is sometimes called a *contravariant vector*. Having an index in the lower position, V_i transforms with the matrix $\Lambda^i{}_{i'}$ inverse to the matrix $\Lambda^{i'}{}_i$ used to transform the vector V^i in Eq. (12.1). In the picture developed in §26 the electric and magnetic fields, $\boldsymbol{E}$ and $\boldsymbol{H}$, appear as covariant vectors in 3–dimensional space—fields of spatial one–forms. Physical force fields are typically covariant vectors, whereas the velocity profiles of matter distributions are ordinary vector fields. In §30 the wave momentum of light will emerge as a one–form, with the particle momentum as a vector.

We can view Eq. (13.1) as lowering the index on the vector V^i using the metric tensor, producing the associated one–form V_i. The *inverse metric tensor* g^{ij} is defined by

$$g^{ij}g_{jk} = \delta^i{}_k\,, \tag{13.3}$$

so that the matrix g^{ij} is the inverse of the matrix g_{ij}. Multiplying across Eq. (13.1) by g^{ki} and contracting on i we find

$$\begin{aligned} g^{ki}V_i &= g^{ki}g_{ij}V^j \\ &= V^k\,, \end{aligned} \tag{13.4}$$

where Eq. (13.3) has been used to obtain the second line. Equation (13.4) can be regarded as raising the index on the one–form V_i using the inverse metric tensor, producing the associated vector V^i. Note that the vector V^i and one–form V_i are the same only in Cartesian coordinates where $g_{ij} = \delta_{ij}$ and $g^{ij} = \delta^{ij}$.

The expression (12.9) for the squared length of a vector is more compactly written using the one–form V_i:

$$|\boldsymbol{V}|^2 = \boldsymbol{V}\cdot\boldsymbol{V} = V_iV^i\,. \tag{13.5}$$

Similarly the scalar product (12.11) of two vectors $\boldsymbol{V}$ and $\boldsymbol{U}$ can be written

$$\boldsymbol{V} \cdot \boldsymbol{U} = g_{ij} V^i U^j = V_i U^i = V^i U_i = V_{i'} U^{i'} = V^{i'} U_{i'} \,, \tag{13.6}$$

where we have used the metric to lower the index on either V^i or U^j (remember Eq. (11.6)) and the coordinate invariance of the scalar product follows from the transformation properties of vectors and one–forms. We can also view Eq. (13.3) as either raising an index on g_{jk} or lowering an index on g^{ij}, giving us the quantities

$$g^i{}_k = g_k{}^i = \delta^i{}_k \,. \tag{13.7}$$

From the position of the indices on the inverse metric tensor g^{ik} we expect it to transform in a vector–like fashion. To verify this we write the vector and one–form in Eq. (13.4) in terms of their transformed values $V^{i'}$ and $V_{j'}$ using Eqs. (12.1) and (13.2):

$$\Lambda^i{}_{i'} V^{i'} = g^{ij} \Lambda^{j'}{}_j V_{j'} \,.$$

We next multiply across by $\Lambda^{k'}{}_i$ and contract on i, using the inverse relationship (10.17):

$$V^{k'} = \Lambda^{k'}{}_i \, g^{ij} \Lambda^{j'}{}_j V_{j'} \,.$$

Substituting $V^{k'} = g^{k'j'} V_{j'}$, which follows from Eq. (13.4), yields

$$g^{k'j'} V_{j'} = \Lambda^{k'}{}_i \, g^{ij} \Lambda^{j'}{}_j V_{j'} \,.$$

Since this must hold for arbitrary one–forms $V_{j'}$ we finally obtain the desired transformation rule (we change the index label k' to i'):

$$g^{i'j'} = \Lambda^{i'}{}_i \Lambda^{j'}{}_j g^{ij} \,. \tag{13.8}$$

Thus the inverse metric tensor does indeed transform similarly to a vector (see Eq. (12.1)), the transformation matrix being $\Lambda^{i'}{}_i$, the inverse to the matrix $\Lambda^i{}_{i'}$ used to transform the lower–index metric tensor in Eq. (11.5).

Exercise 13.1

Find the inverse metric tensor components in spherical and cylindrical polar coordinates.

Solution

The metric tensor in spherical and cylindrical polar coordinates is given by Eqs. (11.17) and (11.21), respectively. Since the components g^{ij} form the matrix inverse of g_{ij} we clearly have for spherical polars

$$g^{ij} = \mathrm{diag}\left(1,\ \frac{1}{r^2},\ \frac{1}{r^2 \sin^2\theta}\right) \tag{13.9}$$

and for cylindrical polars

$$g^{ij} = \mathrm{diag}\left(1,\ \frac{1}{r^2},\ 1\right) . \tag{13.10}$$

It is now an obvious step to the general notion of a tensor, beyond the specific examples of vectors, one–forms and the metric tensor that we have treated in detail.

Tensors can have any number of indices, in upper and lower positions. The transformation rule for a tensor with an arbitrary collection of indices should be clear from the cases (12.1), (13.2), (11.5) and (13.8); we give as an example a four–index tensor:

$$R^{i'}{}_{j'k'l'} = \Lambda^{i'}{}_{i}\Lambda^{j}{}_{j'}\Lambda^{k}{}_{k'}\Lambda^{l}{}_{l'}R^{i}{}_{jkl}\,. \tag{13.11}$$

Note that no thought is necessary in arranging the transformation matrices in Eq. (13.11): the simple requirement that the indices line up correctly on both sides of the equation determines how to transform each index. Tensors can consist of products of other tensors, such as in the simple example

$$T^{ij} = U^{i}W^{j}\,. \tag{13.12}$$

It is clear that Eq. (13.12) adheres to the general transformation law (13.11) because of the transformation rule (12.1) for vectors. One can also raise and lower the indices of a general tensor in complete analogy to Eqs. (13.1) and (13.4); for example

$$R_{ijkl} = g_{im}R^{m}{}_{jkl}\,, \quad R^{i}{}_{j}{}^{k}{}_{l} = g^{km}R^{i}{}_{jml}\,. \tag{13.13}$$

Exercise 13.2
Assuming that $R^{i}{}_{jkl}$ is a tensor, and therefore has the transformation behaviour (13.11), show that R_{ijkl} and $R^{i}{}_{j}{}^{k}{}_{l}$, defined by Eqs. (13.13), also transform in the manner of tensors.

In Eqs. (13.13) new tensors are produced by products of other tensors, as in Eq. (13.12), but with the additional feature of contractions on some of the indices. Contractions always produce new tensors because of their coordinate invariance.

It is straightforward to introduce bases for general tensors, as we have for the particular case of vectors. Consider first a one–form V_i. Strictly speaking, V_i are the *components* of this one–form in the coordinate system x^i, but the one–form itself, like all tensors, is a geometrical object that has an existence independent of whether or not we choose to cover space with a coordinate grid. The situation is no different from that of vectors: a vector $\boldsymbol{U}$ exists independently of any coordinate grid, but when coordinates x^i are chosen, U^i denote its components in the basis $\boldsymbol{e}_i$ of this coordinate system. There must therefore be a *one–form basis* associated with every coordinate system, just as there is a vector basis; the components V_i refer to an expansion of the one–form in this one–form basis. We need a notation for the one–form that does not refer to any coordinate grid, and we choose $\widetilde{\boldsymbol{V}}$, which is to be distinguished from the associated vector $\boldsymbol{V}$. We will in fact be sloppy in the use of this notation when the character of $\boldsymbol{V}$ or $\widetilde{\boldsymbol{V}}$ is obvious from the context; for example, we always denote the electric field strength by $\boldsymbol{E}$, even when it is viewed as a one–form. Denoting the one–form basis induced by the coordinates x^i by $\boldsymbol{\omega}^i$, we thus have, for the one–form $\widetilde{\boldsymbol{V}}$, the expansion

$$\widetilde{\boldsymbol{V}} = V_i\,\boldsymbol{\omega}^i\,. \tag{13.14}$$

Exercise 13.3
Work out how the coordinate one–form basis changes under a coordinate transformation.
Solution
The one–form $\widetilde{\boldsymbol{V}}$ is the same object no matter what basis it is expanded in. Hence, if $\boldsymbol{\omega}^i$ and $\boldsymbol{\omega}^{i'}$ are the one–form bases associated with coordinates x^i and $x^{i'}$, respectively, we have

$$\widetilde{\boldsymbol{V}} = V_{i'}\,\boldsymbol{\omega}^{i'} = V_i\,\boldsymbol{\omega}^i\,. \tag{13.15}$$

Inserting the inverse of Eq. (13.2) in this we find

$$V_{i'}\,\boldsymbol{\omega}^{i'} = \Lambda^{i'}{}_i V_{i'}\,\boldsymbol{\omega}^i\,,$$

and since this must hold for arbitrary $V_{i'}$ we obtain the required transformation rule

$$\boldsymbol{\omega}^{i'} = \Lambda^{i'}{}_i\,\boldsymbol{\omega}^i\,. \tag{13.16}$$

Note that the position of the index on the one–form basis is appropriate to its transformation law (13.16) and the invariance of contractions (see Eq. (13.15)). In Eq. (13.12) we have a tensor defined to be the product of the vectors $\boldsymbol{U}$ and $\boldsymbol{W}$; when writing this *tensor product* in coordinate–free notation the symbol $\otimes$ is used, so that the coordinate–free version of Eq. (13.12) is

$$\boldsymbol{T} = \boldsymbol{U} \otimes \boldsymbol{W}\,, \tag{13.17}$$

where $\boldsymbol{T}$ denotes in a coordinate–free manner the tensor whose components in x^i coordinates are (13.12). Expanding the vectors $\boldsymbol{U}$ and $\boldsymbol{W}$ in the vector basis $\boldsymbol{e}_i$, as in Eq. (12.3), we find

$$\begin{aligned}\boldsymbol{T} &= U^i W^j \boldsymbol{e}_i \otimes \boldsymbol{e}_j \\ &= T^{ij} \boldsymbol{e}_i \otimes \boldsymbol{e}_j\,,\end{aligned} \tag{13.18}$$

which shows how this tensor is expanded in terms of its components T^{ij} in the basis $\boldsymbol{e}_i \otimes \boldsymbol{e}_j$ induced by the coordintes x^i. All tensors whose components have just two indices, both in the upper position, will have similar expansions in the basis $\boldsymbol{e}_i \otimes \boldsymbol{e}_j$. The basis for every tensor is built from appropriate tensor products of the vector and one–form bases; for example a tensor with components $R^i{}_{jkl}$ has the basis expansion

$$\boldsymbol{R} = R^i{}_{jkl} \boldsymbol{e}_i \otimes \boldsymbol{\omega}^j \otimes \boldsymbol{\omega}^k \otimes \boldsymbol{\omega}^l\,. \tag{13.19}$$

Exercise 13.4
Show that the tensors $\boldsymbol{T}$ and $\boldsymbol{R}$ given by Eqs. (13.18) and (13.19) have a similar expansion in every coordinate system.
Solution
From the transformation rules for tensor components, the vector basis and the one–form basis, together with the inverse relation (10.17), one easily finds

$$\boldsymbol{T} = T^{ij} \boldsymbol{e}_i \otimes \boldsymbol{e}_j = T^{i'j'} \boldsymbol{e}_{i'} \otimes \boldsymbol{e}_{j'}\,, \tag{13.20}$$

$$\boldsymbol{R} = R^i{}_{jkl} \boldsymbol{e}_i \otimes \boldsymbol{\omega}^j \otimes \boldsymbol{\omega}^k \otimes \boldsymbol{\omega}^l = R^{i'}{}_{j'k'l'} \boldsymbol{e}_{i'} \otimes \boldsymbol{\omega}^{j'} \otimes \boldsymbol{\omega}^{k'} \otimes \boldsymbol{\omega}'^{l}\,. \tag{13.21}$$

For a more leisurely and rigorous exposition of tensor bases the reader is referred to Schutz [2009] and Misner, Thorne and Wheeler [1973].

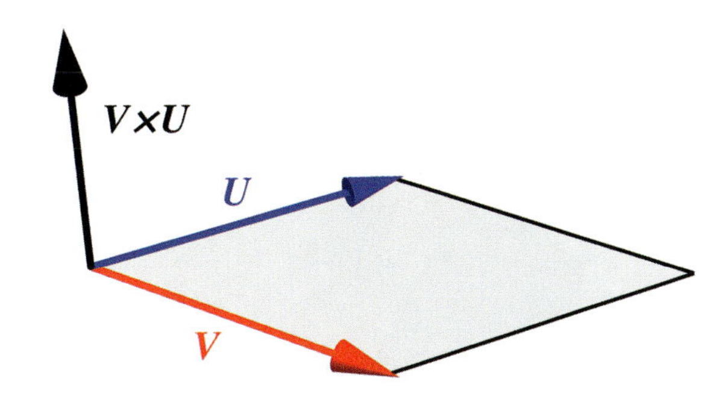

Figure 3.9: The vector product of two vectors $\boldsymbol{V}$ and $\boldsymbol{U}$. The shaded parallelogram has an area equal to the magnitude of the vector product $\boldsymbol{V} \times \boldsymbol{U}$, and the direction of $\boldsymbol{V} \times \boldsymbol{U}$ is perpendicular to $\boldsymbol{U}$ and $\boldsymbol{V}$. The choice of direction of $\boldsymbol{V} \times \boldsymbol{U}$ in the figure, rather than the opposite direction pointing downwards, corresponds to a handedness rule. This means that the tensor used to calculate the vector product, the Levi–Civita tensor, shares this handedness choice (see Eq. (14.2)). Standard vector algebra relates this geometrical construction to the Cartesian expression for $\boldsymbol{V} \times \boldsymbol{U}$; in an arbitrary coordinate system the formula is (14.10).

§14. VECTOR PRODUCTS AND THE LEVI–CIVITA TENSOR

One of the basic operations of vector algebra is the vector product $\boldsymbol{V} \times \boldsymbol{U}$ (Fig. 3.9). Although the rule for evaluating $\boldsymbol{V} \times \boldsymbol{U}$ in Cartesian coordinates is second nature, our goal here is an understanding of vector products as an operation among tensors. Once understood in this way, the simple transformation properties of tensors mean that it will be straightforward to evaluate vector products in any coordinate system.

A key property of $\boldsymbol{V} \times \boldsymbol{U}$ is its antisymmetry, i.e. $\boldsymbol{V} \times \boldsymbol{U} = -\boldsymbol{U} \times \boldsymbol{V}$. This property can be implemented by introducing the *completely antisymmetric symbol*, or *permutation symbol* $[ijk]$ defined by

$$[ijk] = \begin{cases} +1 & , \text{ if } ijk \text{ is an even permutation of } 123\,, \\ -1 & , \text{ if } ijk \text{ is an odd permutation of } 123\,, \\ 0 & , \text{ otherwise}\,. \end{cases} \tag{14.1}$$

This definition means that $[ijk]$ is non–zero only if i, j and k take different values and it is antisymmetric under interchange of two adjacent indices, e.g. $[ijk] = -[ikj]$ (Fig. 3.10). We define the *Levi–Civita tensor* ϵ_{ijk} as the tensor whose components in some particular right–handed Cartesian coordinate system are given by the permutation symbol:

$$\epsilon_{ijk} = [ijk] \qquad (x^i \text{ right–handed Cartesian}). \tag{14.2}$$

This defines a tensor because we can find its components in any other coordinate system, Cartesian or otherwise, by transforming the expression (14.2) according to the general rule (13.11). Accordingly, the Levi–Civita tensor in an arbitrary

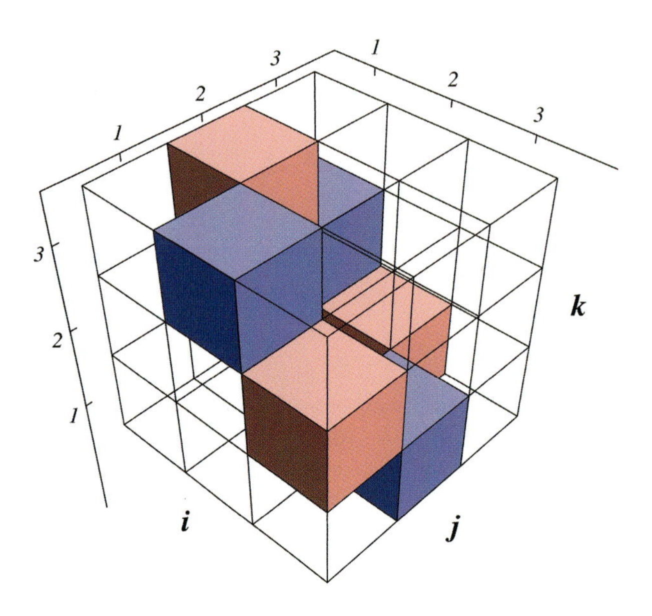

Figure 3.10: Visualization of the permutation symbol (14.1). The axes refer to the three values $\{1, 2, 3\}$ of the three indices i, j and k. The red cubes denote the value $+1$, the blue cubes denote the value -1 and the empty cubes denote the value 0.

coordinate system is

$$\epsilon_{i'j'k'} = \Lambda^i{}_{i'}\Lambda^j{}_{j'}\Lambda^k{}_{k'}[ijk] = \det\mathbf{\Lambda}\,[i'j'k'] \quad (x^i \text{ right–handed Cartesian}), \tag{14.3}$$

where $\mathbf{\Lambda}$ denotes the matrix $\Lambda^i{}_{i'}$, as in (11.9), and in the second equality we have used the Leibniz formula for the determinant of $\Lambda^{l'}{}_l$ (Stoll [1969]). We need to liberate Eq. (14.3) from its reference (through $\det\mathbf{\Lambda}$) to the initial right–handed Cartesian system x^i and write $\epsilon_{i'j'k'}$ in a form involving only quantities in the system $x^{i'}$. This can be done by making use of the square root of Eq. (11.15), which holds here since x^i are Cartesian:

$$\det\mathbf{\Lambda} = \pm\sqrt{g'} \qquad (x^i \text{ right–handed Cartesian}). \tag{14.4}$$

Which sign should we take here? Clearly the sign in question is the sign of $\det\mathbf{\Lambda}$, and this is determined by the issue of *handedness* (Fig. 3.11). If $\det\mathbf{\Lambda}$ is negative the transformation changes the handedness of the coordinate system, so the new system is left–handed (Goldstein, Poole and Safko [2001]). Consider first the case of transformations to another Cartesian system. A general transformation between Cartesian coordinate systems consists of a rotation and a translation, together with possible reflections of the coordinates by means of a sign change (e.g. $x' = -x$). A transformation that changes the sign of one or all three of the coordinates in the right–handed Cartesian system has $\det\mathbf{\Lambda} = -1$ and results in a left–handed Cartesian system (Goldstein, Poole and Safko [2001]). For any transformation between Cartesian systems Eq. (11.10) holds, i.e. the Euclidean metric is preserved

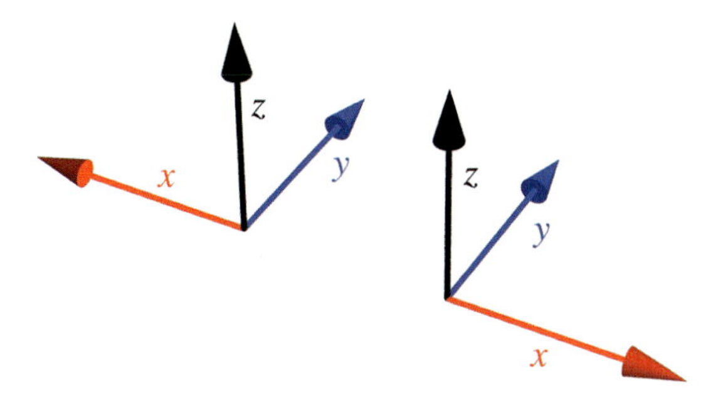

Figure 3.11: Coordinate axes in three dimensions have a handedness. The conventional choice, called right–handedness, is shown on the right in the case of the Cartesian axes x, y, z. A coordinate transformation that flips the sign of one or all three of the coordinates gives the coordinate axes shown on the left; this new triad of axes cannot be superimposed on the original—it has a different handedness (called left–handedness).

$(g_{i'j'} = \delta_{i'j'})$, so $g' = 1$ and (14.4) gives

$$\det\mathbf{\Lambda} = \pm 1 \qquad \text{(Cartesian} \longrightarrow \text{Cartesian)}, \tag{14.5}$$

where the sign is negative if the transformation includes a handedness change. In the case of transformations to an arbitrary system $x^{i'}$ the sign in Eq. (14.4) is similarly determined by whether the new system maintains the right–handedness of the original Cartesian system. Using Eq. (14.4) we can now write the Levi–Civita tensor in arbitrary coordinates (14.3) in terms of the metric; we drop the prime on the indices of the arbitrary coordinate system and obtain

$$\epsilon_{ijk} = \pm\sqrt{g}\,[ijk]\,, \tag{14.6}$$

where it is understood that the plus (minus) sign obtains if the system is right–handed (left–handed). It is now clear that Eq. (14.2) holds in *all* right–handed Cartesian coordinate systems, not just in the one we started with. (Tensors like the Levi–Civita tensor, whose components can undergo a sign change in a coordinate transformation, in addition to the usual tensor transformation law, are sometimes called *pseudo–tensors*.)

Some or all of the indices of ϵ_{ijk} can be raised according to the prescription in Eqs. (13.13); if all are raised we obtain another simple expression to complement Eq. (14.6):

$$\epsilon^{ijk} = g^{il}g^{jm}g^{kn}\epsilon_{lmn} = \pm\sqrt{g}\,g^{il}g^{jm}g^{kn}[lmn] = \pm\sqrt{g}\,\frac{1}{g}[ijk] = \pm\frac{1}{\sqrt{g}}[ijk]\,. \tag{14.7}$$

The third equality in Eq. (14.7) comes from the fact that the determinant of the matrix g^{ij} is $1/g$, since g is the determinant of g_{ij}, the inverse of g^{ij}. Like the permutation symbol (14.1), the Levi–Civita tensor is *completely antisymmetric*; this means that when its components are taken with all indices in the upper or lower position they are antisymmetric under interchange of two adjacent indices, e.g. $\epsilon^{ijk} = -\epsilon^{jik}$.

Exercise 14.1
Determine the handedness of spherical polar and cylindrical polar coordinates, defined in terms of a right–handed Cartesian system by Eqs. (10.1) and (10.2). Write the Levi–Civita tensor in these systems.
Solution
In the case of spherical polars the matrix $\Lambda^{i}{}_{i'}$ appearing in Eq. (14.3) is (10.9); its determinant is

$$\det\mathbf{\Lambda} = r^2 \sin\theta \,,$$

which is positive or zero, since θ has the range $0 \leq \theta \leq \pi$. This implies that the spherical polar basis $\{\boldsymbol{e}_r, \boldsymbol{e}_\theta, \boldsymbol{e}_\phi\}$ forms a right–handed triple, just like the original Cartesian basis $\{\boldsymbol{e}_x, \boldsymbol{e}_y, \boldsymbol{e}_z\}$. From Eqs. (14.6), (14.7) and (11.19) the Levi–Civita tensor in spherical polar coordinates is therefore

$$\epsilon_{ijk} = r^2 \sin\theta \,[ijk]\,, \quad \epsilon^{ijk} = \frac{1}{r^2 \sin\theta}\,[ijk]\,, \quad x^i = \{x^1, x^2, x^3\} = \{r, \theta, \phi\}\,, \quad [123] = [r\theta\phi] = 1\,. \tag{14.8}$$

For cylindrical polars the matrix $\Lambda^{i}{}_{i'}$ is (10.13); its determinant is

$$\det\mathbf{\Lambda} = r\,,$$

which is positive or zero, so the cylindrical polar basis $\{\boldsymbol{e}_r, \boldsymbol{e}_\phi, \boldsymbol{e}_z\}$ also forms a right–handed triple. The Levi–Civita tensor in cylindrical polars is (recall relation (11.23))

$$\epsilon_{ijk} = r\,[ijk]\,, \quad \epsilon^{ijk} = \frac{1}{r}\,[ijk]\,, \quad x^i = \{x^1, x^2, x^3\} = \{r, \phi, z\}\,, \quad [123] = [r\phi z] = 1\,. \tag{14.9}$$

The Levi–Civita tensor is used to write vector products in an arbitrary coordinate system:

$$\boldsymbol{V} \times \boldsymbol{U} = \epsilon^{ijk} V_j U_k \,\boldsymbol{e}_i\,. \tag{14.10}$$

Exercise 14.2
Show that Eq. (14.10) correctly gives the vector product in (right–handed) Cartesian coordinates.
Solution
In right–handed Cartesian coordinates the Levi–Civita tensor is simply the permutation symbol (14.1). Moreover, in Cartesian coordinates the components of the vector V^i and the associated one–form V_i are the same because the metric is δ_{ij}. Equation (14.10) therefore gives in this case

$$\begin{aligned}\boldsymbol{V} \times \boldsymbol{U} &= [xjk] V^j U^k \,\boldsymbol{e}_x + [yjk] V^j U^k \,\boldsymbol{e}_y + [zjk] V^j U^k \,\boldsymbol{e}_z \\ &= (V^y U^z - V^z U^y)\,\boldsymbol{e}_x + (V^z U^x - V^x U^z)\,\boldsymbol{e}_y + (V^x U^y - V^y U^x)\,\boldsymbol{e}_z\,,\end{aligned} \tag{14.11}$$

which is the correct vector–product formula.

The right–hand side of Eq. (14.10) gives the vector product in Cartesian coordinates, and since it is built from tensors by operations (tensor products and contractions) that produce another tensor, it correctly gives the vector $\boldsymbol{V} \times \boldsymbol{U}$ in an arbitrary coordinate system. The components of the vector product can be written in a variety of equivalent forms, for example

$$(\boldsymbol{V} \times \boldsymbol{U})^i = \epsilon^{ijk} V_j U_k = \epsilon^{ij}{}_{k} V_j U^k = \epsilon^{i}{}_{j}{}^{k} V^j U_k = \epsilon^{i}{}_{jk} V^j U^k\,, \tag{14.12}$$

and of course the index i can be lowered on all of these to obtain the associated one–form $(\boldsymbol{V} \times \boldsymbol{U})_i$:

$$(\boldsymbol{V} \times \boldsymbol{U})_i = \epsilon_i{}^{jk} V_j U_k = \epsilon_i{}^j{}_k V_j U^k = \epsilon_{ij}{}^k V^j U_k = \epsilon_{ijk} V^j U^k \,. \tag{14.13}$$

Exercise 14.3
Prove the following important formula:

$$\epsilon^{ijk} \epsilon_{klm} = g^i{}_l g^j{}_m - g^i{}_m g^j{}_l = \delta^i_l \delta^j_m - \delta^i_m \delta^j_l \,. \tag{14.14}$$

Solution
From Eqs. (14.6) and (14.7) follows

$$\epsilon^{ijk} \epsilon_{klm} = [ijk]\,[klm] \,,$$

and one must verify from the definition (14.1) of the permutation symbol that

$$[ijk]\,[klm] = \delta^i_l \delta^j_m - \delta^i_m \delta^j_l \,. \tag{14.15}$$

The result (14.14) now follows from Eq. (13.7). Note that indices can be raised or lowered in Eq. (14.14) to give results such as

$$\epsilon^{ijk} \epsilon_k{}^{lm} = g^{il} g^{jm} - g^{im} g^{jl} \,. \tag{14.16}$$

Exercise 14.4
Prove the following well–known formula of vector algebra:

$$\boldsymbol{A} \times (\boldsymbol{B} \times \boldsymbol{C}) = \boldsymbol{B}(\boldsymbol{A} \cdot \boldsymbol{C}) - \boldsymbol{C}(\boldsymbol{A} \cdot \boldsymbol{B}) \,. \tag{14.17}$$

Solution
We use Eqs. (14.12) and (14.13) to write out the components of the vector $\boldsymbol{A} \times (\boldsymbol{B} \times \boldsymbol{C})$ in terms of the Levi–Civita tensor

$$[\boldsymbol{A} \times (\boldsymbol{B} \times \boldsymbol{C})]^i = \epsilon^{ijk} A_j (\boldsymbol{B} \times \boldsymbol{C})_k = \epsilon^{ijk} A_j \epsilon_{klm} B^l C^m$$

and insert formula (14.14):

$$\begin{aligned} [\boldsymbol{A} \times (\boldsymbol{B} \times \boldsymbol{C})]^i &= (g^i{}_l g^j{}_m - g^i{}_m g^j{}_l) A_j B^l C^m = B^i A_j C^j - C^i A_j B^j \\ &= B^i (\boldsymbol{A} \cdot \boldsymbol{C}) - C^i (\boldsymbol{A} \cdot \boldsymbol{B}) \,, \end{aligned}$$

as required.

Curls are also computed using the Levi–Civita tensor, but they contain a differentiation and we must learn how to differentiate in arbitrary coordinates.

§15. THE COVARIANT DERIVATIVE OF A VECTOR

Consider a scalar field ψ in space that smoothly maps points in space to numbers, for example an electric potential (Fig. 3.12). When a coordinate grid is chosen each point acquires a set of three coordinates x^i and the scalar field becomes a function of these coordinates. We can take the partial derivatives of ψ with respect x^i as follows, where we introduce an ink–saving device:

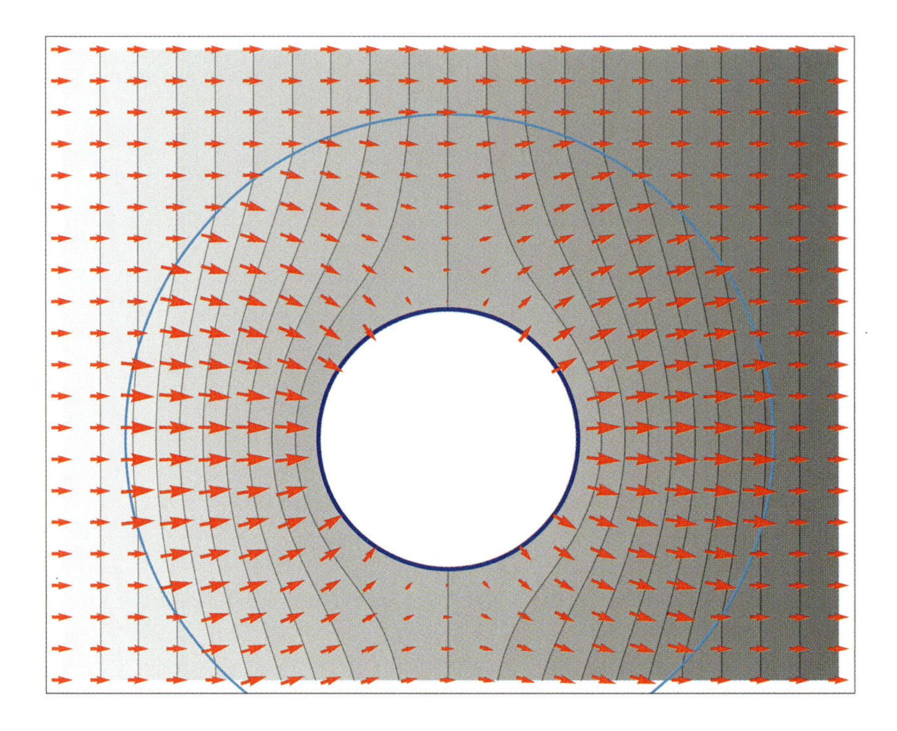

Figure 3.12: A scalar field and its gradient in two dimensions. The contour lines join points where the scalar field has the same value. The arrows depict the gradient of the scalar field; they point in the direction in which the scalar field is increasing and their length is the rate of change. This picture represents the electric potential around an electrostatic cloaking device. The white circle is the boundary of the device and the blue circle is its inner lining; the electric field is excluded from the central white region.

$$\frac{\partial}{\partial x^i}\psi \equiv \partial_i \psi \equiv \psi_{,i}\,. \tag{15.1}$$

Thus a comma means partial differentiation, with the following index giving the coordinate with respect to which the derivative is taken. The definitions (15.1) also explain the common notation ∂_i for the partial derivative operator $\partial/\partial x^i$. In Cartesian coordinates the derivatives (15.1) are of course the components of the gradient vector $\boldsymbol{\nabla}\psi$. It is easy to see, however, from Eqs. (10.7), (10.8) and (13.2), that expression (15.1) transforms as a one–form:

$$\psi_{,i'} = \Lambda^i{}_{i'}\psi_{,i}\,. \tag{15.2}$$

Consistent with the index being in the lower position, the derivatives $\psi_{,i}$ are in fact the components of the one–form associated with the gradient vector, and this distinction can only be ignored in Cartesian coordinates where the vector and the one–form have the same components. In a general coordinate system we must raise the index in Eq. (15.1) using the inverse metric tensor to obtain the components of the gradient vector, i.e.

$$(\boldsymbol{\nabla}\psi)_i = \psi_{,i}\,, \quad (\boldsymbol{\nabla}\psi)^i = g^{ij}\psi_{,j}\,. \tag{15.3}$$

Exercise 15.1
Use Eqs. (15.2) and (13.8) to show that $(\boldsymbol{\nabla}\psi)^i$ in E. (15.3) transform as vector components.

Note from Eq. (15.3) that the components of the gradient vector in coordinate bases are in general more complicated than those of the gradient one–form. It is the gradient one–form $\psi_{,i}$ that is the basic quantity because of the simple tensorial meaning of partial derivatives. For this reason, in differential geometry $\boldsymbol{\nabla}\psi$ is taken to denote the one–form, rather than the associated vector, i.e.

$$\boldsymbol{\nabla}\psi = \psi_{,i}\,\boldsymbol{\omega}^i\,, \tag{15.4}$$

where $\boldsymbol{\omega}^i$ is the one–form basis introduced in §13.

Having incorporated derivatives of scalar fields into the formalism of tensors, we now consider the differentiation of vector fields. As in the case of the scalar field ψ above, with the introduction of a coordinate grid a vector field $\boldsymbol{V}$ becomes a function of the coordinates x^i and so can be differentiated with respect to them. From the definition of a derivative, $\partial\boldsymbol{V}/\partial x^i$ at a general point x^i in space is

$$\left.\frac{\partial}{\partial x^i}\boldsymbol{V}\right|_{x^i} = \lim_{\Delta x^i\to 0}\frac{\boldsymbol{V}|_{x^i+\Delta x^i} - \boldsymbol{V}|_{x^i}}{\Delta x^i} \qquad \text{(no summation over } i\text{!)}, \tag{15.5}$$

Note that differentiation requires the subtraction of two vectors located *at different points*: in (15.5) we must take the value of the vector field $\boldsymbol{V}$ at the point $x^i + \Delta x^i$ and subtract its value at the point x^i. To perform this subtraction we must place the two vectors at the *same* point, so that we must either move the vector $\boldsymbol{V}|_{x^i+\Delta x^i}$ to x^i or move the vector $\boldsymbol{V}|_{x^i}$ to $x^i + \Delta x^i$ (Fig. 3.13). Moving vectors around like this is quite straightforward: we just need to make sure that we do not rotate the vector as we are moving it; if we avoid any rotation it remains the same vector. But this issue is only straightforward because we are in flat, Euclidean space; when we turn to curved spaces, moving vectors around will become a non–trivial operation and hence so will the notion (15.5) of differentiating vector fields.

Exercise 15.2
Show that the derivatives of the Cartesian basis vector fields (12.2) vanish.
Solution
Let $\{\boldsymbol{i},\boldsymbol{j},\boldsymbol{k}\}$ be the triad of basis vector fields in any Cartesian coordinate system. It is trivially obvious that if we take the triad $\{\boldsymbol{i},\boldsymbol{j},\boldsymbol{k}\}|_P$ of vectors located at any point P in space and move it to any other point Q, it will coincide with the triad $\{\boldsymbol{i},\boldsymbol{j},\boldsymbol{k}\}|_Q$ already present at Q. It is therefore clear from the definition (15.5) that

$$\frac{\partial\boldsymbol{i}}{\partial x^i} = 0\,, \quad \frac{\partial\boldsymbol{j}}{\partial x^i} = 0\,, \quad \frac{\partial\boldsymbol{k}}{\partial x^i} = 0 \qquad (x^i \text{ arbitrary}). \tag{15.6}$$

Note that this proof does not specify the coordinates x^i with respect to which we are differentiating the Cartesian basis $\{\boldsymbol{i},\boldsymbol{j},\boldsymbol{k}\}$; the vector fields $\{\boldsymbol{i},\boldsymbol{j},\boldsymbol{k}\}$ can be expressed in terms of any coordinate grid we please and so Eq. (15.6) hold for every x^i.

A vector field $\boldsymbol{V} = V^i \boldsymbol{e}_i$ consists of a sum of products of scalar fields V^i and basis vector fields $\boldsymbol{e}_i$ (Fig. 3.5); we have seen how to differentiate scalars and vectors, so we can differentiate $\boldsymbol{V}$ using the Leibniz rule:

$$\frac{\partial}{\partial x^i} \boldsymbol{V} = \frac{\partial V^j}{\partial x^i} \boldsymbol{e}_j + V^j \frac{\partial \boldsymbol{e}_j}{\partial x^i} . \tag{15.7}$$

The simple relation (15.7) is the most important fact about curvilinear coordinates. In Cartesian coordinates the basis vectors are constant (Exercise 15.2) and to differentiate a vector we need only differentiate its components. The coordinate basis vectors for any other coordinate system, however, change in orientation and magnitude as one moves through space (see Fig. 3.13), so the second term in Eq. (15.7) contributes. As is clear from Eq. (15.5) and Fig. 3.13, the derivative of a vector is itself a vector; thus we can expand $\partial \boldsymbol{e}_j / \partial x^i$ in formula (15.7) in terms of the basis $\boldsymbol{e}_i$, as follows:

$$\frac{\partial \boldsymbol{e}_j}{\partial x^i} = \Gamma^k{}_{ji}\, \boldsymbol{e}_k . \tag{15.8}$$

The quantities $\Gamma^k{}_{ji}$ are called the *Christoffel symbols*; note carefully from definition (15.8) their meaning: $\Gamma^k{}_{ji}$ is the kth component of the derivative of $\boldsymbol{e}_j$ with respect to x^i. Each of the three indices in $\Gamma^k{}_{ji}$ takes three values so there are $3 \times 3 \times 3 = 27$ Christoffel symbols (in three dimensions). In Cartesian coordinates $\Gamma^k{}_{ji}$ vanish, as is

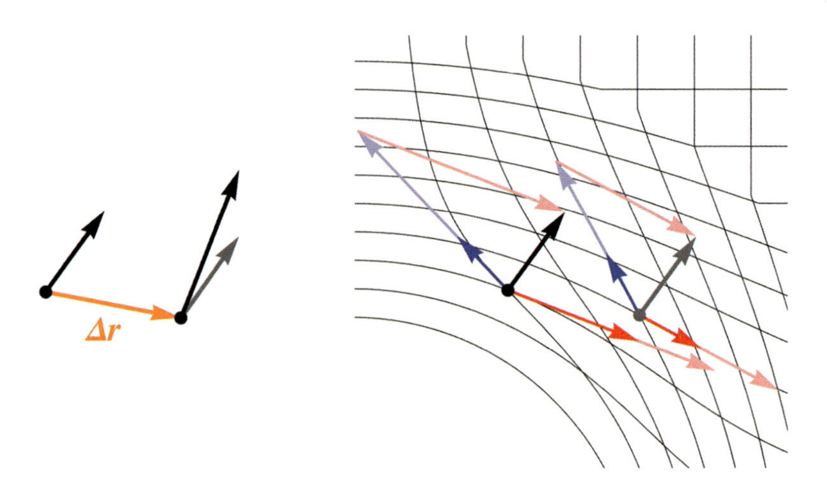

Figure 3.13: Covariant derivative. The derivative (15.5) of a vector field requires the comparison of vectors located at different points. The left picture shows the values of a vector field at two different points (black arrows); to obtain the derivative one arrow must be moved to the other so that the two can be subtracted. In flat space this is a trivial operation; the moved vector (grey arrow) is kept parallel to the original one. As shown in the right picture (see first Fig. 3.5), the basis vectors are in general different at the two points, so that the components of the moved vector (grey arrow) are not those of the vector at the original point (black arrow).

clear from taking x^i in Eq. (15.6) to be the Cartesian coordinate system with basis $\boldsymbol{e}_i = \{\boldsymbol{i}, \boldsymbol{j}, \boldsymbol{k}\}$ and comparing with Eq. (15.8). But in any non–Cartesian system the Christoffel symbols are non–trivial, as in the examples at the end of this Section.

Let us pause to consider how $\Gamma^k{}_{ji}$ change under a coordinate transformation. In the transformed system $x^{i'}$ the Christoffel symbols are defined by

$$\frac{\partial \boldsymbol{e}_{j'}}{\partial x^{i'}} = \Gamma^{k'}{}_{j'i'}\, \boldsymbol{e}_{k'} \,. \tag{15.9}$$

Since we know how basis vectors and partial derivative operators transform, we can deduce from Eqs. (15.8) and (15.9) the transformation law for the Christoffel symbols:

$$\Gamma^{k'}{}_{j'i'} = \Lambda^{k'}{}_k \Lambda^j{}_{j'} \Lambda^i{}_{i'} \Gamma^k{}_{ji} - \Lambda^{k'}{}_{j,i} \Lambda^j{}_{j'} \Lambda^i{}_{i'} \,, \tag{15.10}$$

where the comma in $\Lambda^{k'}{}_{j,i}$ denotes a partial derivative, in accordance with the definition (15.1).

Exercise 15.3

Prove the transformation rule (15.10).

Solution

We insert Eqs. (12.5) and (10.7) into Eq. (15.9):

$$\Lambda^i{}_{i'} \frac{\partial}{\partial x^i}\left(\Lambda^j{}_{j'}\, \boldsymbol{e}_j\right) = \Gamma^{k'}{}_{j'i'} \Lambda^k{}_{k'}\, \boldsymbol{e}_k$$

and expand the derivative on the left–hand side using the Leibniz rule and definition (15.8):

$$\Lambda^i{}_{i'}\left(\frac{\partial \Lambda^j{}_{j'}}{\partial x^i}\, \boldsymbol{e}_j + \Lambda^j{}_{j'} \Gamma^k{}_{ji}\, \boldsymbol{e}_k\right) = \Gamma^{k'}{}_{j'i'} \Lambda^k{}_{k'}\, \boldsymbol{e}_k \,.$$

Now we use the trick of relabeling dummy summation indices; by changing the summation index j in the first term in brackets to k we can move everything to the left–hand side and extract an overall factor of $\boldsymbol{e}_k$:

$$\left(\Lambda^i{}_{i'} \Lambda^k{}_{j',i} + \Lambda^i{}_{i'} \Lambda^j{}_{j'} \Gamma^k{}_{ji} - \Gamma^{k'}{}_{j'i'} \Lambda^k{}_{k'}\right) \boldsymbol{e}_k = 0 \,,$$

where we have employed the shorthand (15.1). Since $\boldsymbol{e}_k$ form a basis they are linearly independent, so we must have

$$\Gamma^{k'}{}_{j'i'} \Lambda^k{}_{k'} = \Lambda^i{}_{i'} \Lambda^k{}_{j',i} + \Lambda^i{}_{i'} \Lambda^j{}_{j'} \Gamma^k{}_{ji} \,.$$

We now multiply across by $\Lambda^{l'}{}_k$ and sum over k, using Eq. (10.17) on the left–hand side:

$$\Gamma^{l'}{}_{j'i'} = \Lambda^{l'}{}_k \Lambda^i{}_{i'} \Lambda^k{}_{j',i} + \Lambda^{l'}{}_k \Lambda^i{}_{i'} \Lambda^j{}_{j'} \Gamma^k{}_{ji} \,. \tag{15.11}$$

This is the transformation law for the Christoffel symbols; we can rewrite it in the form (15.10) by manipulating the factor $\Lambda^{l'}{}_k \Lambda^k{}_{j',i}$ that appears in the first term on the right–hand side of Eq. (15.11) as follows:

$$\begin{aligned}\Lambda^{l'}{}_k \Lambda^k{}_{j',i} &= \left(\Lambda^{l'}{}_k \Lambda^k{}_{j'}\right)_{,i} - \Lambda^{l'}{}_{k,i} \Lambda^k{}_{j'} = \delta^{l'}_{j',i} - \Lambda^{l'}{}_{k,i} \Lambda^k{}_{j'} \\ &= -\Lambda^{l'}{}_{k,i} \Lambda^k{}_{j'} \,,\end{aligned} \tag{15.12}$$

where we used Eq. (10.17) and the fact that $\delta^{l'}_{j'}$ are constants. Inserting Eq. (15.12) in Eq. (15.11) and relabeling some indices we obtain Eq. (15.10).

Note that ${\Gamma^k}_{ji}$ do not obey the transformation law (13.11) of tensor components because of the second term on the right–hand side of Eq. (15.10) that contains only the transformation matrices. The Christoffel symbols do not therefore constitute a tensor. It is not a disaster that we have encountered a non–tensor in differentiating vectors; everything works out fine, as we shall shortly see. First, let us use the transformation rule (15.10) to deduce an important property of the Christoffel symbols: they are symmetric in their lower indices, i.e.

$$ {\Gamma^i}_{jk} = {\Gamma^i}_{kj} . \tag{15.13} $$

Exercise 15.4

Prove the symmetry relation (15.13).

Solution

The Christoffel symbols in any coordinate system can be obtained by transforming from Cartesian coordinates according to the rule (15.10). As noted above, in Cartesian coordinates ${\Gamma^k}_{ji}$ vanish, so for arbitrary $x^{i'}$ we have

$$ {\Gamma^{k'}}_{j'i'} = -{\Lambda^{k'}}_{j,i} {\Lambda^j}_{j'} {\Lambda^i}_{i'} \qquad (x^i \text{ Cartesian}). \tag{15.14} $$

The factor ${\Lambda^{k'}}_{j,i}$ is symmetric in its lower indices, as is seen by writing it explicitly in term of the partial derivatives (10.8):

$$ {\Lambda^{k'}}_{j,i} = \frac{\partial^2 x^{k'}}{\partial x^i \partial x^j} = \frac{\partial^2 x^{k'}}{\partial x^j \partial x^i} = {\Lambda^{k'}}_{i,j} \, . \tag{15.15} $$

We insert Eq. (15.15) in Eq. (15.14), then swap the dummy summation indices i and j, and finally interchange the order of the last two factors:

$$ {\Gamma^{k'}}_{j'i'} = -{\Lambda^{k'}}_{i,j} {\Lambda^j}_{j'} {\Lambda^i}_{i'} = -{\Lambda^{k'}}_{j,i} {\Lambda^i}_{j'} {\Lambda^j}_{i'} = -{\Lambda^{k'}}_{j,i} {\Lambda^j}_{i'} {\Lambda^i}_{j'} \qquad (x^i \text{ Cartesian}). \tag{15.16} $$

Equation (15.14) shows that the last quantity in Eq. (15.16) is ${\Gamma^{k'}}_{i'j'}$ so in the coordinate system $x^{i'}$ we have shown that ${\Gamma^{k'}}_{j'i'} = {\Gamma^{k'}}_{i'j'}$; but this coordinate system is arbitrary, therefore Eq. (15.13) holds for any coordinates.

Alternatively, we can use the expression (12.6) for the components of the basis $\boldsymbol{e}_{i'}$ in the basis $\boldsymbol{e}_i$, where we choose the latter to be the Cartesian basis $\{\boldsymbol{i}, \boldsymbol{j}, \boldsymbol{k}\}$. When expressed in a Cartesian basis, the derivative (15.9) of $\boldsymbol{e}_{i'}$ is obtained by just differentiating the components (12.6), since the derivatives of the Cartesian basis vanish (Exercise 15.2):

$$ \frac{\partial \mathbf{e}_{i'}}{\partial x^{j'}} = \frac{\partial^2 x^i}{\partial x^{j'} \partial x^{i'}} \mathbf{e}_i \qquad (x^i \text{ Cartesian}). \tag{15.17} $$

The commutativity of partial derivatives then shows that $\partial \boldsymbol{e}_{i'} / \partial x^{j'} = \partial \boldsymbol{e}_{j'} / \partial x^{i'}$, and applying this last result to Eq. (15.9) gives ${\Gamma^{k'}}_{j'i'} = {\Gamma^{k'}}_{i'j'}$.

Although in general there are 27 Christoffel symbols (in three dimensions), due to the symmetry (15.13) at most 18 of them are independent.

Returning to the vector derivative (15.7), we insert the Christoffel symbols by means of Eq. (15.8):

$$ \begin{aligned} \frac{\partial}{\partial x^i} \mathbf{V} &= {V^j}_{,i} \, \mathbf{e}_j + {\Gamma^k}_{ji} V^j \, \mathbf{e}_k \\ &= \left({V^j}_{,i} + {\Gamma^j}_{ki} V^k \right) \mathbf{e}_j \\ &\equiv {V^j}_{;i} \, \mathbf{e}_j \, , \end{aligned} \tag{15.18} $$

where the second line is obtained by relabeling summation indices and in the last line we have defined the quantities

$$V^j{}_{;i} = V^j{}_{,i} + \Gamma^j{}_{ki} V^k \, . \tag{15.19}$$

It is clear from Eq. (15.18) that $V^j{}_{;i}$ are the components of the derivative of $\boldsymbol{V}$ with respect to x^i. We have seen that the derivative of a scalar (a zero–index tensor) gives a one–form (a one–index tensor), so we would expect the derivative of a vector, specified by the quantities $V^j{}_{;i}$, to be a two–index tensor. Neither of the two terms on the right–hand side of Eq. (15.19) separately constitutes a tensor, but one can show that $V^j{}_{;i}$ do in fact form a two–index tensor by inspecting their behaviour under a coordinate transformation.

Exercise 15.5
Prove that $V^j{}_{;i}$ obey the transformation rule

$$V^{j'}{}_{;i'} = \Lambda^{j'}{}_j \Lambda^i{}_{i'} V^j{}_{;i} \, , \tag{15.20}$$

and are thus the components of a tensor.
Solution
The transformation rule for $V^j{}_{;i}$ follows from the known transformation rules of everything on the right–hand side of Eq. (15.19). In the transformed coordinates $x^{i'}$, Eq. (15.19) reads

$$V^{j'}{}_{;i'} = V^{j'}{}_{,i'} + \Gamma^{j'}{}_{k'i'} V^{k'} \, . \tag{15.21}$$

We substitute Eqs. (10.7), (12.1) and (15.10) into Eq. (15.21):

$$\begin{aligned}
V^{j'}{}_{;i'} &= \left(\Lambda^{j'}{}_j V^j\right)_{,i} \Lambda^i{}_{i'} + \left(\Lambda^{j'}{}_j \Lambda^k{}_{k'} \Lambda^i{}_{i'} \Gamma^j{}_{ki} - \Lambda^{j'}{}_{k,i} \Lambda^k{}_{k'} \Lambda^i{}_{i'}\right) \Lambda^{k'}{}_l V^l \\
\text{[using Eq. (10.17)]} \quad &= \left(\Lambda^{j'}{}_{j,i} V^j + \Lambda^{j'}{}_j V^j{}_{,i}\right) \Lambda^i{}_{i'} + \left(\Lambda^{j'}{}_j \Gamma^j{}_{ki} - \Lambda^{j'}{}_{k,i}\right) \Lambda^i{}_{i'} V^k \\
\text{[relabeling a summation index]} \quad &= \left(\Lambda^{j'}{}_{j,i} V^j + \Lambda^{j'}{}_j V^j{}_{,i}\right) \Lambda^i{}_{i'} + \left(\Lambda^{j'}{}_j \Gamma^j{}_{ki} V^k - \Lambda^{j'}{}_{j,i} V^j\right) \Lambda^i{}_{i'} \\
&= \Lambda^{j'}{}_j \Lambda^i{}_{i'} \left(V^j{}_{,i} + \Gamma^j{}_{ki} V^k\right) \\
&= \Lambda^{j'}{}_j \Lambda^i{}_{i'} V^j{}_{;i} \, ,
\end{aligned}$$

as required.

The tensor with components $V^j{}_{;i}$ is called the *covariant derivative* of $\boldsymbol{V}$. The semi–colon in the definition (15.19) thus denotes covariant differentiation of the vector field, in which the variation of both the vector components and of the basis are properly accounted for, whereas the comma in Eq. (15.19) denotes differentiation of the vector components only. "Covariant derivative" therefore just means "correct derivative"!

Note that the gradient of a scalar field is a special case of covariant differentiation where there is no need to consider the basis vectors; in terms of the semicolon notation for covariant derivatives we thus have for a scalar

$$\psi_{;i} = \psi_{,i} \, . \tag{15.22}$$

The notation $\boldsymbol{\nabla}\psi$ expresses the covariant derivative of a scalar field ψ without any reference to a coordinate grid and there is a similar coordinate–free notation for the covariant derivative of a vector field $\boldsymbol{V}$, namely $\boldsymbol{\nabla V}$. If a coordinate grid is chosen we can expand the tensor $\boldsymbol{\nabla V}$ in terms of its components $V^j{}_{;i}$ in the manner described in §13:

$$\boldsymbol{\nabla V} = V^j{}_{;i}\, \boldsymbol{e}_j \otimes \boldsymbol{\omega}^i\,. \tag{15.23}$$

Thus in proper tensor notation the covariant derivatives (15.8) of the basis vectors are ($\boldsymbol{e}_i = \delta_i^j \boldsymbol{e}_j$ have components δ_i^j and we apply Eqs. (15.23) and (15.19))

$$\boldsymbol{\nabla e}_i = \Gamma^j{}_{ik}\, \boldsymbol{e}_j \otimes \boldsymbol{\omega}^k\,. \tag{15.24}$$

An alternative notation for the tensor components of a covariant derivative is to use the symbol ∇_i to represent the derivative operation and the extra index it adds to the object being differentiated, for example

$$\nabla_i\psi \equiv \psi_{,i} = \psi_{;i}\,, \quad \nabla_i V^j \equiv V^j{}_{;i}\,. \tag{15.25}$$

In standard vector calculus one is confined to scalar and vector fields, and so one does not encounter the two–index tensor $\boldsymbol{\nabla V}$ in its full glory. Rather, covariant derivatives are manipulated in various ways so that the resulting object is always either a scalar or a vector. We shall perform these manipulations in a proper tensor fashion in §17, but first we need to see how the results of this Section determine the covariant derivatives of all tensors.

Exercise 15.6

Compute the components of $\boldsymbol{\nabla V}$ (where $\boldsymbol{V}$ is a general vector field) in spherical polar coordinates.

Solution

We first require the Christoffel symbols in spherical polar coordinates. These can be easily found from Eqs. (15.8), (12.7) and (15.6), with the result

$$\begin{gathered}
\Gamma^\theta{}_{r\theta} = \frac{1}{r}\,, \quad \Gamma^\phi{}_{r\phi} = \frac{1}{r}\,, \quad \Gamma^\theta{}_{\theta r} = \frac{1}{r}\,, \quad \Gamma^r{}_{\theta\theta} = -r\,, \quad \Gamma^\phi{}_{\theta\phi} = \cot\theta\,, \\
\Gamma^\phi{}_{\phi r} = \frac{1}{r}\,, \quad \Gamma^\phi{}_{\phi\theta} = \cot\theta\,, \quad \Gamma^r{}_{\phi\phi} = -r\sin^2\theta\,, \quad \Gamma^\theta{}_{\phi\phi} = -\sin\theta\cos\theta\,,
\end{gathered} \tag{15.26}$$

all the other Christoffel symbols vanishing. Note that the symmetry (15.13) is evident. (Alternatively, one can transform from Cartesian coordinates, in which $\Gamma^i{}_{jk}$ are zero, using Eqs. (15.10) and (10.9)–(10.11), but the recipe (15.8) presents the easier path.)

The components $V^j{}_{;i}$ of $\boldsymbol{\nabla V}$ are now obtained from Eq. (15.19):

$$\begin{gathered}
V^r{}_{;r} = V^r{}_{,r}\,, \quad V^r{}_{;\theta} = V^r{}_{,\theta} - rV^\theta\,, \quad V^r{}_{;\phi} = V^r{}_{,\phi} - r\sin^2\theta\, V^\phi\,, \\
V^\theta{}_{;r} = V^\theta{}_{,r} + \frac{V^\theta}{r}\,, \quad V^\theta{}_{;\theta} = V^\theta{}_{,\theta} + \frac{V^r}{r}\,, \quad V^\theta{}_{;\phi} = V^\theta{}_{,\phi} - \sin\theta\cos\theta\, V^\phi\,, \\
V^\phi{}_{;r} = V^\phi{}_{,r} + \frac{V^\phi}{r}\,, \quad V^\phi{}_{;\theta} = V^\phi{}_{,\theta} + \cot\theta\, V^\phi\,, \quad V^\phi{}_{;\phi} = V^\phi{}_{,\phi} + \frac{V^r}{r} + \cot\theta\, V^\theta\,.
\end{gathered} \tag{15.27}$$

Exercise 15.7

Repeat Exercise 15.6 for the case of cylindrical polar coordinates.

Solution
The Christoffel symbols in this case are found from Eqs. (15.8), (12.8) and (15.6):

$$\Gamma^{\phi}{}_{r\phi} = \frac{1}{r}, \quad \Gamma^{\phi}{}_{\phi r} = \frac{1}{r}, \quad \Gamma^{r}{}_{\phi\phi} = -r, \tag{15.28}$$

all the other Christoffel symbols vanishing. The components $V^{j}{}_{;i}$ of $\boldsymbol{\nabla V}$ are given by Eq. (15.19):

$$\begin{gathered} V^{r}{}_{;r} = V^{r}{}_{,r}, \qquad V^{r}{}_{;\phi} = V^{r}{}_{,\phi} - rV^{\phi}, \qquad V^{r}{}_{;z} = V^{r}{}_{,z}, \\ V^{\phi}{}_{;r} = V^{\phi}{}_{,r} + \frac{V^{\phi}}{r}, \qquad V^{\phi}{}_{;\phi} = V^{\phi}{}_{,\phi} + \frac{V^{r}}{r}, \qquad V^{\phi}{}_{;z} = V^{\phi}{}_{,z}, \\ V^{z}{}_{;r} = V^{z}{}_{,r}, \qquad V^{z}{}_{;\phi} = V^{z}{}_{,\phi}, \qquad V^{z}{}_{;z} = V^{z}{}_{,z}. \end{gathered} \tag{15.29}$$

---

§16. COVARIANT DERIVATIVES OF TENSORS AND OF THE METRIC

Our knowledge of how to differentiate scalars and vectors leads directly to the expressions for the covariant derivatives of tensors in general. Consider the scalar product (13.6), written in terms of a one–form and a vector: U_iV^i. Since this is a scalar field, the covariant derivative is the ordinary partial derivative, as in Eq. (15.22):

$$(U_iV^i)_{;j} = (U_iV^i)_{,j} = U_{i,j}V^i + U_iV^i{}_{,j}, \tag{16.1}$$

where we have employed the Leibniz rule for the partial derivative. Let us rewrite this equation in terms of the covariant derivative of the vector V^i, by substituting for the partial derivative $V^i{}_{,j}$ using Eq. (15.19):

$$(U_iV^i)_{,j} = U_{i,j}V^i + U_iV^i{}_{;j} - U_i\Gamma^i{}_{kj}V^k = U_iV^i{}_{;j} + (U_{i,j} - \Gamma^k{}_{ij}U_k)V^i, \tag{16.2}$$

where the second equality is obtained by relabeling summation indices. The first quantity in expression (16.2) is a tensor, the gradient one–form of the scalar U_iV^i, so the final expression is also a tensor. Now everything in the final expression, except the quantity in brackets, has already been shown to be a tensor; therefore the quantity in brackets is also a tensor—it is the covariant derivative of the one–form U_i:

$$U_{i;j} \equiv \nabla_j U_i \equiv U_{i,j} - \Gamma^k{}_{ij}U_k. \tag{16.3}$$

In definition (16.3) we employ both the semi–colon notation for covariant–derivative components as well the ∇_j notation. Compare Eq. (16.3) carefully with Eq. (15.19): note the difference in the sign of the Γ–term and how the Christoffel symbols are contracted on the vector and one–form, respectively. Just as in the case of the vector covariant derivative (15.19), neither of the two terms in Eq. (16.3) separately forms a tensor, but when they are combined as shown the result is the tensor $U_{i;j}$.

Exercise 16.1
Show directly that $U_{i;j}$, given by definition (16.3), form a tensor by proving the tensor transformation rule

$$U_{i';j'} = \Lambda^i{}_{i'}\Lambda^j{}_{j'}U_{i;j}. \tag{16.4}$$

Solution
The transformation behaviour is established as in the case of the vector covariant derivative in Exercise 15.5.

We see from Eqs. (16.1)–(16.3) that the covariant derivative obeys the Leibniz rule:

$$(U_i V^i)_{;j} = U_i V^i{}_{;j} + U_{i;j} V^i \,. \tag{16.5}$$

One can deduce the expression for the covariant derivative of any tensor by constructing a scalar from it with vectors and one–forms and applying the procedure used above to find the one–form derivative. For example, to deduce the covariant derivative of a tensor $A_j{}^i$, we form the scalar $A_j{}^i U_i V^j$ and compute as follows:

$$\begin{aligned}(A_j{}^i U_i V^j)_{;k} &= (A_j{}^i U_i V^j)_{,k} = A_j{}^i{}_{,k} U_i V^j + A_j{}^i U_{i,k} V^j + A_j{}^i U_i V^j{}_{,k} \\ &= \left(A_j{}^i{}_{,k} - \Gamma^l{}_{jk} A_l{}^i + \Gamma^i{}_{lk} A_j{}^l\right) U_i V^j + A_j{}^i U_{i;k} V^j + A_j{}^i U_i V^j{}_{;k} \,, \end{aligned} \tag{16.6}$$

where we obtain the second line by using Eqs. (15.19) and (16.3). The right–hand side of Eq. (16.6) is a tensor (since $(A_j{}^i U_i V^j)_{;k}$ is a tensor), so the sum in brackets must be a tensor; it is the covariant derivative of $A_j{}^i$:

$$A_j{}^i{}_{;k} = \nabla_k A_j{}^i = A_j{}^i{}_{,k} - \Gamma^m{}_{jk} A_m{}^i + \Gamma^i{}_{mk} A_j{}^m \,. \tag{16.7}$$

In Eq. (16.6) we have again an example of the general property of the covariant derivative that it obeys the Leibniz rule:

$$(A_j{}^i U_i V^j)_{;k} = A_j{}^i{}_{;k} U_i V^j + A_j{}^i U_{i;k} V^j + A_j{}^i U_i V^j{}_{;k} \,. \tag{16.8}$$

A consideration of the general procedure used to derive Eq. (16.7) reveals how to write the covariant derivative of any tensor. There is a partial derivative of the tensor components and a sum of Γ–terms, one for each index of the tensor; for each upper index the Γ–term is exactly as in the vector derivative (15.19), whereas for each lower index the Γ–term is exactly as in the one–form derivative (16.3).

Exercise 16.2
Write down the covariant derivative of the 4–index tensor $S^{ij}{}_{kl}$.
Solution

$$S^{ij}{}_{kl;m} = S^{ij}{}_{kl,m} + \Gamma^i{}_{nm} S^{nj}{}_{kl} + \Gamma^j{}_{nm} S^{in}{}_{kl} - \Gamma^n{}_{km} S^{ij}{}_{nl} - \Gamma^n{}_{lm} S^{ij}{}_{kn} \,. \tag{16.9}$$

An important case is the covariant derivative of the metric tensor; applying the above recipe we obtain for the covariant derivative of g_{ij} and of the inverse metric g^{ij}:

$$g_{ij;k} = g_{ij,k} - \Gamma^l{}_{ik} g_{lj} - \Gamma^l{}_{jk} g_{il} \,, \tag{16.10}$$

$$g^{ij}{}_{;k} = g^{ij}{}_{,k} + \Gamma^i{}_{lk} g^{lj} + \Gamma^j{}_{lk} g^{il} \,. \tag{16.11}$$

A highly significant property of the metric tensor now emerges if we consider the fact that, as a tensor, it transforms as

$$g_{i'j';k'} = \Lambda^i{}_{i'} \Lambda^j{}_{j'} \Lambda^k{}_{k'} g_{ij;k} \,. \tag{16.12}$$

It is clear from Eq. (16.10) that the covariant derivative of the metric vanishes in Cartesian coordinates, where $g_{ij} = \delta_{ij}$ and the Christoffel symbols are all zero. But Eq. (16.12) shows that if $g_{ij;k}$ is zero in one coordinate system it is zero in all coordinate systems,* so we have

$$g_{ij;k} = 0 \,. \tag{16.13}$$

Similar reasoning applied to relation (16.11) shows that

$$g^{ij}{}_{;k} = 0 \,. \tag{16.14}$$

Exercise 16.3
Verify the property (16.13) explicitly for spherical polar coordinates.
Solution
This is a matter of substituting the spherical polar metric (11.17) and Christoffel symbols (15.26) into Eq. (16.10). Most components of $g_{ij;k}$ are trivially seen to vanish; we give one example where all three terms in Eq. (16.10) give a non–zero contribution to the overall vanishing $g_{ij;k}$:

$$g_{\phi\phi;\theta} = g_{\phi\phi,\theta} - \Gamma^l{}_{\phi\theta} g_{l\phi} - \Gamma^l{}_{\phi\theta} g_{\phi l} = g_{\phi\phi,\theta} - 2\Gamma^\phi{}_{\phi\theta} g_{\phi\phi} = 2r^2 \sin\theta\cos\theta - 2\cot\theta\, r^2 \sin^2\theta = 0 \,.$$

The reader may have noticed a consistency issue between covariant derivatives and the index raising and lowering operations (13.13). By using Eqs. (13.13) to lower an index on the vector covariant derivative $V^i{}_{;j}$ we obtain $V_{i;j}$, which for consistency with Eq. (16.3) must be the covariant derivative of the one–form $V_i = g_{ij}V^j$ associated with the vector V^i. It is the vanishing of the covariant derivative of the metric that ensures consistency here; using Eq. (16.13) we easily see that the covariant derivative of the one–form $V_i = g_{ij}V^j$ is indeed the vector covariant derivative $V^i{}_{;j}$ with the first index lowered:

$$V_{i;j} = (g_{ik}V^k)_{;j} = g_{ik;j}V^k + g_{ik}V^i{}_{;j} = g_{ik}V^i{}_{;j} \,, \tag{16.15}$$

where the Leibniz rule for covariant derivatives is also used. Similarly, property (16.14) ensures that taking the covariant derivative of the vector $V^i = g^{ik}V_k$ is equivalent to raising the first index on the one–form covariant derivative $V_{i;j}$:

$$V^i{}_{;j} = (g^{ik}V_k)_{;j} = g^{ik}{}_{;j}V_k + g^{ik}V_{i;j} = g^{ik}V_{i;j} \,. \tag{16.16}$$

As covariant derivatives are tensors we can also raise the index associated with the derivative itself, for example

$$V_i{}^{;j} = g^{jk}V_{i;k} \qquad \text{or} \qquad \nabla^j V_i = g^{jk}\nabla_k V_i \,. \tag{16.17}$$

Equation (16.17) defines what it means to have a covariant–derivative index in the upper position.

---

*This is a general property of tensors as a consequence of their transformation rule (13.11): if a tensor field vanishes in one coordinate system it vanishes in all of them. Similarly, if a tensor field is zero at some point P in a particular coordinate system, then it is zero at P in all coordinate systems.

Exercise 16.4
Show that

$$r_{i;j} = g_{ij}\,, \tag{16.18}$$

where r^i is the position vector defined in Exercise 12.5.
Solution
Since Eq. (16.18) is an equation involving only tensors, if it holds in one coordinate system the transformation rule (13.11) of tensors ensures that it holds in all coordinates. We consider Cartesian coordinates, in which the position vector has components $r^i = \{x, y, z\}$ (see Exercise 12.5). The metric in Cartesian coordinates is δ_{ij}, so the one–form components r_i are equal to r^i, and the Christoffel symbols vanish; hence the covariant derivative $r_{i;j}$ in these coordinates is equal to the partial derivatives $r^i{}_{,j}$. Equation 16.18 therefore reads in this case $r^i{}_{,j} = \delta_{ij}$, which is indeed true since $r^i = \{x, y, z\}$. This proves the general result (16.18), but the reader may find it instructive to verify it in the case of spherical polar coordinates. In spherical polars the components of r^i are $r^i = \{r, 0, 0\}$ (see Exercise 12.5) and Eq. (16.18) follows from the metric (11.17) and the Christoffel symbols (15.26).

Exercise 16.5
Show that the Levi–Civita tensor has zero covariant derivative, i.e.

$$\epsilon_{ijk;l} = 0\,. \tag{16.19}$$

Solution
As in the case of the metric in Eq. (16.13), the result follows easily from considering a Cartesian grid, where the Christoffel symbols vanish and the covariant derivative is a partial derivative. We recall from §14 that in Cartesian coordinates the components ϵ_{ijk} are constants, equal to 1, −1, or 0. Hence Eq. (16.19) holds in Cartesian systems and since it is a tensor equation it holds in all coordinates. Bear in mind that the indices in Eq. (16.19) can be raised and lowered, so that, for example, it implies $\epsilon^{ijk}{}_{;l} = 0$.

It is very important that Eqs. (16.13) and (16.10) serve to determine the Christoffel symbols in terms of the metric tensor. To see this, we insert Eq. (16.13) into Eq. (16.10) and write it three times, with different permutations of the indices:

$$\begin{aligned} g_{ij,k} &= \Gamma^l{}_{ik} g_{lj} + \Gamma^l{}_{jk} g_{il}\,, \\ g_{ik,j} &= \Gamma^l{}_{ij} g_{lk} + \Gamma^l{}_{kj} g_{il}\,, \\ -g_{jk,i} &= -\Gamma^l{}_{ji} g_{lk} - \Gamma^l{}_{ki} g_{jl}\,. \end{aligned} \tag{16.20}$$

In view of the symmetry (11.6) of the metric tensor and the symmetry (15.13) of the Christoffel symbols, the sum of the last three equations gives

$$2g_{il}\Gamma^l{}_{jk} = g_{ij,k} + g_{ik,j} - g_{jk,i}\,. \tag{16.21}$$

Multiplying across with the inverse metric tensor g^{mi} and contracting on i we finally obtain (after relabeling some indices)

$$\Gamma^i{}_{jk} = \frac{1}{2} g^{il}\left(g_{lj,k} + g_{lk,j} - g_{jk,l}\right). \tag{16.22}$$

Equation (16.22) represents the most economic way of computing the Christoffel symbols in general.

Exercise 16.6
Re–calculate the spherical polar Christoffel symbols (15.26) using the recipe (16.22) and the spherical polar metric (11.17), (13.9).

---

§17. Divergence, curl and Laplacian

In this Section we will see how the familiar derivative operations of vector calculus are constructed from the more general, but less widely known, covariant derivative operation. This is not a matter of simply restating commonplace concepts; the understanding of the divergence, curl and Laplacian as tensor operations brings to bear a powerful calculational machinery that greatly simplifies the handling of curvilinear coordinates. Moreover, the tensor machinery is essential for treating non–Euclidean spaces, which is the summit to which our mathematical considerations will carry us.

The divergence of a vector field $\boldsymbol{V}$, which we denote as usual by $\boldsymbol{\nabla}\cdot\boldsymbol{V}$, is a scalar field obtained by contracting the two indices on the covariant derivative tensor $\boldsymbol{\nabla V}$:

$$\boldsymbol{\nabla}\cdot\boldsymbol{V} = \nabla_i V^i = V^i{}_{;i}\,. \tag{17.1}$$

The coordinate–independence of contractions has been noted before (in this case it ensures $V^i{}_{;i} = V^{i'}{}_{;i'}$) so that formula (17.1) defines the same scalar field whatever coordinates are used to evaluate it. In Cartesian coordinates the familiar divergence of vector calculus is $V^i{}_{,i}$, which (only) in these coordinates is equal to $V^i{}_{;i}$. Thus formula (17.1) expresses the divergence in a form valid in all coordinates.

Exercise 17.1
Write the expressions for the divergence (17.1) in spherical polar and cylindrical polar coordinates. Use the results for the covariant derivative $\boldsymbol{\nabla V}$ in these coordinates (see Exercises 15.6 and 15.7).
Solution
For spherical polar coordinates Eqs. (17.1) and (15.27) give

$$\begin{aligned}\boldsymbol{\nabla}\cdot\boldsymbol{V} &= V^r{}_{;r} + V^\theta{}_{;\theta} + V^\phi{}_{;\phi} = V^r{}_{,r} + V^\theta{}_{,\theta} + V^\phi{}_{,\phi} + \frac{2}{r}V^r + \cot\theta\, V^\theta \\ &= \frac{1}{r^2}\frac{\partial}{\partial r}\left(r^2 V^r\right) + \frac{1}{\sin\theta}\frac{\partial}{\partial\theta}\left(\sin\theta\, V^\theta\right) + \frac{\partial}{\partial\phi}V^\phi\,.\end{aligned} \tag{17.2}$$

Remember that in Eq. (17.2) we are using the coordinate basis (12.7). To find the expression in the orthonormal, non–coordinate basis (12.15) requires an obvious rescaling of the vector components. We denote the components of $\boldsymbol{V}$ in the orthonormal basis (12.15) by $\{\hat{V}^r, \hat{V}^\theta, \hat{V}^\phi\}$; it is clear from (12.15) that $V^r = \hat{V}^r$, $V^\theta = \hat{V}^\theta/r$ and $V^\phi = \hat{V}^\phi/(r\sin\theta)$, so (17.2) is

$$\boldsymbol{\nabla}\cdot\boldsymbol{V} = \frac{1}{r^2}\frac{\partial}{\partial r}\left(r^2\hat{V}^r\right) + \frac{1}{r\sin\theta}\frac{\partial}{\partial\theta}\left(\sin\theta\,\hat{V}^\theta\right) + \frac{1}{r\sin\theta}\frac{\partial}{\partial\phi}\hat{V}^\phi\,, \tag{17.3}$$

which is the expression found in the electromagnetism textbooks. Note from Eqs. (17.2) and (17.3) that the result is simpler in the coordinate basis than in the orthonormal basis.

For cylindrical polar coordinates we obtain from Eqs. (17.1) and (15.29)

$$\nabla \cdot \boldsymbol{V} = V^r{}_{;r} + V^\phi{}_{;\phi} + V^z{}_{;z} = V^r{}_{,r} + V^\phi{}_{,\phi} + V^z{}_{,z} + \frac{1}{r} V^r$$

$$= \frac{1}{r}\frac{\partial}{\partial r}\left(rV^r\right) + \frac{\partial}{\partial \theta} V^\theta + \frac{\partial}{\partial \phi} V^\phi . \tag{17.4}$$

This is the result for the coordinate basis (12.8). In the orthonormal basis (12.17) we denote the components of $\boldsymbol{V}$ by $\{\hat{V}^r, \hat{V}^\phi, \hat{V}^z\}$; it is clear from (12.17) that $V^r = \hat{V}^r$, $V^\phi = \hat{V}^\phi / r$ and $V^z = \hat{V}^z$, and Eq. (17.4) is

$$\nabla \cdot \boldsymbol{V} = \frac{1}{r}\frac{\partial}{\partial r}\left(r\hat{V}^r\right) + \frac{1}{r}\frac{\partial}{\partial \theta}\hat{V}^\theta + \frac{\partial}{\partial \phi}\hat{V}^\phi . \tag{17.5}$$

We can utilize expression (16.22) for the Christoffel symbols to derive a very simple formula for the divergence of a vector in arbitrary coordinates. From Eqs. (17.1) and (15.19) the divergence is

$$\nabla \cdot \boldsymbol{V} = V^i{}_{;i} = V^i{}_{,i} + \Gamma^i{}_{ji} V^j , \tag{17.6}$$

and inserting expression (16.22) gives

$$\nabla \cdot \boldsymbol{V} = V^i{}_{,i} + \frac{1}{2} g^{il}\left(g_{lj,i} - g_{ji,l}\right) V^j + \frac{1}{2} g^{il} g_{li,j} V^j . \tag{17.7}$$

The second term on the right–hand side of Eq. (17.7) vanishes, as is seen by relabeling dummy indices and employing the symmetry of the metric (and its inverse):

$$\frac{1}{2} g^{il}\left(g_{lj,i} - g_{ji,l}\right) = \frac{1}{2} g^{il} g_{lj,i} - \frac{1}{2} g^{li} g_{jl,i} = 0 . \tag{17.8}$$

A useful general result of "index gymnastics" is revealed by Eq. (17.8). Note that $g_{lj,i} - g_{ji,l}$ is antisymmetric (i.e. it changes sign) under interchange of i and l, whereas g^{il} is symmetric under this interchange. Examination of Eq. (17.8) shows that it is this symmetry/antisymmetry contrast in the summed indices that implies the result of the summations is zero. It is therefore generally the case that when a symmetric pair of indices is summed with an anti–symmetric pair, as in Eq. (17.8), the result vanishes (whether or not the objects involved are tensors).

Returning to Eq. (17.7), we employ the simplification (17.8) and the symmetry of the metric:

$$\nabla \cdot \mathbf{V} = V^i{}_{,i} + \frac{1}{2} g^{il} g_{il,j} V^j . \tag{17.9}$$

Now, the term involving the metric in Eq. (17.9) can be written in terms of the determinant g of g_{ij} using the following identity:

$$g^{il} g_{il,j} = \frac{2}{\sqrt{g}}\left(\sqrt{g}\right)_{,j} . \tag{17.10}$$

Exercise 17.2
Prove Eq. (17.10).

Solution
By means of cofactor expansion, we express the determinant g in terms of the cofactors C_{ij} of the matrix g_{ij}:

$$g = g_{1i} C_{1i} \,, \tag{17.11}$$

where we have chosen to use the first row of g_{ij} and sum over i. The inverse matrix g^{ij} is given by Cramer's rule

$$g^{ij} = \frac{1}{g} C_{ij} \,, \tag{17.12}$$

where the mismatch in the positions of the indices on the two sides of the equation should cause no alarm since the cofactors C_{ij} do not separately form a tensor. Use of Cramer's rule (17.12) in Eq. (17.11) gives

$$g = g g_{1i} \, g^{1i} \,. \tag{17.13}$$

It is clear from Eq. (17.13) that the partial derivative $\partial g/\partial g_{1j}$ of g with respect to any of the three metric components g_{1j} is gg^{1j}. But in the cofactor expansion (17.11) we could have chosen the second or third row of the matrix g_{ij}, instead of the first row; consideration of these two alternatives shows that the partial derivative $\partial g/\partial g_{ij}$ with respect to any metric component g_{ij} is

$$\frac{\partial g}{\partial g_{ij}} = g g^{ij} \,. \tag{17.14}$$

We now use the chain rule to write $g_{,k}$, the partial derivative with respect to the coordinates, in terms of formula (17.14):

$$g_{,k} = \frac{\partial g}{\partial g_{ij}} g_{ij,k} = g g^{ij} g_{ij,k} \,. \tag{17.15}$$

The required result (17.10) follows from Eq. (17.15) since $\left(\sqrt{g}\right)_{,j} = g_{,j}/(2\sqrt{g})$.

Insertion of identity (17.10) in Eq. (17.9) leads to the simplified formula for the divergence

$$\begin{aligned} \boldsymbol{\nabla} \cdot \boldsymbol{V} &= V^i_{\ ,i} + \frac{1}{\sqrt{g}} \left(\sqrt{g}\right)_{,i} V^i \\ &= \frac{1}{\sqrt{g}} \left(\sqrt{g}\, V^i\right)_{,i} = \frac{1}{\sqrt{g}} \partial_i \sqrt{g}\, V^i \,. \end{aligned} \tag{17.16}$$

The advantage of this formula compared to Eq. (17.6) is clear: there are no Christoffel symbols, just a sum of partial derivatives, so it is much easier to evaluate.

Exercise 17.3
Show that the divergence formula (17.16) gives the previous results (17.2) and (17.4) for spherical and cylindrical polar coordinates, respectively.
Solution
All that is required is the determinant g of the metric, which in spherical polars is (11.19) and in cylindrical polars is (11.23); one can then see without any calculation that Eq. (17.16) gives Eqs. (17.2) and (17.4).

It is non–trivial that the combination of partial derivatives and the metric appearing in formula (17.16) gives the same scalar field in every coordinate system; our derivation of course proves that this is the case. A further advantage of the divergence formula (17.16) is that it allows us to easily see that Gauss's theorem (Fig. 3.14)

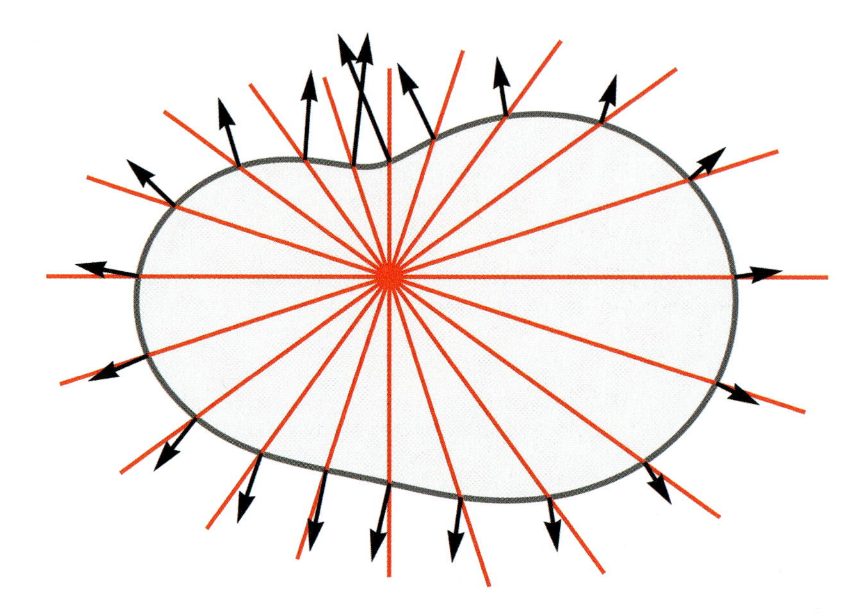

Figure 3.14: Gauss's theorem. The red lines indicate a vector field emerging from a source point. The integral of the divergence of the vector field over the shaded volume is equal to the integral of the flux of the vector field over the boundary of the volume. The black arrows show the projection of the vector field on to the boundary surface, which gives the flux. In this example the length of the vector field fall off as $1/r^2$, where r is the distance from the source; the integral of the flux is then independent of the surface enclosing the source.

holds in arbitrary coordinates. If we integrate the divergence over a volume Ω using the volume element (11.16) and Eq. (17.16) we obtain

$$\begin{aligned} \int_\Omega V^i{}_{;i}\sqrt{g}\,\mathrm{d}^3x &= \int_\Omega \left(\sqrt{g}\,V^i\right)_{,i}\mathrm{d}^3x \\ &= \oint_{\partial\Omega} V^i n_i \sqrt{g}\,\mathrm{d}^2x\,, \end{aligned} \tag{17.17}$$

where the last step follows from the familiar Gauss's theorem involving partial derivatives. The unit vector n^i is the normal to the boundary surface $\partial\Omega$ of the volume Ω; we see that in a general coordinate system the surface element is $\sqrt{g}\,\mathrm{d}^2x$.

Another derivative operation familiar from vector calculus is the curl of a vector. Like the vector product (14.10), the curl $\boldsymbol{\nabla}\times\boldsymbol{V}$ is formed using the Levi–Civita tensor; it is defined as

$$\boldsymbol{\nabla}\times\boldsymbol{V} = \epsilon^{ijk}V_{k;j}\,\boldsymbol{e}_i\,. \tag{17.18}$$

One easily confirms that this is the familiar curl operation by evaluating it in Cartesian coordinates where the Levi–Civita tensor is the permutation symbol and $V_{k;j}$

are partial derivatives $V_{k,j}$. Since it is constructed from tensors, definition (17.18) is the correct expression for the curl in arbitrary coordinates.

A simplification occurs in the general formula (17.18). Expanding the covariant derivative in Eq. (17.18) using Eq. (16.3) we encounter the factor $\epsilon^{ijk}\Gamma^l{}_{kj}$, in which the summed indices are anti–symmetric on ϵ^{ijk} but symmetric on $\Gamma^l{}_{kj}$ (recall Eq. (15.13) and the complete antisymmetry of ϵ^{ijk}). The general rule discussed after Eq. (17.8) therefore applies and shows that $\epsilon^{ijk}\Gamma^l{}_{kj} = 0$. The covariant derivative in Eq. (17.18) is thus equivalent to a partial derivative because its indices are contracted with the antisymmetric Levi–Civita tensor. For the components of the curl we can therefore use the expressions

$$(\boldsymbol{\nabla} \times \boldsymbol{V})^i = \epsilon^{ijk} V_{k,j} \,, \quad (\boldsymbol{\nabla} \times \boldsymbol{V})_i = \epsilon_i{}^{jk} V_{k,j} \,, \tag{17.19}$$

which are easier to evaluate than Eq. (17.18) since they do not depend on the Christoffel symbols. Although we know that Eqs. (17.19) are tensors because we obtained them from the definition (17.18), this fact is somewhat obscured in Eqs. (17.19) by the presence of the non–tensor $V_{k,j}$.

The Laplacian is an operation involving a double covariant derivative. Multiple covariant derivatives are denoted in a coordinate–free manner by successive application of the symbol $\boldsymbol{\nabla}$: thus $\boldsymbol{\nabla\nabla V}$ denotes the covariant derivative of $\boldsymbol{\nabla V}$ (where for definiteness we take $\boldsymbol{V}$ to be a vector field). The components of the tensor $\boldsymbol{\nabla\nabla V}$ are denoted by $V^i{}_{;j;k} = \nabla_k \nabla_j V^i$, and the second covariant derivative is evaluated in terms of the Christoffel symbols according to the general rule derived in §16:

$$V^i{}_{;j;k} = V^i{}_{;j,k} + \Gamma^i{}_{lk} V^l{}_{;j} - \Gamma^l{}_{jk} V^i{}_{;l} \,. \tag{17.20}$$

The Laplacian operation is defined by taking two covariant derivatives and then contracting the resulting tensor on the two indices produced by the differentiations. Denoting the operator in question by $\boldsymbol{\nabla}^2$ we thus obtain for the Laplacian of $\boldsymbol{V}$ the vector:

$$\boldsymbol{\nabla}^2 \boldsymbol{V} = V^i{}_{;j}{}^{;j} \, \boldsymbol{e}_i = \nabla_j \nabla^j V^i \, \boldsymbol{e}_i \,. \tag{17.21}$$

Note that the rule (13.13) for raising indices and the symmetry of the metric imply $V^i{}_{;j}{}^{;j} = g^{jk} V^i{}_{;j;k} = V^{i;j}{}_{;j}$, or, in terms of the ∇ notation, $\nabla_j \nabla^j V^i = \nabla^j \nabla_j V^i$.

Consider the simplest case, the Laplacian of a scalar field ψ; this is the scalar field

$$\boldsymbol{\nabla}^2 \psi = \psi_{,i}{}^{;i} \,. \tag{17.22}$$

By writing the contraction in Eq. (17.22) in terms of the metric and using the fact that the metric has zero covariant derivative, we can easily show that $\boldsymbol{\nabla}^2\psi$ is $(\boldsymbol{\nabla}\psi)^i{}_{;i}$, the divergence of the gradient vector $(\boldsymbol{\nabla}\psi)^i = g^{ij}\psi_{,j}$:

$$\begin{aligned} \boldsymbol{\nabla}^2 \psi &= g^{ij} \psi_{,j;i} = (g^{ij}\psi_{,j})_{;i} = (\boldsymbol{\nabla}\psi)^i{}_{;i} \\ &= \frac{1}{\sqrt{g}} \left(\sqrt{g}\, g^{ij} \psi_{,j} \right)_{,i} \,. \end{aligned} \tag{17.23}$$

In writing the final result (17.23) we have applied formula (17.16) for the divergence to the case of the gradient vector $(\boldsymbol{\nabla}\psi)^i = g^{ij}\psi_{,j}$. For spherical polar coordinates we use Eqs. (11.19) and (13.9) in formula (17.23) and easily obtain the well–known result

$$\nabla^2\psi = \frac{1}{r^2}\frac{\partial}{\partial r}\left(r^2\frac{\partial\psi}{\partial r}\right) + \frac{1}{r^2\sin\theta}\frac{\partial}{\partial\theta}\left(\sin\theta\frac{\partial\psi}{\partial\theta}\right) + \frac{1}{r^2\sin^2\theta}\frac{\partial^2\psi}{\partial\phi^2}\,. \tag{17.24}$$

Any reader who has had the misfortune of having to work out this expression without access to the simple formula (17.23) should now appreciate the power of the machinery we have presented in this Chapter. Note that since $\nabla^2\psi$ is a scalar, the result (17.24) is independent of whether a coordinate or orthonormal basis is used.

Exercise 17.4
Write the Laplacian of the scalar field ψ in cylindrical polar coordinates.
Solution
Use of Eqs. (11.23) and (13.10) in formula (17.23) immediately gives

$$\nabla^2\psi = \frac{1}{r}\frac{\partial}{\partial r}\left(r\frac{\partial\psi}{\partial r}\right) + \frac{1}{r^2}\frac{\partial^2\psi}{\partial\phi^2} + \frac{\partial\psi}{\partial z}\,. \tag{17.25}$$

As a further salutary example, consider the monochromatic wave equation, the Helmholtz equation, for the electric field $\boldsymbol{E}$:

$$\nabla^2\boldsymbol{E} + \frac{\omega^2}{c^2}\boldsymbol{E} = 0\,. \tag{17.26}$$

We take the electric field $\boldsymbol{E}$ to be a one–form, so that its components are E_i. The meaning of the first term in Eq. (17.26) as a covariant–derivative operation has been explained above; if, as is usually the case, the object $\nabla^2\boldsymbol{E}$ is introduced without such an explanation, then this conventional notation is rather treacherous for the student. When the wave equation (17.26) is considered in curvilinear coordinates, for example to find the radiation modes in a waveguide, the student is apt to think that the components $(\nabla^2\boldsymbol{E})_i$ of the one–form $\nabla^2\boldsymbol{E}$ are the Laplacians $\nabla^2(E_i)$ of the electric field components E_i—this is *not* true. As shown in Eq. (17.21), the components of the one–form $\nabla^2\boldsymbol{E}$ are $E_{i;j}{}^{;j} = E_i{}^{;j}{}_{;j}$, whereas the Laplacians of the components E_i are $(g^{jk}E_{i,j})_{;k}$. In fact the three Laplacians $(g^{jk}E^i{}_{,j})_{;k}$ are not even the components of a one–form, because $E_{i,j}$ do not form a tensor. This potential error in the interpretation of Eq. (17.26) is probably best forestalled by writing it in component form using the semi–colon notation:

$$E_i{}^{;j}{}_{;j} + \frac{\omega^2}{c^2}E_i = 0\,. \tag{17.27}$$

Only in Cartesian coordinates is Eq. (17.27) three separate scalar wave equations for the components E_i; in all other coordinate systems it gives coupled equations for E_i. Tensor analysis provides the machinery to extract the content of Eq. (17.26) in a systematic manner.

Exercise 17.5
Find the Helmholtz equation for the electric field in cylindrical polar coordinates.
Solution
Using Eq. (16.3) and the Christoffel symbols (15.28) in cylindrical polars, we obtain the components of the covariant derivative $E_{i;j}$ in cylindrical polar coordinates:

$$\begin{aligned}
&E_{r;r} = E_{r,r}\,, \qquad E_{r;\phi} = E_{r,\phi} - \frac{E_\phi}{r}\,, \qquad E_{r;z} = E_{r,z}\,,\\
&E_{\phi;r} = E_{\phi,r} - \frac{E_\phi}{r}\,, \qquad E_{\phi;\phi} = E_{\phi,\phi} + rE_r\,, \qquad E_{\phi;z} = E_{\phi,z}\,,\\
&E_{z;r} = E_{z,r}\,, \qquad E_{z;\phi} = E_{z,\phi}\,, \qquad E_{z;z} = E_{z,z}\,.
\end{aligned} \tag{17.28}$$

The one–form $E_i{}^{;j}{}_{;j} = (\boldsymbol{\nabla}^2 \boldsymbol{E})_i$ can be written in terms of $E_{i;j}$ as follows:

$$E_i{}^{;j}{}_{;j} = g^{jk} E_{i;j;k} = g^{jk}\left(E_{i;j,k} - \Gamma^l{}_{ik} E_{l;j} - \Gamma^l{}_{jk} E_{i;l}\right).$$

We use this to write the three components of the one–form $E_i{}^{;j}{}_{;j}$; for each component we first substitute for the inverse metric (13.10) and Christoffel symbols (15.28) in cylindrical polars and then we insert the values of $E_{i;j}$ found in Eqs. (17.28):

$$\begin{aligned}
E_r{}^{;j}{}_{;j} = (\boldsymbol{\nabla}^2 \boldsymbol{E})_r &= E_{r;r,r} + \frac{1}{r^2} E_{r;\phi,\phi} + E_{r;z,z} - \frac{1}{r^3} E_{\phi;\phi} + \frac{1}{r} E_{r;r}\\
&= E_{r,r,r} + \frac{1}{r^2}\left(E_{r,\phi,\phi} - \frac{1}{r} E_{\phi,\phi}\right) + E_{r,z,z} - \frac{1}{r^3}\left(E_{\phi,\phi} + rE_r\right) + \frac{1}{r} E_{r,r}\\
&= \boldsymbol{\nabla}^2 (E_r) - \frac{2}{r^3} E_{\phi,\phi} - \frac{1}{r^2} E_r\,,
\end{aligned} \tag{17.29}$$

$$\begin{aligned}
E_\phi{}^{;j}{}_{;j} = (\boldsymbol{\nabla}^2 \boldsymbol{E})_\phi &= E_{\phi;r,r} + \frac{1}{r^2} E_{\phi;\phi,\phi} + E_{\phi;z,z} - \frac{1}{r} E_{\phi;r} + \frac{1}{r} E_{r;\phi} + \frac{1}{r} E_{\phi;r}\\
&= E_{\phi,r,r} + \frac{1}{r^2} E_\phi - \frac{1}{r} E_{\phi,r} + \frac{1}{r^2}\left(E_{\phi,\phi,\phi} + rE_{r,\phi}\right) + E_{\phi,z,z}\\
&\quad + \frac{1}{r}\left(E_{r,\phi} - \frac{1}{r} E_\phi\right)\\
&= \boldsymbol{\nabla}^2 (E_\phi) - \frac{2}{r} E_{\phi,r} + \frac{2}{r} E_{r,\phi}\,,
\end{aligned} \tag{17.30}$$

$$\begin{aligned}
E_z{}^{;j}{}_{;j} = (\boldsymbol{\nabla}^2 \boldsymbol{E})_z &= E_{z;r,r} + \frac{1}{r^2} E_{z;\phi,\phi} + E_{z;z,z} + \frac{1}{r} E_{z;r}\\
&= E_{z;r,r} + \frac{1}{r^2} E_{z;\phi,\phi} + E_{z;z,z} + \frac{1}{r} E_{z,r}\\
&= \boldsymbol{\nabla}^2 (E^z)\,.
\end{aligned} \tag{17.31}$$

The final expressions (17.29)–(17.31) for $(\boldsymbol{\nabla}^2 \boldsymbol{E})_i$ have been written in terms of the Laplacians $\boldsymbol{\nabla}^2(E_i)$ of the components E_i, given by Eq. (17.25). Note that only in the case of the z–component (17.31) is $(\boldsymbol{\nabla}^2 \boldsymbol{E})_i = \boldsymbol{\nabla}^2(E_i)$. The three components (17.27) of the Helmholtz equation (17.26) are thus

$$\nabla^2(E_r) - \frac{2}{r^3}E_{\phi,\phi} + \left(\frac{\omega^2}{c^2} - \frac{1}{r^2}\right)E_r = 0\,, \tag{17.32}$$

$$\nabla^2(E_\phi) - \frac{2}{r}E_{\phi,r} + \frac{2}{r}E_{r,\phi} + \frac{\omega^2}{c^2}E_\phi = 0\,, \tag{17.33}$$

$$\nabla^2(E_z) + \frac{\omega^2}{c^2}E_z = 0\,. \tag{17.34}$$

We see that E_z is not coupled to the other two components of $\boldsymbol{E}$ and it obeys the scalar Helmholtz equation. Thus a particular solution exists with $E_r = E_\phi = 0$ and is obtained by solving the familiar scalar wave equation (17.34). Since the magnetic field $\boldsymbol{H}$ obeys the same equation (17.26) one can similarly obtain a solution with $H_r = H_\phi = 0$.

---

§18. Curvature

We have now become acquainted with the powerful mathematics of arbitrary coordinates in Euclidean space. But there is an extra bonus for our efforts—this machinery is precisely what is required to deal with *curved* spaces. In this Section we show how tensor analysis is used to define and quantify curvature.

Our interest here is in spaces that are, in some sense we need to make precise, stretched or distorted versions of the Euclidean space on which we have so far performed tensor analysis. We can still measure distances between points in such spaces and put coordinate grids on them, so we still have a line element

$$\mathrm{d}s^2 = g_{ij}\,\mathrm{d}x^i\mathrm{d}x^j\,, \tag{18.1}$$

defining a metric tensor g_{ij}. As in Euclidean space, the metric is symmetric,

$$g_{ij} = g_{ji}\,, \tag{18.2}$$

because (relabeling indices) $g_{ij}\mathrm{d}x^i\mathrm{d}x^j = g_{ji}\mathrm{d}x^j\mathrm{d}x^i = g_{ji}\mathrm{d}x^i\mathrm{d}x^j$, so g_{ji} quantifies distance exactly like g_{ij}. Since lengths in curved space are measured by the metric tensor exactly as in flat space, the volume element in curved space has the same expression (11.16) in terms of the metric as in flat space:

$$\mathrm{d}V = \sqrt{g}\,\mathrm{d}^3x\,. \tag{18.3}$$

The stretching and distortion of Euclidean space alter the distances between points and therefore change the line element and metric tensor from their Euclidean forms. Thus, a notion of curvature for such spaces is naturally induced by the metric tensor, as follows: if there exist coordinates on the space in which the line element has the Euclidean form (11.1) then the space is *flat*, otherwise it is *curved*.

Curvature of a 3–dimensional space is difficult to visualize, but a familiar curved 2–dimensional space is the surface of a sphere (Fig. 3.15). We can obtain the line element of the sphere from the line element (11.18) of 3–dimensional Euclidean space written in spherical polar coordinates (Fig. 3.1). In these coordinates a sphere of

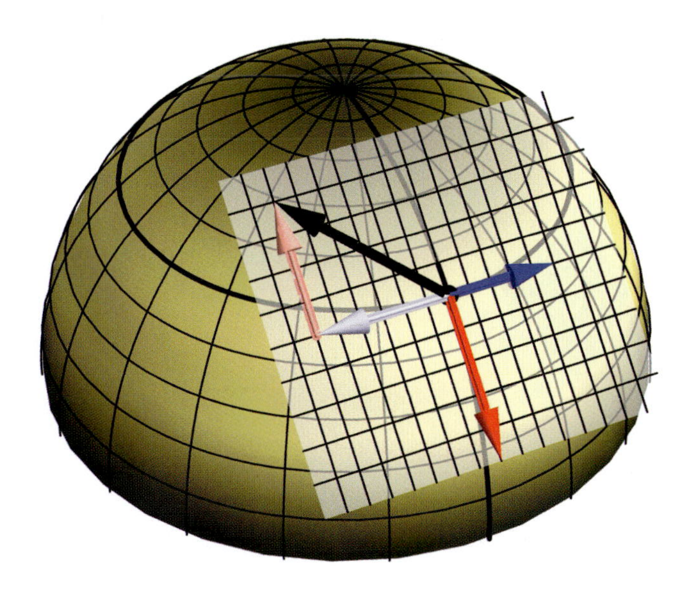

Figure 3.15: The surface of a sphere is a curved space. The picture shows the coordinate lines on the sphere given by the $\{\theta, \phi\}$ coordinates of the spherical polar system, used in the line element (18.4). The dark red and dark blue arrows show the basis vectors $\boldsymbol{e}_\theta$ and $\boldsymbol{e}_\phi$, respectively, at one point. These basis vectors are tangent to the coordinate lines, as in flat space, but note here how they "live" in a space tangent to the curved space (sphere) at each point. The black arrow is a general vector on the sphere, which can be expanded in terms of the basis vectors just as in flat space (see Fig. 3.5).

radius a is given by $r = a$; the radius is constant so $\mathrm{d}r = \mathrm{d}a = 0$ and Eq. (11.18) shows that the line element of the sphere is

$$\mathrm{d}s^2 = a^2(\mathrm{d}\theta^2 + \sin^2\theta\,\mathrm{d}\phi^2)\,. \tag{18.4}$$

There is no transformation to coordinates $\{x^1, x^2\}$ in which this line element takes the Euclidean form

$$\mathrm{d}s^2 = (\mathrm{d}x^1)^2 + (\mathrm{d}x^2)^2\,. \tag{18.5}$$

We will prove that no such transformation exists by a systematic method later on. The obstacle that prevents the transformation of Eq. (18.4) to the Euclidean (18.5) is not that the sphere is a closed space with a finite area; we can consider any non–infinitesimal patch of the sphere, ignoring its global structure, and we would still be unable to find a coordinate transformation to Eq. (18.5) in this patch. The crucial fact that makes the sphere a curved space is that we cannot form a patch of the sphere from a flat piece of paper without stretching the paper and thereby changing its geometry from that of Euclidean space.

Consider, in contrast to the sphere, the surface of a cylinder (Fig. 3.16). A cylinder can be formed by rolling up a flat piece of paper, so this space must be flat. The line element of a cylinder of radius a is obtained from the 3–dimensional

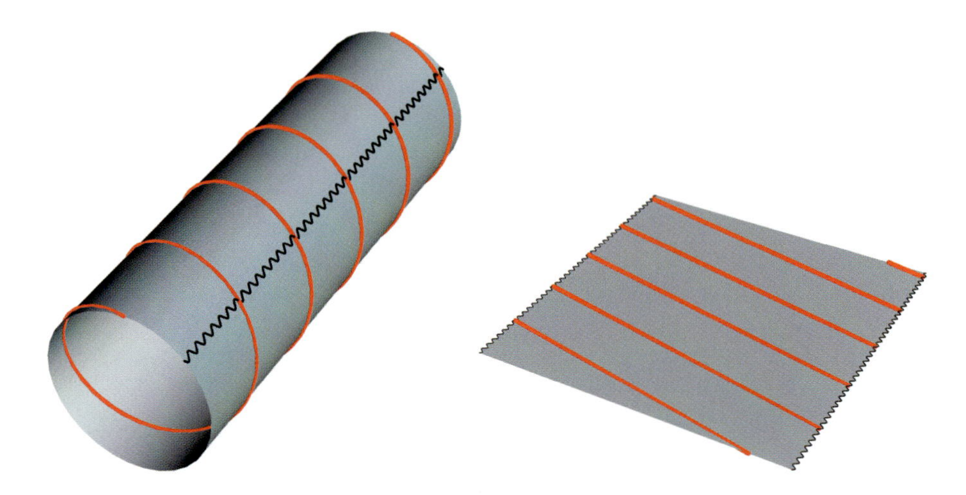

Figure 3.16: The surface of a cylinder is a rolled–up plane. It therefore has the same intrinsic geometry as the plane—Euclidean geometry—which means the cylinder is an intrinsically flat space. The straight red lines on the plane join points by the shortest possible routes; this is of course also true of these lines on the cylinder, even though here the lines are no longer straight when viewed from the ambient 3–dimensional space (the red lines are geodesics on the cylinder, to be discussed in the next Section).

Euclidean line element (11.22) in cylindrical polar coordinates by setting $r = a$ and $\mathrm{d}r = \mathrm{d}a = 0$:

$$\mathrm{d}s^2 = a^2\,\mathrm{d}\phi^2 + \mathrm{d}z^2 = (\mathrm{d}(a\phi))^2 + \mathrm{d}z^2\,. \tag{18.6}$$

This is of the Eulidean form (18.5), proving that the cylinder is flat. It is not important here that the coordinate ϕ in (18.6) is periodic; curvature is a local property of a space, in contrast to its topology. A cylinder and a plane have different topologies but they have the same curvature, namely zero.

The reader may object that the surface of a cylinder is obviously curved, and therefore the above definition of curvature is inadequate. What the example of the cylinder reveals, in fact, is a second notion of curvature, qualitatively different from that captured by our previous definition, and one that requires the space in question to be embedded in a higher dimensional space. The first kind of curvature, defined above in terms of the metric tensor, is called *intrinsic curvature* because it is revealed by distances between points in the space and these are determined by measurements confined to the space itself. The second kind of curvature is called *extrinsic curvature* and this cannot be determined by measurements confined to the space itself, as is clear from the example of the cylinder which has a Euclidean metric and therefore has the intrinsic geometry of flat space. The cylinder has zero intrinsic curvature, but it has a non–zero extrinsic curvature if it is viewed as embedded in a 3–dimensional space. In simple terms, the extrinsic curvature has to do with the deviation of lines on the cylinder surface from straight lines in the ambient 3–dimensional space. A simple example of an intrinsically flat but extrinsically curved 2–dimensional space

Figure 3.17: A wavy sheet. This surface can be formed from a flat plane without stretching or compressing it, so it has the same intrinsic Euclidean geometry as a flat plane. Its intrinsic curvature is zero, but when imbedded in in 3–dimensional Euclidean space as shown in the figure it has a non–zero extrinsic curvature.

that, unlike the cylinder, has a trivial topology is shown in Fig. 3.17. This space is obtained by "making waves" on a plane; it is clear that this can be done without stretching the plane so the resulting space still has the Euclidean metric and is intrinsically flat, though an extrinsic curvature can be defined with regard to the ambient 3–dimensional space in Fig. 3.17. Beings confined to the surface in Fig. 3.17 would find that the theorems of Euclid describe exactly the geometry in which they reside, and they would have no way of deducing exactly how they are embedded in a possible higher–dimensional world. On the other hand, beings confined to the surface of a sphere would experience a non–Euclidean geometry, and this is the meaning of intrinsic curvature. We will not describe here how to quantify extrinsic curvature and instead refer the reader to Misner, Thorne and Wheeler [1973]. Our focus in the remainder of this Chapter will be intrinsic curvature.

Although human beings are 3–dimensional creatures who exist in 3–dimensional space, for the most part we have found ourselves restricted to the surface of a curved 2–dimensional space, the Earth. In §22 we will discuss an example where the intrinsic curvature (and rotation) of the Earth becomes directly apparent—Foucault's pendulum.

§19. GEODESICS

How can we quantify the (intrinsic) curvature of a space? We first need to generalize the notion of a straight line to the case of curved spaces like the sphere. The key property of a straight line joining two points in flat space is that it is the shortest path between those points. In a curved space we can still construct the shortest line between two points and this is called a *geodesic*. For a sphere, the geodesics are the

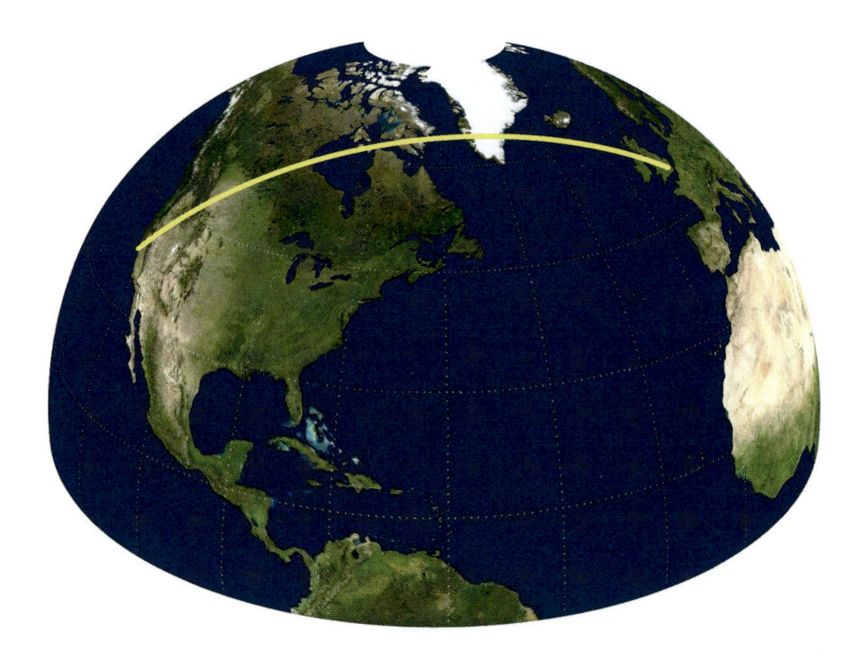

Figure 3.18: Geodesic. The shortest line on Earth joining London and San Francisco. This is a geodesic on the Earth's surface; all such geodesics on a sphere are portions of great circles, circles whose centre is the centre of the sphere. (The figure was created by inverting the Mercator projection of Fig. 2.40).

great circles, i.e. the circles whose centres are the centre of the sphere (Fig. 3.18). A curve $x^i(\xi)$ with parameter ξ that joins a point P, with coordinates $x^i(\xi_1)$, to a point Q, with coordinates $x^i(\xi_2)$, has a length $s(P,Q)$ given by the integrated line element (11.11). For the curve $x^i(\xi)$ to be a geodesic the length (11.11) must be a minimum, so the change $\delta s(P,Q)$ in the length must be zero when we make an infinitesimal change $x^i(\xi) \rightarrow x^i(\xi)+\delta x^i(\xi)$ in the curve joining P and Q (maintaining the parameter values ξ_1 and ξ_2 at the endpoints). This is another example of the variational calculus problem discussed in §2. In the language of classical mechanics, $s(P,Q)$ in Eq. (11.11) plays the role of the action while the "Lagrangian" is

$$L = \sqrt{g_{ij}\dot{x}^i\dot{x}^j}\,, \quad \dot{x}^i \equiv \frac{\mathrm{d}x^i(\xi)}{\mathrm{d}\xi}\,. \tag{19.1}$$

The geodesic is the curve $x^i(\xi)$ that minimizes the length $s(P,Q)$. As shown in §2, for this extremal curve the Lagrangian satisfies the Euler–Lagrange equations (2.6), which in this case are the three equations

$$\frac{\mathrm{d}}{\mathrm{d}\xi}\frac{\partial L}{\partial \dot{x}^i} - \frac{\partial L}{\partial x^i} = 0\,, \tag{19.2}$$

with L given by Eq. (19.1). Bearing in mind that g_{ij} in Eq. (19.1) are functions of $x^i(\xi)$, we obtain explicitly from the Euler–Lagrange equations (19.2)

$$\begin{aligned} 0 &= \frac{\mathrm{d}}{\mathrm{d}\xi}\left(\frac{1}{L} g_{ij}\dot{x}^j\right) - \frac{1}{2L}\, g_{jk,i}\, \dot{x}^j \dot{x}^k \\ &= L\left[\frac{1}{L}\frac{\mathrm{d}}{\mathrm{d}\xi}\left(g_{ij}\frac{1}{L}\frac{\mathrm{d}x^j}{\mathrm{d}\xi}\right) - \frac{1}{2} g_{jk,i}\frac{1}{L}\frac{\mathrm{d}x^j}{\mathrm{d}\xi}\frac{1}{L}\frac{\mathrm{d}x^k}{\mathrm{d}\xi}\right] . \end{aligned} \tag{19.3}$$

Note from Eqs. (11.3) and (19.1) that $L = \mathrm{d}s/\mathrm{d}\xi$, where s is the length along the curve; we can therefore replace $L\,\mathrm{d}\xi$ in the geodesic equation (19.3) by $\mathrm{d}s$ and we obtain

$$\begin{aligned} 0 &= \frac{\mathrm{d}}{\mathrm{d}s}\left(g_{ij}\frac{\mathrm{d}x^j}{\mathrm{d}s}\right) - \frac{1}{2} g_{jk,i}\frac{\mathrm{d}x^j}{\mathrm{d}s}\frac{\mathrm{d}x^k}{\mathrm{d}s} \\ &= g_{ij,k}\frac{\mathrm{d}x^k}{\mathrm{d}s}\frac{\mathrm{d}x^j}{\mathrm{d}s} + g_{ij}\frac{\mathrm{d}^2x^j}{\mathrm{d}s^2} - \frac{1}{2} g_{jk,i}\frac{\mathrm{d}x^j}{\mathrm{d}s}\frac{\mathrm{d}x^k}{\mathrm{d}s} . \end{aligned} \tag{19.4}$$

In moving from Eq. (19.3) to Eq. (19.4) we have effectively changed the parameterization of the geodesic from ξ to the length s. We get the same equation if ξ is proportional to s with constant proportionality factor, but for other parameterizations the geodesic equation assumes the more complicated form (19.3). A final simplification of the geodesic equation is obtained by writing it in terms of the Christoffel symbols. To achieve this we re–express the first metric–dependent term in Eq. (19.4) as follows:

$$g_{ij,k}\frac{\mathrm{d}x^k}{\mathrm{d}s}\frac{\mathrm{d}x^j}{\mathrm{d}s} = \frac{1}{2}\left(g_{ij,k}\frac{\mathrm{d}x^k}{\mathrm{d}s}\frac{\mathrm{d}x^j}{\mathrm{d}s} + g_{ik,j}\frac{\mathrm{d}x^j}{\mathrm{d}s}\frac{\mathrm{d}x^k}{\mathrm{d}s}\right) = \frac{1}{2}\Big(g_{ij,k} + g_{ik,j}\Big)\frac{\mathrm{d}x^j}{\mathrm{d}s}\frac{\mathrm{d}x^k}{\mathrm{d}s} . \tag{19.5}$$

The identity (19.5) exhibits another trick of index gymnastics: the contraction of $g_{ij,k}$ with the symmetric pair of indices on $(\mathrm{d}x^k/\mathrm{d}s)(\mathrm{d}x^j/\mathrm{d}s)$ picks out the part of $g_{ij,k}$ that is symmetric in those indices. Use of Eq. (19.5) in Eq. (19.4) yields

$$g_{ij}\frac{\mathrm{d}^2x^j}{\mathrm{d}s^2} + \frac{1}{2}\Big(g_{ij,k} + g_{ik,j} - g_{jk,i}\Big)\frac{\mathrm{d}x^j}{\mathrm{d}s}\frac{\mathrm{d}x^k}{\mathrm{d}s} = 0 , \tag{19.6}$$

and multiplication by g^{li} with a contraction on i produces in the second term the expression (16.22) for the Christoffel symbols. The significance of Eq. (16.22) in curved space is however not yet clear, since it was derived using properties of *flat* space, in particular Eq. (16.13), deduced by considering flat space in Cartesian coordinates. In what follows we shall show that the covariant derivative computed using Christoffel symbols given by Eq. (16.22) does indeed generalize appropriately from flat space to the curved spaces we are considering. For now, let us introduce the curved–space Christoffel symbols by defining them to be still given by the previous flat–space expression

$$\Gamma^i{}_{jk} = \frac{1}{2} g^{il}\Big(g_{lj,k} + g_{lk,j} - g_{jk,l}\Big) , \tag{19.7}$$

which means they maintain the symmetry (15.13):

$$\Gamma^i{}_{jk} = \Gamma^i{}_{kj}\,. \tag{19.8}$$

Then we obtain from Eq. (19.6) in the manner described the simple geodesic equation

$$\frac{\mathrm{d}^2 x^i(s)}{\mathrm{d}s^2} + \Gamma^i{}_{jk}\frac{\mathrm{d}x^j(s)}{\mathrm{d}s}\frac{\mathrm{d}x^k(s)}{\mathrm{d}s} = 0\,. \tag{19.9}$$

Equation (19.9) determines a unique geodesic curve once an initial position $x^i(s_0)$ and tangent vector (direction of the geodesic) $\mathrm{d}x^i(s)/\mathrm{d}s|_{s_0}$ are specified.

In Euclidean space the solutions of the geodesic equation (19.9) are straight lines; this is easily seen by choosing Cartesian coordinates, which have $\Gamma^i{}_{jk} = 0$, so the solution is $\boldsymbol{a} + s\boldsymbol{b}$, where $\boldsymbol{a}$ and $\boldsymbol{b}$ are constant vectors. For the sphere, the geodesic equation (19.9) describes a great circle, the shortest path between points in this curved space.

Exercise 19.1

Prove that the geodesics of a sphere are the great circles.

Solution

Equation (19.9) in this case must be understood in a 2–dimensional version; we will write it in the coordinates $\{\theta, \phi\}$ used in the line element (18.4) of the sphere of radius a. First, we require the Christoffel symbols in these coordinates; they are easily obtained from the Christoffel symbols (15.26) of flat 3–dimensional space in spherical polar coordinates by noting that for the sphere the dimension associated with the coordinate r is removed so one need only drop the $\Gamma^i{}_{jk}$ in Eqs. (15.26) that have r appearing in their indices. Alternatively, one can of course directly calculate the Christoffel symbols using Eq. (16.22) and the metric tensor of the sphere given by Eq. (18.4). In either case one finds

$$\Gamma^\phi{}_{\theta\phi} = \Gamma^\phi{}_{\phi\theta} = \cot\theta\,, \quad \Gamma^\theta{}_{\phi\phi} = -\sin\theta\cos\theta\,, \tag{19.10}$$

all the other Christoffel symbols vanishing. The geodesic equation (19.9) is thus

$$\frac{\mathrm{d}^2\theta(s)}{\mathrm{d}s^2} - \sin[\theta(s)]\cos[\theta(s)]\,\frac{\mathrm{d}\phi(s)}{\mathrm{d}s}\frac{\mathrm{d}\phi(s)}{\mathrm{d}s} = 0\,, \tag{19.11}$$

$$\frac{\mathrm{d}^2\phi(s)}{\mathrm{d}s^2} + 2\cot[\theta(s)]\,\frac{\mathrm{d}\phi(s)}{\mathrm{d}s}\frac{\mathrm{d}\theta(s)}{\mathrm{d}s} = 0\,. \tag{19.12}$$

Consider an arbitrary great circle. We can choose our coordinates $\{\theta, \phi\}$ so that this great circle is given by $\theta = \pi/2$ (the coordinate grid can be rotated so that the chosen great circle coincides with $\theta = \pi/2$). The curve $\theta = \pi/2$ is parametrized in terms of its length s by the equations (recall a is the radius of the sphere)

$$\theta(s) = \pi/2\,, \quad \phi(s) = \phi_0 + s/a\,, \quad (\phi_0 \text{ a constant}). \tag{19.13}$$

Now Eqs. (19.13) are clearly a solution of Eqs. (19.11) and (19.12), so the great circle is a geodesic. But the chosen great circle was arbitrary, hence all great circles are geodesics. Conversely, the great circles supply all possible initial conditions $\{\theta(s_0), \phi(s_0)\}$ and $\{\mathrm{d}\theta(s)/\mathrm{d}s|_{s_0}, \mathrm{d}\phi(s)/\mathrm{d}s|_{s_0}\}$ for the differential equations (19.11) and (19.12), and these initial conditions give unique solutions.

As noted above, the vector $\mathrm{d}x^i(s)/\mathrm{d}s$ appearing in the geodesic equation (19.9) is the tangent vector to the geodesic curve $x^i(s)$. It is clear from Eqs. (11.3) and (12.9)

that the choice of length s as the parameter of the curve means that $\mathrm{d}x^i(s)/\mathrm{d}s$ is a unit vector. Denoting this tangent vector by U^i and using the chain–rule relation

$$\frac{\mathrm{d}}{\mathrm{d}s} = \frac{\mathrm{d}x^i(s)}{\mathrm{d}s}\frac{\partial}{\partial x^i(s)} = U^i\frac{\partial}{\partial x^i(s)}\,, \qquad U^i \equiv \frac{\mathrm{d}x^i(s)}{\mathrm{d}s}\,, \tag{19.14}$$

we can write Eq. (19.9) in a very enlightening form:

$$\frac{\mathrm{d}U^i}{\mathrm{d}s} + \Gamma^i{}_{jk}U^jU^k = U^k\left(\frac{\partial U^i}{\partial x^k(s)} + \Gamma^i{}_{jk}U^j\right) = 0\,. \tag{19.15}$$

Through our definition of the curved–space Christoffel symbols (19.7) we have generalized the covariant derivative to curved space, and the expression in brackets in Eq. (19.15) will be recognized as the covariant derivative of the tangent vector U^i. We thus obtain from Eq. (19.15) the following equivalent form of the geodesic equation:

$$U^kU^i{}_{;k} = U^k\,\nabla_kU^i = 0\,, \quad U^i \equiv \frac{\mathrm{d}x^i(s)}{\mathrm{d}s}\,. \tag{19.16}$$

To understand what Eq. (19.16) is saying, consider first its meaning in flat space. In conventional vector–calculus notation, one would write Eq. (19.16) in flat space as $(\boldsymbol{U}\cdot\boldsymbol{\nabla})\boldsymbol{U} = 0$, which is the statement that the derivative of the tangent vector $\boldsymbol{U}$ in the direction $\boldsymbol{U}$ is zero; in other words, as one follows the tangent vector along the curve it does not change, so the tangent vectors at all points on the curve are parallel and of equal length. This clearly implies that the curve is a straight line, showing explicitly in the flat–space case that Eq. (19.16) describes a geodesic. In a general (flat or curved) space, $U^kU^i{}_{;k}$ is the covariant derivative of the tangent vector $\boldsymbol{U}$ in the direction $\boldsymbol{U}$; it quantifies the rate of change of the tangent vector as one moves along the curve. If $U^kU^i{}_{;k}$ vanishes this means that $\boldsymbol{U}$ stays "parallel" to itself and the curve is "straight" as far as is possible in a general curved space, so it is entirely reasonable that the geodesic equation should turn out to be equivalent to Eq. (19.16). But this conclusion that geodesics, which we defined as the shortest paths between points, also obey Eq. (19.16) required the generalization of the covariant derivative to curved spaces through the definition (19.7) of the curved–space Christoffel symbols. This way of generalizing the covariant derivative is therefore wholly appropriate.*

§20. PARALLEL TRANSPORT AND COVARIANT DERIVATIVES

The condition (19.16) defines the notion of the tangent vectors at all points on a curve being "parallel", in so far as this is possible in a curved space. For this reason

*As stated at the beginning of this Section, our concern is with curved spaces that, when "smoothed out", become Euclidean space with its familiar properties. For such spaces one is led, as above, to a covariant derivative determined by the metric tensor through Eq. (19.7). These spaces are known as *Riemannian manifolds*. It is perfectly legitimate however to consider somewhat more abstract spaces that have a covariant derivative operation not given by (19.7); indeed, one can have a covariant derivative in a space that does not possess a metric tensor (see for example Hawking and Ellis [1973]).

geodesics, which all obey Eq. (19.16), are said to *parallel transport* their tangent vectors. This concept of parallel transport has an obvious generalization: if at every point the covariant derivative of a vector field $\boldsymbol{V}$ along another vector field $\boldsymbol{W}$ is zero, then $\boldsymbol{V}$ is said to be parallel transported along $\boldsymbol{W}$:

$$W^k V^i_{\ ;k} = W^k \nabla_k V^i = 0 \quad \Longleftrightarrow \quad \boldsymbol{V} \text{ parallel transported along } \boldsymbol{W}. \tag{20.1}$$

In terms of a curve $x^i(\xi)$ to which $\boldsymbol{W}$ is tangent, so that $W^i = \mathrm{d}x^i(\xi)/\mathrm{d}\xi$, the parallel transport condition (20.1) can be written out using the chain rule (19.14) applied to W^i:

$$W^k \left(\frac{\partial V^i}{\partial x^k(\xi)} + \Gamma^i_{\ jk} V^j \right) = \frac{\mathrm{d}V^i}{\mathrm{d}\xi} + \Gamma^i_{\ jk} \frac{\mathrm{d}x^k(\xi)}{\mathrm{d}\xi} V^j = 0\,. \tag{20.2}$$

We see from Eq. (20.2) that the covariant derivative of $\boldsymbol{V}$ along the curve $x^i(\xi)$ has the obvious expression $\mathrm{d}V^i/\mathrm{d}\xi$ in Cartesian coordinates in flat space; the general expression for this covariant derivative is often denoted by $\mathrm{D}V^i/\mathrm{D}\xi$, so that Eq. (20.2) becomes

$$\frac{\mathrm{D}V^i}{\mathrm{D}\xi} \equiv \frac{\mathrm{d}V^i}{\mathrm{d}\xi} + \Gamma^i_{\ jk} \frac{\mathrm{d}x^k(\xi)}{\mathrm{d}\xi} V^j = 0 \quad \Longleftrightarrow \quad \boldsymbol{V} \text{ parallel transported along } x^i(\xi). \tag{20.3}$$

In the notation of Eq. (20.3), we see from Eqs. (19.15) and (19.16) that the geodesic equation is

$$\frac{\mathrm{D}}{\mathrm{D}s} \left(\frac{\mathrm{d}x^i(s)}{\mathrm{d}s} \right) = 0 \quad \Longleftrightarrow \quad x^i(s) \text{ a geodesic.} \tag{20.4}$$

As was alluded to after Eq. (15.5), the rate of change of a vector field involves a comparison of vectors located at different points and this comparison is not straightforward in curved space. From the definition (20.1) of parallel transport we see that the vanishing of $V^i_{\ ;k}$ implies the parallel transport of $\boldsymbol{V}$ along the basis vectors $\boldsymbol{e}_k$. This shows that $V^i_{\ ;k}$ measures how the vector field $\boldsymbol{V}$ changes as one moves along $\boldsymbol{e}_k$; thus, for the covariant derivative in curved space, the values of $\boldsymbol{V}$ at neighbouring points are compared by parallel transport of one to the other in order to obtain the rate of change $V^i_{\ ;k}$.

Note that in a curved space there is no coordinate system in which the Christoffel symbols (19.7) vanish, since by definition the metric–tensor components in curved space are always non–Euclidean and therefore vary from point to point.* The Christoffel symbols in curved space perform the same function as in flat space: they quantify through Eq. (15.24) the variation of the coordinate basis as one moves around the space (Fig. 3.15), and this rate of change of the basis must be included in differentiating tensors. The difference in curved space is that there is no basis

*If g_{ij} were constant everywhere one could find a coordinate system in which the metric is Euclidean, as follows. Calculate the three eigenvectors of the matrix g_{ij}; these vectors are orthogonal since g_{ij} is symmetric, and they are constant vectors if g_{ij} are constant. Rescale the eigenvectors to have unit length and thereby obtain a Cartesian basis of orthonormal vectors. The metric components in a Cartesian basis are the Euclidean δ_{ij}.

that "stays the same" at all points, in contrast to the Cartesian basis in flat space, so the Christoffel symbols are always non–trivial. An interesting property that we shall not prove here is that it is always possible to find a coordinate system such that *at any one point* in a curved space (1) the metric g_{ij} is equal to δ_{ij} and (2) the first–order partial derivatives $g_{ij,k}$ are zero. This clearly means that in these coordinates the Christoffel symbols (19.7) vanish *at the chosen point*; but if one moves away from this point g_{ij} is no longer equal to δ_{ij} and the partial derivatives $g_{ij,k}$ no longer vanish, giving non–zero $\Gamma^i{}_{jk}$. This property of Riemannian manifolds is called *local flatness*; we refer the reader to Schutz [2009] or Misner, Thorne and Wheeler [1973] for the proof.

The transformation rule (15.10) holds also for the Christoffel symbols in curved space since in all cases they are given in terms of the metric tensor by Eq. (19.7) and the transformation behaviour of the right–hand side of this equation is the same in flat or curved space. In §16 we derived the expression (16.22) for the Christoffel symbols from the vanishing (16.13) in flat space of the covariant derivative of the metric tensor. In curved space we have taken Eq. (19.7) as the fundamental equations of the Christoffel symbols and we can reverse the derivation in §16 to obtain also for curved spaces the general result

$$g_{ij;k} = 0\,. \tag{20.5}$$

Exercise 20.1

Derive Eq. (20.5) from Eq. (19.7).

Solution

Multiplication across Eq. (19.7) with g_{mi} and contraction on i gives

$$g_{mi}\Gamma^i{}_{jk} = \frac{1}{2}\big(g_{mj,k} + g_{mk,j} - g_{jk,m}\big). \tag{20.6}$$

We rewrite Eq. (20.6) with a different permutation of indices:

$$g_{ji}\Gamma^i{}_{mk} = \frac{1}{2}\big(g_{jm,k} + g_{jk,m} - g_{mk,j}\big) \tag{20.7}$$

and add Eq. (20.6) to Eq. (20.7), with use of Eq. (18.2):

$$g_{jm,k} = g_{im}\Gamma^i{}_{jk} + g_{ji}\Gamma^i{}_{mk}\,. \tag{20.8}$$

This is Eq. (20.5).

Exercise 20.2

The Levi–Civita tensor in curved space is defined by Eq. (14.6), which implies Eq. (14.7):

$$\epsilon_{ijk} = \pm\sqrt{g}\,[ijk]\,, \quad \epsilon^{ijk} = \pm\frac{1}{\sqrt{g}}\,[ijk]\,. \tag{20.9}$$

In Exercise 16.5 it was shown that in flat space the Levi–Civita tensor has zero covariant derivative:

$$\epsilon_{ijk;l} = 0 \quad\Longrightarrow\quad \epsilon^{ijk}{}_{;l} = 0\,. \tag{20.10}$$

Show that this holds also in curved space.

Solution
We write out $\epsilon_{ijk;l}$ in terms of the Christoffel symbols:

$$\epsilon_{ijk;l} = \epsilon_{ijk,l} - \Gamma^m{}_{il}\epsilon_{mjk} - \Gamma^m{}_{jl}\epsilon_{imk} - \Gamma^m{}_{kl}\epsilon_{ijm} \,. \tag{20.11}$$

By using the fact that ϵ_{ijk} is completely antisymmetric it is easy to show from Eq. (20.11) that $\epsilon_{ijk;l}$ is completely antisymmetric in its first three indices. This means that $\epsilon_{ijk;l}$ is zero unless the first three indices have different values, hence every non–zero component of $\epsilon_{ijk;l}$ can be obtained from $\epsilon_{123;l}$ by interchanging indices (e.g. $\epsilon_{132;l} = -\epsilon_{123;l}$). From Eqs. (20.11) and (20.9) we find

$$\begin{aligned}\epsilon_{123;l} &= \epsilon_{123,l} - \Gamma^m{}_{1l}\epsilon_{m23} - \Gamma^m{}_{2l}\epsilon_{1m3} - \Gamma^m{}_{3l}\epsilon_{12m} \\ &= \pm\left[(\sqrt{g})_{,l} - \Gamma^1{}_{1l}\sqrt{g} - \Gamma^2{}_{2l}\sqrt{g} - \Gamma^3{}_{3l}\sqrt{g}\right][123] \\ &= \pm\left[(\sqrt{g})_{,l} - \Gamma^i{}_{il}\sqrt{g}\right][123] \,. \end{aligned} \tag{20.12}$$

Equation (19.7) gives

$$\Gamma^i{}_{il} = \frac{1}{2}g^{ik}g_{ki,l} + \frac{1}{2}g^{ik}(g_{kl,i} - g_{il,k}) = \frac{1}{2}g^{ik}g_{ki,l} = \frac{(\sqrt{g})_{,l}}{\sqrt{g}} \,, \tag{20.13}$$

where the second equality follows from Eq. (17.8) and the third equality is Eq. (17.10). Insertion of Eq. (20.13) into Eq. (20.12) gives the first of Eqs. (20.10), and since indices on the tensor $\epsilon_{ijk;l}$ can be raised it follows that $\epsilon^{ijk}{}_{;l} = 0$.

---

§21. Geodesic deviation and the Riemann tensor

The geodesics carry all the information about the curvature of a space and so their behaviour can be used to quantify the amount of curvature. In flat space two geodesics (straight lines) either have a constant separation (parallel lines) or the separation distance between them changes at a constant rate, i.e. it changes linearly with distance along the lines. The second derivative of the separation with respect to distance along the geodesics is therefore zero in flat space. In curved space by contrast, this second derivative does not vanish, a phenomenon called *geodesic deviation*, and this is used to measure the curvature. As an example, consider again the surface of a sphere (Fig. 3.19). We see clearly that the separation of any two geodesics (great circles) does not change linearly with distance along the geodesics: the separation increases from zero to a maximum and then decreases to zero again. At the points where the tangent vectors to the geodesics are drawn in Fig. 3.19, the separation is a maximum and the tangent vectors are parallel as viewed in the three–dimensional space in which the sphere is embedded. The geodesics are parallel at these points, inasmuch as a notion of parallelism can be introduced on the sphere, but these "parallel lines" meet! Thus the postulates of Euclidean geometry do not hold on the sphere: it is a non–Euclidean space and this fact is revealed by geodesic deviation. As another example, consider the geodesics on the surface of a cylinder (Fig. 3.16). These are drawn by simply cutting open the cylinder into a flat sheet, drawing straight lines on the sheet and rolling it up again to reform the cylinder. Clearly the geodesics behave just as in the plane: there is no geodesic deviation

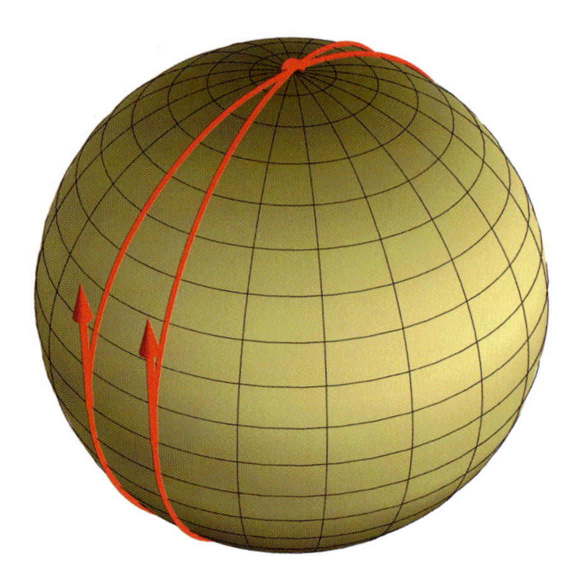

Figure 3.19: Two geodesics (great circles) on the sphere. The separation of the geodesics as a function of distance from a meeting point is not linear; as one moves along, the geodesics have an "acceleration" towards each other. Tangent vectors to the geodesics are drawn at the points where the separation has reached a maximum. These vectors are parallel in 3–dimensional space.

(second derivative of separation with respect to distance along geodesics is zero), showing again that a cylinder is a flat space.

How are we to compute the geodesic deviation? We need to consider a family of geodesics $x^i(s,\xi)$, where the parameter ξ labels (continuously) the geodesics in the family and for fixed ξ the parameter s is the distance along a geodesic. Note that ξ is not the distance between geodesics since two fixed values of ξ determine two geodesics that may be moving apart or approaching each other. For the purposes of our calculation the parameters s and ξ need only have an arbitrarily small range, so that $x^i(s,\xi)$ defines an arbitrarily small 2–dimensional surface in the space (Fig. 3.20). This means that we can assume no two geodesics in the family $x^i(s,\xi)$ intersect, so that $\{s,\xi\}$ labels distinct points on the small surface $x^i(s,\xi)$. Consider for some s and ξ the points $x^i(s,\xi)$ and $x^i(s,\xi+\mathrm{d}\xi)$; these are points on two infinitesimally separated geodesics with the same parameter value s. Since $x^i(s,\xi+\mathrm{d}\xi) = x^i(s,\xi) + [\partial x^i(s,\xi)/\partial\xi]\mathrm{d}\xi$ the vector joining these two points is given by

$$V^i = \frac{\partial x^i(s,\xi)}{\partial\xi}\,. \tag{21.1}$$

This vector gives the separation between the neighbouring geodesics $x^i(s,\xi)$ and $x^i(s,\xi+\mathrm{d}\xi)$. Now what we require for the geodesic deviation is the rate of change of $\boldsymbol{V}$ as s increases, specifically its second covariant derivative along the geodesic

$x^i(s,\xi)$ (ξ fixed); in the notation introduced in the previous Section this second covariant derivative is

$$\frac{\mathrm{D}}{\mathrm{D}s}\left(\frac{\mathrm{D}V^i}{\mathrm{D}s}\right). \tag{21.2}$$

For fixed ξ the geodesic $x^i(s,\xi)$ has a tangent vector

$$U^i = \frac{\partial x^i(s,\xi)}{\partial s} \tag{21.3}$$

that obeys the geodesic equation (19.16): $U^j U^i{}_{;j} = 0$. The vectors (21.1) and (21.3) have the following important property that results from them being tangent vectors induced by the coordinate grid $\{s,\xi\}$ on the surface $x^i(s,\xi)$: the covariant derivative of $\boldsymbol{V}$ along $\boldsymbol{U}$ is equal to the covariant derivative of $\boldsymbol{U}$ along $\boldsymbol{V}$. This property is easily proved from the definitions (21.1) and (21.3), and the commutativity of partial derivatives:

$$\begin{aligned} U^j V^i{}_{;j} &= \frac{\mathrm{D}}{\mathrm{D}s}V^i = \frac{\partial V^i}{\partial s} + \Gamma^i{}_{kj}\frac{\partial x^j}{\partial s}V^k \\ &= \frac{\partial^2 x^i}{\partial s\,\partial\xi} + \Gamma^i{}_{kj}\frac{\partial x^j}{\partial s}\frac{\partial x^k}{\partial\xi} \\ &= \frac{\partial^2 x^i}{\partial\xi\,\partial s} + \Gamma^i{}_{kj}\frac{\partial x^k}{\partial\xi}\frac{\partial x^j}{\partial s} \\ &= \frac{\mathrm{D}}{\mathrm{D}\xi}U^i = V^j U^i{}_{;j}\,. \end{aligned} \tag{21.4}$$

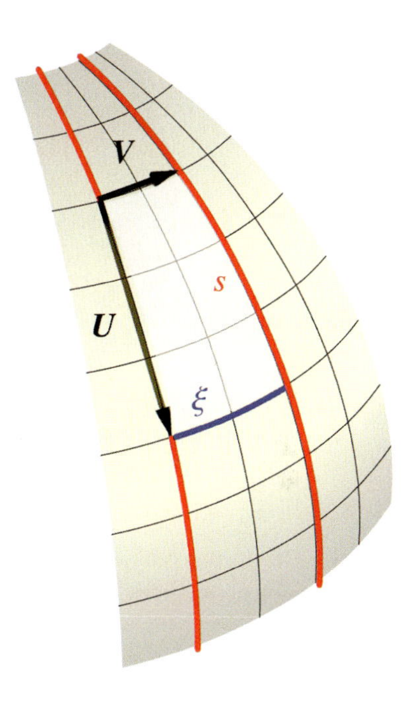

Figure 3.20: A family of geodesics, all parametrized by their length s, and labeled by a continuous parameter ξ. The set of numbers $\{s,\xi\}$ are coordinates for a 2–dimensional patch of the space, provided the patch is small enough so that no two geodesics in the family intersect. The vector field $\boldsymbol{U}$ on the patch is defined by having a value at each point given by the tangent vector to the geodesic through that point. The vector field $\boldsymbol{V}$ represents the separation between neighbouring geodesics; in general this separation varies with s.

(Note that the property (21.4) does not hold in general for two arbitrary vectors.) We now evaluate the geodesic deviation (21.2), wherein we make use of the result (21.4), the Leibniz rule for covariant derivatives, and the geodesic equation $U^j U^i{}_{;j} = 0$:

$$\begin{aligned} \frac{\mathrm{D}}{\mathrm{D}s}\left(\frac{\mathrm{D}V^i}{\mathrm{D}s}\right) &= U^k\left(U^l V^i{}_{;l}\right)_{;k} \\ &= U^k\left(V^l U^i{}_{;l}\right)_{;k} \\ &= U^k V^l{}_{;k} U^i{}_{;l} + U^k V^l U^i{}_{;l;k} \\ &= V^k U^l{}_{;k} U^i{}_{;l} + U^k V^l U^i{}_{;l;k} = V^k\left[\left(U^l U^i{}_{;l}\right)_{;k} - U^l U^i{}_{;l;k}\right] + U^k V^l U^i{}_{;l;k} \\ &= -V^k U^l U^i{}_{;l;k} + U^k V^l U^i{}_{;l;k} = -V^l U^k U^i{}_{;k;l} + U^k V^l U^i{}_{;l;k} \\ &= U^k V^l\left(U^i{}_{;l;k} - U^i{}_{;k;l}\right). \end{aligned} \tag{21.5}$$

In Eq. (21.5) the geodesic deviation is expressed in terms of a commutator $U^i{}_{;l;k} - U^i{}_{;k;l}$ of covariant derivatives. We know that *partial* derivatives commute, but the question of commutation of covariant derivatives is not something we have addressed up to now. Since geodesic deviation is caused by curvature and is quantified by a commutator of covariant derivatives, it must be the case that this commutator vanishes in flat space but not in curved spaces. Let us compute the commutator $U^i{}_{;l;k} - U^i{}_{;k;l}$ in terms of the Christoffel symbols; first we write out $U^i{}_{;l;k}$:

$$\begin{aligned} U^i{}_{;l;k} &= U^i{}_{;l,k} + \Gamma^i{}_{jk} U^j{}_{;l} - \Gamma^j{}_{lk} U^i{}_{;j} \\ &= \left(U^i{}_{,l} + \Gamma^i{}_{ml} U^m\right)_{,k} + \Gamma^i{}_{jk}\left(U^j{}_{,l} + \Gamma^j{}_{ml} U^m\right) - \Gamma^j{}_{lk}\left(U^i{}_{,j} + \Gamma^i{}_{mj} U^m\right) \\ &= U^i{}_{,l,k} + \Gamma^i{}_{ml,k} U^m + \Gamma^i{}_{ml} U^m{}_{,k} + \Gamma^i{}_{jk}\left(U^j{}_{,l} + \Gamma^j{}_{ml} U^m\right) - \Gamma^j{}_{lk}\left(U^i{}_{,j} + \Gamma^i{}_{mj} U^m\right) \end{aligned} \tag{21.6}$$

and then we subtract from this the same quantity with the indices l and k interchanged. Several cancelations occur in this subtraction: partial derivatives commute so $U^i{}_{,l,k} - U^i{}_{,k,l} = 0$, the symmetry (19.8) of the Christoffel symbols shows that the last term in Eq. (21.6) is cancelled by the corresponding term in $U^i{}_{;k;l}$, and the sum $\Gamma^i{}_{ml} U^m{}_{,k} + \Gamma^i{}_{jk} U^j{}_{,l}$ that appears in Eq. (21.6) is symmetric in l and k and so is also cancelled by subtracting $U^i{}_{;k;l}$. We thus find

$$\begin{aligned} U^i{}_{;l;k} - U^i{}_{;k;l} &= \Gamma^i{}_{ml,k} U^m - \Gamma^i{}_{mk,l} U^m + \Gamma^i{}_{jk} \Gamma^j{}_{ml} U^m - \Gamma^i{}_{jl} \Gamma^j{}_{mk} U^m \\ &= \left(\Gamma^i{}_{jl,k} - \Gamma^i{}_{jk,l} + \Gamma^i{}_{mk} \Gamma^m{}_{jl} - \Gamma^i{}_{ml} \Gamma^m{}_{jk}\right) U^j , \end{aligned} \tag{21.7}$$

where we performed some index relabeling in writing the second line. No particular properties of the vector $\boldsymbol{U}$ were used in deriving Eq. (21.7) so it holds for all vector fields. Thus covariant derivatives fail to commute for any non–zero vector field if the quantity in brackets in Eq. (21.7) does not vanish. This quantity depends solely on the metric tensor, and therefore solely on the geometry of the space. It quantifies the geodesic deviation (21.5), which is the signature of curvature, and it is a tensor since when contracted with the vector U^i it gives the tensor $U^i{}_{;l;k} - U^i{}_{;k;l}$. It is one of the most important objects in geometry, the *Riemann curvature tensor**

$$R^i{}_{jkl} \equiv \Gamma^i{}_{jl,k} - \Gamma^i{}_{jk,l} + \Gamma^i{}_{mk} \Gamma^m{}_{jl} - \Gamma^i{}_{ml} \Gamma^m{}_{jk} . \tag{21.8}$$

One can also verify explicitly using the transformation rule (15.10) for the Christoffel symbols that $R^i{}_{jkl}$ transforms as a tensor. In terms of the Riemann tensor (21.8) the commutator (21.7) of covariant derivatives is

$$U^i{}_{;l;k} - U^i{}_{;k;l} = R^i{}_{jkl} U^j , \tag{21.9}$$

*In 1854 Bernhard Riemann gave his first university lecture. His subject was the foundations of geometry. In his lecture, Riemann developed the notion of a curved space of an arbitrary number of dimensions, explained how beings in such a space could measure its curvature, and defined the curvature tensor. He went on in the same lecture to argue that the actual geometry of physical space was a matter of empirical observation and that it may not be Euclidean, or 3–dimensional. It took 60 years for physicists to fully understand the importance of Riemann's insights.

or, written in terms of the operator (15.25),

$$(\nabla_k \nabla_l - \nabla_l \nabla_k) U^i = R^i{}_{jkl} U^j \,. \tag{21.10}$$

We obtain for the geodesic deviation (21.5):

$$\frac{\mathrm{D}}{\mathrm{D}s}\left(\frac{\mathrm{D}V^i}{\mathrm{D}s}\right) = R^i{}_{jkl} U^j U^k V^l \,. \tag{21.11}$$

We introduced the idea of geodesic deviation by noting that it does not occur in flat space, and this accords with the result (21.11). In flat space we can by definition introduce Cartesian coordinates with vanishing Christoffel symbols and so the Riemann tensor (21.8) is zero in these coordinates, implying it is zero in all coordinates, and the geodesic deviation (21.11) vanishes. Thus flat space implies zero Riemann tensor. What about the converse? Does vanishing Riemann tensor imply flat space? Yes, as the following argument shows. If $R^i{}_{jkl} = 0$ there is no geodesic deviation (21.11) and so in the 2–dimensional surface $x^i(s, \xi)$ defined above we can choose the geodesics labelled by ξ to have fixed separations that do not change as s changes. We can therefore choose ξ to be the distance between the geodesics, which stays fixed as we move along them. But this means that $\{s, \xi\}$ can be chosen to be a Cartesian coordinate grid on the surface $x^i(s, \xi)$, and since this surface is arbitrary it is not difficult to see that this implies we can construct a Cartesian coordinate grid in the full 3–dimensional space. The space is therefore flat by definition. We have thus established

$$R^i{}_{jkl} = 0 \qquad \Longleftrightarrow \qquad \text{space is flat.} \tag{21.12}$$

The Riemann tensor quantifies curvature, and a simple geometrical effect of curvature is geodesic deviation which occurs if and only if the space is curved.

We mentioned on p. 135 that one can always tailor coordinates to any one point in a space such that the partial derivatives $g_{ij,k}$ are zero at that point. In general, however, it is not possible to also make the second–order partial derivatives $g_{ij,k,l}$ vanish at the point in question. This is in line with our findings about curvature since we see from Eqs. (21.8) and (19.7) that the Riemann tensor depends on $g_{ij,k,l}$ as well as $g_{ij,k}$. If coordinates existed in which both $g_{ij,k}$ and $g_{ij,k,l}$ were zero at one arbitrary point, the Riemann tensor would vanish at that point, in all coordinate systems, and since the chosen point is arbitrary the Riemann tensor would vanish throughout every space and there would be no such thing as curvature.

Two other curvature quantities are obtained by contracting indices on $R^i{}_{jkl}$. The 2–index *Ricci tensor* is defined by

$$R_{ij} = R^k{}_{ikj} \,, \tag{21.13}$$

and contraction of the indices of the Ricci tensor gives the *curvature scalar*, denoted by R:

$$R = g^{ij} R_{ij} = R^i{}_i = R^{ij}{}_{ij} \,. \tag{21.14}$$

On the face of it there seems to be a large amount of information necessary to specify curvature since the 4–index tensor $R^i{}_{jkl}$ has $3^4 = 81$ components in three dimensions. The number of *independent* components of the Riemann tensor is, however, significantly less, since there are many relations between them. One of these relations ($R^i{}_{jkl} = -R^i{}_{jlk}$) is immediately apparent from definition (21.8), but to find all of them we need to lower the upper index on $R^i{}_{jkl}$ with the metric and rearrange the constituent metric derivatives in R_{ijkl} to obtain

$$R_{ijkl} = \frac{1}{2}\left(g_{il,j,k} - g_{ik,j,l} + g_{jk,i,l} - g_{jl,i,k}\right) + g_{mn}\left(\Gamma^m{}_{il}\Gamma^n{}_{jk} - \Gamma^m{}_{ik}\Gamma^n{}_{jl}\right) . \tag{21.15}$$

Exercise 21.1
Derive Eq. (21.15).
Solution
Lowering the upper index in definition (21.8) we obtain

$$R_{ijkl} = g_{in}R^n{}_{jkl} = g_{in}\Gamma^n{}_{jl,k} - g_{in}\Gamma^n{}_{jk,l} + g_{in}\Gamma^n{}_{mk}\Gamma^m{}_{jl} - g_{in}\Gamma^n{}_{ml}\Gamma^m{}_{jk} . \tag{21.16}$$

Writing out the first term in Eq. (21.16) using Eq. (19.7) we find after some simplification and use of $g_{in}g^{nm}{}_{,k} = (g_{in}g^{nm})_{,k} - g_{in,k}g^{nm} = -g^{nm}g_{in,k}$ the result

$$g_{in}\Gamma^n{}_{jl,k} = \frac{1}{2}\left(g_{ij,l,k} + g_{il,j,k} - g_{jl,i,k}\right) - \frac{1}{2}g^{nm}g_{in,k}\left(g_{mj,l} + g_{ml,j} - g_{jl,m}\right) , \tag{21.17}$$

while for the third term in Eq. (21.16) use of Eq. (19.7) gives

$$\begin{aligned} g_{in}\Gamma^n{}_{mk}\Gamma^m{}_{jl} &= \frac{1}{4}g^{nm}\left(g_{im,k} + g_{ik,m} - g_{mk,i}\right)\left(g_{nj,l} + g_{nl,j} - g_{jl,n}\right) \\ &= \frac{1}{4}g^{nm}g_{im,k}\left(g_{nj,l} + g_{nl,j} - g_{jl,n}\right) + \frac{1}{4}g^{nm}\left(g_{ik,m} - g_{mk,i}\right)\left(g_{nj,l} + g_{nl,j} - g_{jl,n}\right) . \end{aligned} \tag{21.18}$$

The symmetry of the metric and index relabeling shows that the second term in Eq. (21.17) and the first term in Eq. (21.18) differ only by a factor of $-1/2$; adding these two equations therefore gives

$$\begin{aligned} g_{in}\Gamma^n{}_{jl,k} + g_{in}\Gamma^n{}_{mk}\Gamma^m{}_{jl} =& \frac{1}{2}\left(g_{ij,l,k} + g_{il,j,k} - g_{jl,i,k}\right) \\ &+ \frac{1}{4}g^{nm}\left(-g_{im,k} + g_{ik,m} - g_{mk,i}\right)\left(g_{nj,l} + g_{nl,j} - g_{jl,n}\right) \\ =& \frac{1}{2}\left(g_{ij,l,k} + g_{il,j,k} - g_{jl,i,k}\right) - g_{mn}\Gamma^m{}_{ik}\Gamma^n{}_{jl} . \end{aligned} \tag{21.19}$$

To obtain Eq. (21.16) we subtract from Eq. (21.19) the same expression with k and l interchanged; the term $g_{ij,l,k}$ is symmetric in k and l and so is cancelled in this subtraction, and the result is (21.15).

From the right–hand side of expression (21.15) one easily verifies the following symmetry properties of the Riemann tensor components:

$$R_{ijkl} = -R_{ijlk} , \quad R_{ijkl} = -R_{jikl} , \quad R_{ijkl} = R_{klij} , \tag{21.20}$$

$$R_{ijkl} + R_{iklj} + R_{iljk} = 0 . \tag{21.21}$$

The third of Eqs. (21.20) shows that the Ricci tensor (21.13) is symmetric:

$$R_{ij} = R_{ji} \, . \tag{21.22}$$

The general results derived here for curvature at no point required a restriction to spaces of three dimensions, and they are valid for an arbitrary number d of dimensions, with the indices understood to range over d values. The Riemann tensor in d dimensions has d^4 components, but because of the symmetry relations (21.20) and (21.21) the number of *independent* components is

$$\frac{d^2(d^2-1)}{12} \, . \tag{21.23}$$

Thus in three dimensions only 6 of the 81 components of the Riemann tensor are independent.

The lowest dimensional space with intrinsic curvature is 2–dimensional, and in this case the number of independent components (21.23) of the Riemann tensor is just one. But the curvature scalar (21.14) provides one quantity to specify the curvature, so it must be possible in two dimensions to write the Riemann tensor in terms of the curvature scalar. This is indeed the case, and with capital Latin letters $A, B, C, \ldots$ labeling the indices in two dimensions the relation is

$$R^{AB}{}_{CD} = \frac{1}{2} R \left(\delta^A_C \, \delta^B_D - \delta^A_D \, \delta^B_C \right) \, . \tag{21.24}$$

The form of the quantity in brackets in Eq. (21.24) ensures that the Riemann tensor has the symmetries (21.20)–(21.21) (when lowering indices in Eq. (21.24) recall that the Kroneker delta is the metric tensor with a raised index). It is easy to see by contracting indices on $R^{AB}{}_{CD}$ that Eq. (21.24) is consistent with the definition (21.14) of R. As noted above, in three dimensions the Riemann tensor has 6 independent components. Now this is exactly the number of independent components of the Ricci tensor in three dimensions, because of the symmetry (21.22). It must therefore be possible in this case to express $R^i{}_{jkl}$ in terms of R_{ij}. The relation also contains the contraction of R_{ij}, the curvature scalar R, and can be written as

$$R^{ij}{}_{kl} = \delta^i_k R^j{}_l - \delta^j_k R^i{}_l - \delta^i_l R^j{}_k + \delta^j_l R^i{}_k + \frac{1}{4} R \left(\delta^i_k \delta^j_l - \delta^j_k \delta^i_l - \delta^i_l \delta^j_k + \delta^j_l \delta^i_k \right) \, . \tag{21.25}$$

One can check by contracting indices on $R^{ij}{}_{kl}$ that formula (21.25) is consistent with the definitions (21.13) and (21.14) of R_{ij} and R. In the case of four or more dimensions the Riemann tensor has more independent components than the Ricci tensor and so cannot be constructed from it. A physically important example of such a curved space is the 4–dimensional space–time briefly described in §25.

Exercise 21.2
Compute the Riemann tensor of the sphere.
Solution
The sphere is a 2–dimensional space so in this case the indices in Eq. (21.8) take only two values.

As before we choose the coordinates $\{\theta, \phi\}$ and substitute the Christoffel symbols (19.10) in these coordinates into definition (21.8); the non–vanishing components of the Riemann tensor are

$$\begin{aligned} R^{\phi}{}_{\theta\theta\phi} &= -R^{\phi}{}_{\theta\phi\theta} = -\csc^2\theta + \cot^2\theta = -1\,, \\ R^{\theta}{}_{\phi\theta\phi} &= -R^{\theta}{}_{\phi\phi\theta} = -\cos(2\theta) + \cos^2\theta = \sin^2\theta\,. \end{aligned} \tag{21.26}$$

It is easy to check from the metric revealed by the line element (18.4) that the Riemann tensor (21.26) can be written

$$R^{A}{}_{BCD} = \frac{1}{a^2}\left(\delta^{A}{}_{C}\, g_{BD} - \delta^{A}{}_{D}\, g_{BC}\right), \tag{21.27}$$

or, raising an index,

$$R^{AB}{}_{CD} = \frac{1}{a^2}\left(\delta^{A}_{C}\,\delta^{B}_{D} - \delta^{A}_{D}\,\delta^{B}_{C}\right). \tag{21.28}$$

Equations (21.27) and (21.28) are tensor equations and so they correctly express the Riemann tensor in *any* coordinate system on the sphere. We obtain from Eq. (21.27) the Ricci tensor (21.13), which in turn gives the curvature scalar (21.14):

$$R_{AB} = R^{C}{}_{ACB} = \frac{1}{a^2} g_{BD}\,, \quad R = g^{AB} R_{AB} = \frac{2}{a^2}\,, \tag{21.29}$$

in agreement with the curvature (9.32) of Maxwell's fish–eye, the optical implementation of the geometry of the 2–sphere. We note from the value $2/a^2$ of the curvature scalar that Eq. (21.28) is a case of the general result (21.24) valid for all 2–dimensional spaces. The curvature of any 2–dimensional space is specified by R, and in the case of the sphere this scalar is a constant—the sphere is a space of constant curvature. The curvature decreases as the radius a increases and goes to zero as $a \to \infty$, in which limit the sphere becomes a flat surface.

There is one other important identity satisfied by the Riemann tensor, one that involves its covariant derivative:

$$R^{i}{}_{jkl;m} + R^{i}{}_{jlm;k} + R^{i}{}_{jmk;l} = 0\,. \tag{21.30}$$

These relations are called the *Bianchi identities*. They can be verified by expanding the covariant derivatives and the Riemann tensor in Eq. (21.30) in terms of the Christoffel symbols, although they are usually established by a more clever argument. We refer the reader to Misner, Thorne and Wheeler [1973] for a detailed description of the Bianchi identities and their exact geometrical meaning.

Exercise 21.3

Equation (21.9) shows that in a curved space covariant derivatives of a vector field do not commute. Show that two covariant derivatives of a scalar field f always commute, even in curved space, i.e.

$$f_{,i;j} = f_{,j;i}\,, \tag{21.31}$$

where we use the fact that the first covariant derivative of f is a partial derivative (Eq. (15.22)).

Solution

Evaluating the left–hand side of Eq. (21.31) we obtain

$$f_{,i;j} = f_{,i,j} - \Gamma^{k}{}_{ij} f_{,k}\,, \tag{21.32}$$

and since partial derivatives commute and the Christoffel symbols have the symmetry (19.8) this can be rewritten as

$$f_{,i;j} = f_{,j,i} - \Gamma^{k}{}_{ji} f_{,k}\,, \tag{21.33}$$

which is the right–hand side of Eq. (21.31).

Exercise 21.4
The commutator of covariant derivatives of a tensor with more than one index satisfies a relation similar to the vector result (21.9). For a 2–index tensor T_{ij} prove that

$$T_{ij;k;l} - T_{ij;l;k} = R_{imlk}T^m{}_j + R_{jmlk}T_i{}^m\,. \tag{21.34}$$

Solution
This is matter of evaluating both sides of Eq. (21.34) in terms of Christoffel symbols.

---

§22. Parallel transport around a closed loop

In addition to geodesic deviation, curvature also manifests itself in the non–triviality of parallel transport of vectors in a curved space. According to definition (20.3), a vector moving along a curve by parallel transport stays parallel to its direction at previous points on the curve. This operation leads to qualitatively different behaviour in a curved space compared to flat space. As an example, consider the parallel transport of a vector around a closed loop on the sphere (Fig. 3.21). Upon returning to its starting point the vector is rotated relative to its starting value. It is trivial to see that after parallel transport around any closed loop in flat space a vector coincides with its initial value. The change in a vector after parallel transport around closed loops happens if and only if the space is curved, and this rotation must therefore be determined by the Riemann tensor.

Let us compute, in an arbitrary space, the effect on a vector of parallel transport around a small closed loop. As shown in Fig. 3.22, we take the loop to have four corners: A with coordinates x^i, B with coordinates $x^i + \delta_1 x^i$, C with coordinates $x^i + \delta_1 x^i + \delta_2 x^i$ and D with coordinates $x^i + \delta_2 x^i$. We consider an arbitrary vector $\boldsymbol{V}$, located at A, and parallel transport it around the loop. Throughout the calculation we content ourselves with being accurate to lowest order in the small quantities $\delta_1 x^i$ and $\delta_2 x^i$; thus the results become exact for an infinitesimally small loop.

Parallel transport of $\boldsymbol{V}$ from A to B is achieved by ensuring that the covariant derivative of $\boldsymbol{V}$ is zero along this branch of the loop. Thus at A, the covariant derivative $V^i_{;j}$ in the direction $\delta_1 x^i$ must vanish:

$$V^i_{;j}\big|_A\, \delta_1 x^j = V^i_{,j}\big|_A\, \delta_1 x^j + \Gamma^i{}_{kj} V^k\big|_A\, \delta_1 x^j = 0\,. \tag{22.1}$$

Now for the resulting components of $\boldsymbol{V}$ at B we have by Taylor expansion in first order

$$V^i\big|_B = V^i\big|_A + V^i_{,j}\big|_A\, \delta_1 x^j = V^i\big|_A - \Gamma^i{}_{kj} V^k\big|_A\, \delta_1 x^j\,, \tag{22.2}$$

with use of Eq. (22.1). Similarly, parallel transport of $V^i|_B$ from B to C in the direction $\delta_2 x^i$ gives

$$V^i\big|_C = V^i\big|_B - \Gamma^i{}_{kj} V^k\big|_B\, \delta_2 x^j\,, \tag{22.3}$$

parallel transport of $V^i|_C$ from C to D in the direction $-\delta_1 x^i$ gives

$$V^i\big|_D = V^i\big|_C + \Gamma^i{}_{kj} V^k\big|_C\, \delta_1 x^j\,, \tag{22.4}$$

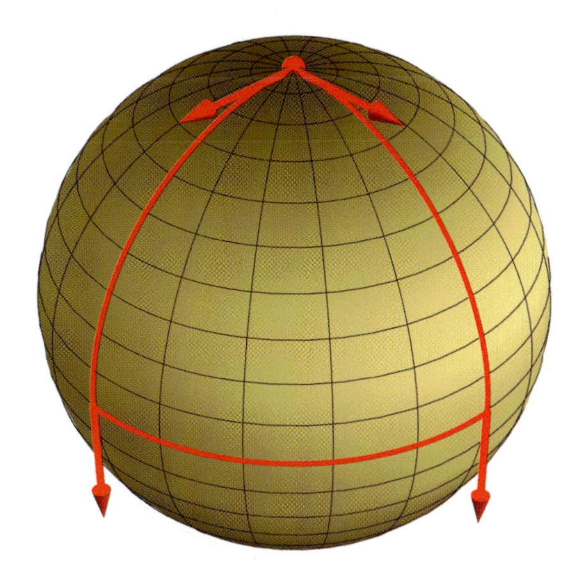

Figure 3.21: Parallel transport around a closed loop on the sphere. A vector at the North Pole is parallel transported to the equator along the geodesic to which it is tangent. As a geodesic parallel transports its tangent vector, the vector is still tangent to the geodesic at the equator. The vector is then parallel transported a quarter of the distance around the equator (also a geodesic). Parallel transport preserves the vector, as much as the curvature allows, so it stays at the same angle to the equator. Finally the vector is parallel transported back to the North Pole along a geodesic. On return to the North Pole the vector is rotated by 90 degrees with respect to the initial vector. This rotation is completely specified by the Riemann curvature tensor. Note that on a plane (zero Riemann tensor) this kind of operation does not rotate the vector.

and finally parallel transport of $V^i\big|_D$ from D to A in the direction $-\delta_2 x^i$ gives

$$V^i\big|_{A(\text{final})} = V^i\big|_D + \Gamma^i{}_{kj}V^k\big|_D\ \delta_2 x^j\,. \tag{22.5}$$

We subtract the value of the vector when it returns to A form its original value at A to obtain the change δV^i; from Eqs. (22.2)–(22.5) this is

$$\begin{aligned}\delta V^i &\equiv V^i\big|_{A(\text{final})} - V^i\big|_A \\ &= \Gamma^i{}_{kj}V^k\big|_D\ \delta_2 x^j + \Gamma^i{}_{kj}V^k\big|_C\ \delta_1 x^j - \Gamma^i{}_{kj}V^k\big|_B\ \delta_2 x^j - \Gamma^i{}_{kj}V^k\big|_A\ \delta_1 x^j\,. \end{aligned}\tag{22.6}$$

Now from the coordinates of the four corner points it is clear that (again, to lowest order in each of $\delta_1 x^i$ and $\delta_2 x^i$) by Taylor expansion

$$\Gamma^i{}_{kj}V^k\big|_B\ \delta_2 x^j = \Gamma^i{}_{kj}V^k\big|_A\ \delta_2 x^j + \left(\Gamma^i{}_{kj}V^k\right)_{,l}\Big|_A\ \delta_1 x^l\,\delta_2 x^j\,, \tag{22.7}$$

$$\Gamma^i{}_{kj}V^k\big|_C\ \delta_1 x^j = \Gamma^i{}_{kj}V^k\big|_A\ \delta_1 x^j + \left(\Gamma^i{}_{kj}V^k\right)_{,l}\Big|_A\ \delta_2 x^l\,\delta_1 x^j\,, \tag{22.8}$$

$$\Gamma^i{}_{kj}V^k\big|_D\ \delta_2 x^j = \Gamma^i{}_{kj}V^k\big|_A\ \delta_2 x^j\,. \tag{22.9}$$

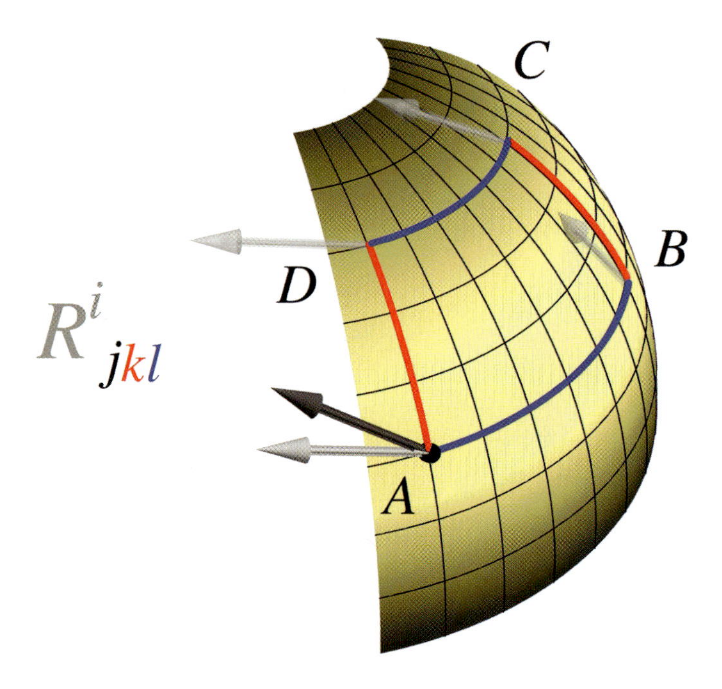

Figure 3.22: Visualization of the Riemann curvature tensor. Parallel transport around a closed loop. The black vector at point A is parallel transported around the closed path $ABCD$; the light–coloured vectors at these four points on the loop are the result of the parallel transport. Upon returning to A the vector has changed. The colours on the indices of the Riemann–tensor components show how the difference between the two vectors at A is calculated: the first index is that of the vector difference, the black index is contracted with the initial vector, and the vectors representing the red and blue parts of the path are contracted with the red and blue indices. The resulting formula (22.12) is exact for an infinitesimal loop.

Substitution of Eqs. (22.7)–(22.9) into Eq. (22.6) yields (after index relabeling)

$$\delta V^i = \left[\left(\Gamma^i{}_{kl} V^k\right)_{,j}\Big|_A - \left(\Gamma^i{}_{kj} V^k\right)_{,l}\Big|_A \right] \delta_1 x^l \, \delta_2 x^j \, . \tag{22.10}$$

We expand the partial derivatives in Eq. (22.10) and eliminate the partial derivatives of V^i at A using Eq. (22.1); this produces (after index relabeling)

$$\delta V^i = \left(\Gamma^i{}_{kl,j} - \Gamma^i{}_{ml}\Gamma^m{}_{kj} - \Gamma^i{}_{kj,l} + \Gamma^i{}_{mj}\Gamma^m{}_{kl}\right) V^k \, \delta_1 x^l \, \delta_2 x^j , \tag{22.11}$$

or, in terms of the definition (21.8) of the Riemann tensor

$$\delta V^i = R^i{}_{kjl} V^k \, \delta_1 x^l \, \delta_2 x^j \, . \tag{22.12}$$

Equation (22.12) establishes a beautiful visualization of intrinsic curvature and the Riemann tensor (Fig. 3.22): curvature is quantified by the change in a vector when parallel transported around an infinitesimal closed loop, for which Eq. (22.12) is exact. The effect of parallel transport around a loop of arbitrary size is also determined by the Riemann tensor since it can be obtained from Eq. (22.12) by an integration procedure over the surface enclosed by the loop. It is not difficult to see that an equivalent statement of our conclusion for closed loops is that in a curved space the result of parallel transporting a vector from one point to another depends on the path taken.

A physical example of parallel transport around a closed loop in a curved space is provided by *Foucault's pendulum* (Hart, Miller and Mills [1987]). This is a pendulum free of any (appreciable) non–gravitational disturbances that oscillates long enough for the Earth's rotation to be visible in the pendulum's motion. In practice this means the pendulum must be long and heavy; Jean Bernard Léon Foucault's famous 1851 pendulum in the Pantheon in Paris was 67 metres long with a 28 kilogram weight on the end. What is observed is a steady rotation of the line in which the pendulum oscillates (Fig. 3.23), and the rate of this rotation depends on the latitude on Earth at which the pendulum is located. An exact treatment of this problem

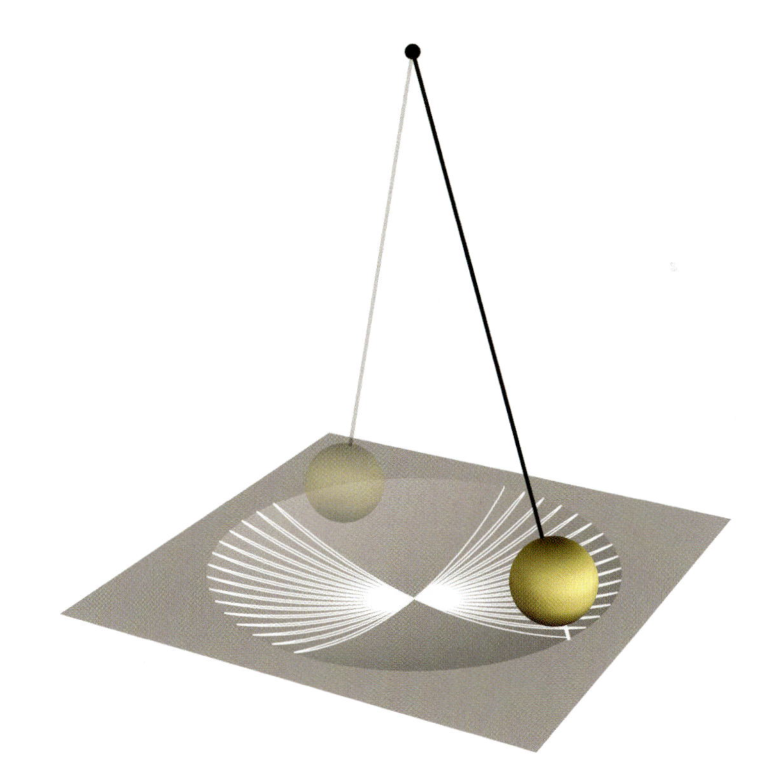

Figure 3.23: Foucault's pendulum. Left undisturbed a pendulum is seen to experience a rotation of its line of oscillation (chalk traces). This rotation is caused by the curvature and rotation of the Earth. To a very good approximation, the line of oscillation is parallel transported by the Earth's rotation.

presents a tedious exercise in Newtonian mechanics, but the following considerations lead to the standard, and very accurate, formula for the period of rotation of the pendulum's line of oscillation. Viewed from an inertial frame we can consider the rotation of the Earth as moving the pendulum around the surface of a sphere along a fixed line of latitude, with the pendulum returning to its starting point on the sphere after 24 hours. The only force on the pendulum is the radial force of gravity, so there is no torque acting to rotate its line of oscillation. This means the line of oscillation should try to remain parallel to itself as the pendulum is transported around the curve of fixed latitude. We can represent the line of oscillation as a vector tangent to the sphere, and we conclude that this vector must be parallel transported around a closed loop of fixed latitude with every complete rotation of the Earth.

Now we must calculate. We choose as previously the spherical polar coordinates $x^A = \{\theta, \phi\}$ on the sphere, so that a line of latitude is given by θ = constant (note that this measures latitude from the North Pole of the sphere whereas the angle of latitude is conventionally measured from the equator). An arbitrary line $x^A(s) = \{\theta(s), \phi(s)\}$ of latitude is parametrized by its length s as follows:

$$\theta(s) = \theta_0 \,, \quad \phi(s) = \phi_0 + \frac{s}{a \sin\theta_0} \,, \quad (\theta_0 \text{ a constant}), \tag{22.13}$$

where a, as before, is the radius of the sphere, in this case the Earth. Note that motion along $x^A(s)$ in the direction of increasing s is motion from west to east, the direction of the Earth's rotation. Let the vector on the sphere specifying the line of oscillation of the pendulum be V^A (of course $-V^A$ is also on this line). The vector V^A is parallel transported along $x^A(s)$, so it obeys Eq. (20.3), which in this 2–dimensional case is

$$\frac{\mathrm{d}V^A}{\mathrm{d}s} + \Gamma^A{}_{BC} \frac{\mathrm{d}x^C(s)}{\mathrm{d}s} V^B = 0 \,. \tag{22.14}$$

We write out Eq. (22.14) using Eq. (22.13) and the Christoffel symbols (19.10):

$$\frac{\mathrm{d}V^\theta}{\mathrm{d}s} - \frac{1}{a}\cos\theta_0\, V^\phi = 0 \,, \qquad \frac{\mathrm{d}V^\phi}{\mathrm{d}s} + \frac{\cos\theta_0}{a\sin^2\theta_0} V^\theta = 0 \,. \tag{22.15}$$

Taking the derivatives of Eqs. (22.15) with respect to s we see that both V^θ and V^ϕ satisfy the same second–order equation:

$$\frac{\mathrm{d}^2 V^A}{\mathrm{d}s^2} + \frac{1}{a^2}\cot^2\theta_0\, V^A = 0 \,. \tag{22.16}$$

This last equation has the general solution

$$V^A = C_1^A \cos(ks) + C_2^A \sin(ks) \,, \quad k \equiv \frac{1}{a}\cot\theta_0 \,, \tag{22.17}$$

where C_1^A and C_2^A are arbitrary constant vectors. Substituting Eq. (22.17) back into Eqs. (22.15) removes two of the four arbitrary constants C_1^A, C_2^A. In terms of the value $\{V_0^\theta, V_0^\phi\}$ of V^A at $s = 0$ the final solution is

$$\begin{aligned} V^\theta &= V_0^\theta \cos(ks) + V_0^\phi \sin\theta_0 \sin(ks) \,, \\ V^\phi &= V_0^\phi \cos(ks) - V_0^\theta \csc\theta_0 \sin(ks) \,. \end{aligned} \tag{22.18}$$

Equations (22.18) describe a rotation of the vector V^A, and thus of the pendulum's line of oscillation, at a constant rate with respect to distance s along the line of latitude. The Earth's rotation causes a daily complete loop around the line of latitude, a distance

$$s_{\text{day}} = 2\pi a \sin\theta_0 \,. \tag{22.19}$$

But this is not the distance required for the pendulum's line of oscillation to complete a rotation of 360°. This latter distance is reached when the phase ks in Eq. (22.18) changes by 2π, giving us the Foucault distance

$$s_{\text{Foucault}} = \frac{2\pi}{k} = 2\pi a \tan\theta_0 \,. \tag{22.20}$$

Since the Earth rotates at a constant angular velocity the distance s in Eqs. (22.18) is proportional to time; thus s_{day} is proportional to the period of rotation T_{day} of the Earth (23.934 hours to be exact) while s_{Foucault} is proportional to the period of rotation T_{Foucault} of the pendulum's line of oscillation. We therefore obtain the Foucault equation

$$T_{\text{Foucault}} = \frac{T_{\text{day}}}{\cos\theta_0} \,. \tag{22.21}$$

(If the angle of latitude θ_0 is measured from the equator, rather than from the North Pole, the cosine in Eq. (22.21) becomes a sine.)

A positive value of k corresponds to a clockwise rotation of V^A in Eq. (22.18) when viewed from above the surface of the sphere, so the definition (22.17) of k shows that for latitudes in the northern hemisphere ($0 \leq \theta_0 < \pi/2$) the pendulum's line of oscillation rotates clockwise whereas in the southern hemisphere ($\pi/2 < \theta_0 \leq \pi$) the rotation is counterclockwise. The change in the sign of T_{Foucault} in Eq. (22.21) as θ_0 passes through the equator $\theta_0 = \pi/2$ corresponds to this reversal in the direction of rotation of the line of oscillation. At the equator itself T_{Foucault} becomes infinite, i.e. the line does not rotate at all. Away from the equator the line rotates with a period longer than T_{day} except at the North and South Poles where $|T_{\text{Foucault}}| = T_{\text{day}}$. There is of course no parallel transport at the Poles, but we have obtained the result in these two cases because the behaviour of the pendulum must change continuously as the Poles are approached and the loop of latitude contracts to a point. The reader should be able to see that the results at the Poles and the Equator are obvious without calculation.

Exercise 22.1
Show that the length of the vector (22.18) is independent of s and so does not change as the vector is parallel transported along the line of latitude. (Remember to use the metric tensor of the sphere.) This proves that the vector is just rotated, without being stretched or compressed.

From Eqs. (22.18) and (22.19) we see that during one day the pendulum rotates through the angle

$$\phi_{\text{day}} = ks_{\text{day}} = 2\pi\cos\theta_0 \,. \tag{22.22}$$

Foucault's pendulum acts like a clock whose behaviour depends on latitude (Fig. 3.24). The angle (22.22) is positive in the northern hemisphere and negative in the southern, but the amount of daily rotation $|\phi_{\text{day}}|$ lags behind that of the Earth (2π), except at the Poles. In the northern hemisphere we can express this lag as

$$\phi_{\text{day}} - 2\pi = 2\pi(\cos\theta_0 - 1) = -\int_0^{2\pi}\int_0^{\theta_0} \sin\theta \, d\theta \, d\phi , \tag{22.23}$$

where the last expression is the upper area enclosed by the line of latitude on the unit sphere. Clearly, the same result holds for the lag $|\phi_{\text{day}}| - 2\pi$ at the corresponding line of latitude on the southern hemisphere in terms of the lower area it encloses on the unit sphere. This area–dependent lapse would not exist without curvature and rotation: Foucault's pendulum strikingly demonstrates both the rotation and the curvature of the Earth.*

*Foucault's pendulum belongs to a wide class of phenomena caused by parallel transport in non–Euclidean parametric spaces. Such a phase lapse is called the *geometric phase* or *Berry phase* (Berry [1984], Nakahara [2003]).

Figure 3.24: Foucault's pendulum is a clock whose rate depends on latitude. The picture shows the daily clockwise rotation of the pendulum at four locations in the northern hemisphere: these are, in order of increasing latitude, Singapore, Mecca, Paris and the North Pole.

§23. CONFORMALLY FLAT SPACES

We now apply the mathematics of curved spaces to the line element that was introduced at the start of the book in connection with Fermat's principle, see Eq. (1.1). In any number of dimensions, there is an obvious subset of curved spaces whose Riemann tensor is determined by just one scalar field; these are the spaces whose metric tensor components in some coordinate system x^i take the form

$$g_{ij} = n^2\, \delta_{ij}\,, \tag{23.1}$$

where n is a scalar function ($n = n(x^i)$). The metric (23.1) differs from the flat–space metric in Cartesian coordinates by the factor n^2 and it is therefore clear that the curvature of the space is determined solely by this scalar function. After a general coordinate transformation the metric tensor components will of course no longer have the simple form (23.1); what is required is that there exists a coordinate system x^i in which Eq. (23.1) is true.

Spaces whose metric can be written as in Eq. (23.1) are called *conformally flat*, because they differ from flat space by a *conformal transformation*. A conformal transformation is a change of the metric by a scalar factor:

$$g_{ij} \rightarrow \Omega^2\, g_{ij}\,. \tag{23.2}$$

The significance of a conformal transformation is that it changes lengths in a space but not angles. A general angle θ between two directions along vectors $\boldsymbol{U}$ and $\boldsymbol{V}$ satisfies

$$\cos\theta = \frac{\boldsymbol{U}\cdot\boldsymbol{V}}{|\boldsymbol{U}||\boldsymbol{V}|} = \frac{g_{ij}U^iV^j}{\sqrt{g_{ij}U^iU^j g_{kl}V^kV^l}}\,, \tag{23.3}$$

and this angle is clearly invariant under a conformal transformation (23.2). It is essential to understand that a conformal transformation is in general *not* equivalent to an active coordinate transformation: the scalar function Ω in the new metric (23.2) in general gives a new geometry that is not related to the original one by a coordinate transformation. This difference between conformal and coordinate transformations is clear from the example of conformally flat spaces defined by the metric (23.1); these spaces are in general curved (see Eq. (23.4) below) but they are changed into flat space by a conformal transformation, a feat no coordinate transformation can achieve. There is however a subset of conformal transformations that *are* equivalent to active coordinate transformations; a trivial example is a constant coordinate rescaling $x^i \rightarrow ax^i$, which gives a conformal transformation (23.2) of the metric with $\Omega = a^2$. In §5 we considered a special class of conformal transformations of flat space in two dimensions that were also coordinate transformations; these transformations were given by analytic functions.

Although we are using latin indices for 3–dimensional quantities, Eq. (23.1) defines a conformally flat space in d dimensions when the indices are understood to range over d values. The Riemann tensor components in coordinates where the metric takes the form (23.1) are

$$R^i{}_{jkl} = \frac{1}{n}\Big(n_{,j,k}\delta_{il} - n_{,j,l}\delta_{ik} - n_{,i,k}\delta_{jl} + n_{,i,l}\delta_{jk}\Big) + \frac{1}{n^2}\Big[n_{,j}\Big(n_{,l}\delta_{ik} - n_{,k}\delta_{il}\Big) + n_{,i}\Big(n_{,k}\delta_{jl} - n_{,l}\delta_{jk}\Big) - n_{,m}n_{,m}\Big(\delta_{ik}\delta_{jl} - \delta_{il}\delta_{jk}\Big)\Big], \tag{23.4}$$

and this result holds in any number of dimensions. Note that the indices in Eq. (23.4) do not line up on the left– and right–hand sides and it contains a summation that is not a contraction of an upper and lower index. This is because Eq. (23.4) is not a tensor equation: it holds only in coordinates in which the metric of the conformally flat space has components (23.1). This is in contrast to results such as (21.27) for the sphere, which is a tensor equation and so holds in every coordinate system.

Exercise 23.1
Derive the Riemann tensor (23.4) from the metric (23.1).
Solution
The inverse of the metric (23.1) is $g^{ij} = n^{-2}\delta_{ij}$, where we do not bother writing the indices in the upper position on the Kroneker delta since the result (23.4) will not be a tensor equation. The metric and its inverse give the following Christoffel symbols (19.7):

$$\Gamma^i{}_{jk} = \frac{1}{n}\left(n_{,k}\delta_{ij} + n_{,j}\delta_{ik} - n_{,i}\delta_{jk}\right). \tag{23.5}$$

These must be substituted into definition (21.8) and after straightforward simplifications the result is (23.4).

As discussed in §21, in three dimensions all the information about the curvature is in fact contained in the Ricci tensor. Contracting indices in the d–dimensional result (23.4) to obtain the Ricci tensor (21.13) brings in the trace δ_{ii}, which is equal to the dimensionality d of the space; in three dimensions ($\delta_{ii} = 3$) we easily find

$$R_{ij} = -\frac{1}{n}\Big(n_{,i,j} + n_{,k,k}\delta_{ij}\Big) + \frac{1}{n^2}\Big(n_{,i}n_{,j} - n_{,k}n_{,k}\delta_{ij}\Big), \tag{23.6}$$

giving the curvature scalar

$$R = g^{ij}R_{ij} = n^{-2}R_{ii} = -\frac{4\,n_{,i,i}}{n^3} - \frac{2\,n_{,i}n_{,i}}{n^4}. \tag{23.7}$$

We introduced conformally flat spaces as a restricted class whose curvature is determined by one scalar function. But recall from §21 that the curvature of every 2–dimensional space is determined by the curvature scalar R, a single scalar function. It should therefore be the case that in two dimensions the metric of every space can be written in the form (23.1), with n providing enough freedom to determine the curvature. This is in fact true: *all 2–dimensional spaces are conformally flat*, i.e. there exist coordinates in which the metric components are

$$g_{AB} = n^2\,\delta_{AB}, \tag{23.8}$$

where as before capital letters are used for indices in two dimensions. As every curved 2–dimensional geometry can be determined by a single function, it follows that to create such geometries for light requires materials whose optical properties are determined by only one function; such materials are isotropic. Implementation of all geometries in planar integrated optics can thus be achieved with isotropic media.

Exercise 23.2
Prove that all 2–dimensional spaces are conformally flat.
Solution
The line element of an arbitrary 2–dimensional space in an arbitrary coordinate system x^A is

$$\mathrm{d}s^2 = g_{AB}\,\mathrm{d}x^A\,\mathrm{d}x^B = g_{11}(\mathrm{d}x^1)^2 + 2g_{12}\,\mathrm{d}x^1\mathrm{d}x^2 + g_{22}(\mathrm{d}x^2)^2\,, \tag{23.9}$$

since $g_{21} = g_{12}$. The determinant of the metric in Eq. (23.9) is $g = g_{11}g_{22} - g_{12}^2$ and it is easy to verify that Eq. (23.9) can be written as

$$\mathrm{d}s^2 = \left(\sqrt{g_{11}}\,\mathrm{d}x^1 + \frac{g_{12} + \mathrm{i}\sqrt{g}}{\sqrt{g_{11}}}\,\mathrm{d}x^2\right)\left(\sqrt{g_{11}}\,\mathrm{d}x^1 + \frac{g_{12} - \mathrm{i}\sqrt{g}}{\sqrt{g_{11}}}\,\mathrm{d}x^2\right). \tag{23.10}$$

Although we have introduced complex quantities into the expression (23.10) for the line element, $\mathrm{d}s^2$ is of course real. The two factors in the product in Eq. (23.10) are complex conjugates of each other, and they look like the differential $\mathrm{d}f$ of a function of the coordinates x^A:

$$\mathrm{d}f(x^1, x^2) = f_{,1}\,\mathrm{d}x^1 + f_{,2}\,\mathrm{d}x^2\,. \tag{23.11}$$

But the coefficients of the differentials in Eq. (23.11) have the property $(f_{,1})_{,2} = (f_{,2})_{,1}$, because partial derivatives commute, whereas the corresponding coefficients in Eq. (23.10) will in general not have this property. The two factors in Eq. (23.10) are thus in general not the differentials of functions of x^A. The theory of differential equations (Riley, Hobson and Bence [2006]), however, shows that there exists an integrating factor $h(x^1, x^2)$ such that, when it is multiplied by the first factor in Eq. (23.10) the result *is* the differential of a function $f(x^1, x^2)$:

$$h\left(\sqrt{g_{11}}\,\mathrm{d}x^1 + \frac{g_{12} + \mathrm{i}\sqrt{g}}{\sqrt{g_{11}}}\,\mathrm{d}x^2\right) = \mathrm{d}f\,. \tag{23.12}$$

Similarly, an integrating factor exists for the second factor in Eq. (23.10), and complex conjugation of Eq. (23.12) shows that it is h^*:

$$h^*\left(\sqrt{g_{11}}\,\mathrm{d}x^1 + \frac{g_{12} - \mathrm{i}\sqrt{g}}{\sqrt{g_{11}}}\,\mathrm{d}x^2\right) = \mathrm{d}f^*\,. \tag{23.13}$$

Denoting the real and imaginary parts of f by $x^{1'}$ and $x^{2'}$, respectively ($f = x^{1'} + \mathrm{i}x^{2'}$), we obtain by substitution of Eqs. (23.12) and (23.13) into Eq. (23.10)

$$\begin{aligned}\mathrm{d}s^2 &= \frac{1}{|h|^2}\left(\mathrm{d}x^{1'} + \mathrm{i}\,\mathrm{d}x^{2'}\right)\left(\mathrm{d}x^{1'} - \mathrm{i}\,\mathrm{d}x^{2'}\right) = \frac{1}{|h|^2}\left[\left(\mathrm{d}x^{1'}\right)^2 + \left(\mathrm{d}x^{2'}\right)^2\right]\\ &= \frac{1}{|h|^2}\,\delta_{A'B'}\,\mathrm{d}x^{A'}\,\mathrm{d}x^{B'}\,.\end{aligned} \tag{23.14}$$

We have thus shown that coordinates $x^{A'}$ exist in which the metric takes the conformally flat form (23.8), as required.

Exercise 23.3
The sphere, being 2–dimensional, is conformally flat. Find a coordinate system in which its metric tensor components take the form (23.8).

Solution
We have in fact already performed this exercise in §9. Recall that according to Fermat's principle, Maxwell's fish–eye in 2–dimensions has the conformally flat line element (9.28), which we showed is that of a sphere in stereographic coordinates. Here we find another coordinate system in which the metric components take the form (23.8).

We rewrite the metric (18.4) of the sphere in the coordinates $\{\theta, \phi\}$ as

$$\mathrm{d}s^2 = a^2 \sin^2\theta \left(\frac{1}{\sin^2\theta}\,\mathrm{d}\theta^2 + \mathrm{d}\phi^2 \right) . \tag{23.15}$$

Our goal will be achieved if we transform the coordinate θ to a new coordinate u defined by

$$\frac{1}{\sin\theta}\,\mathrm{d}\theta = \mathrm{d}u\,. \tag{23.16}$$

Integrating this gives us the coordinate transformation up to an additive constant which we drop:

$$u = \ln\left(\tan\frac{\theta}{2}\right) \qquad \Longrightarrow \qquad \theta = 2\arctan(\mathrm{e}^u)\,. \tag{23.17}$$

Since θ has the range $0 \leq \theta \leq \pi$, the range of u is $-\infty \leq u \leq \infty$. From Eqs. (23.17) follows

$$\sin\theta = \operatorname{sech} u\,, \tag{23.18}$$

and using this and Eq. (23.16) in Eq. (23.15) gives the line element with the desired form of the metric:

$$\mathrm{d}s^2 = a^2 \operatorname{sech}^2 u \left(\mathrm{d}u^2 + \mathrm{d}\phi^2\right) . \tag{23.19}$$

The transformation (23.17) defines the Mercator projection of the sphere—see Exercise 9.10.

In any 2–dimensional space we can choose coordinates in which the metric has the obviously conformally flat form (23.8), and the Riemann tensor in these coordinates is given by Eq. (23.4) with indices ranging over two values. Contracting the Riemann tensor to obtain the Ricci tensor (21.13) we use the trace $\delta_{AA} = 2$ and find

$$R_{AB} = \frac{1}{n^2}\Big(n_{,C}\,n_{,C} - n\,n_{,C,C}\Big)\delta_{AB}\,, \tag{23.20}$$

which gives the curvature scalar (5.12)

$$R = g^{AB} R_{AB} = n^{-2} R_{AA} = -\frac{2\,n_{,A,A}}{n^3} + \frac{2\,n_{,A}n_{,A}}{n^4}\,. \tag{23.21}$$

Exercise 23.4
Equation (23.19) gives the line element of the sphere in coordinates $\{u, \phi\}$; the metric in these coordinates is thus (23.8) with $n = a\operatorname{sech} u$. Compute the curvature scalar (23.21) in the coordinates $\{u, \phi\}$ and find the previous result $R = 2/a^2$ obtained in Exercise 21.2 using the coordinates $\{\theta, \phi\}$.

Recall the statement of Fermat's principle in §1. In the geometrical language of this Chapter, Eqs. (1.1) and (1.2) state that light propagates in a conformally flat space in which the conformal factor n^2 in Eq. (23.1) is the square of the refractive index. In general the conformally flat space experienced by light is curved. In §5 we used conformal transformations of flat space in two dimensions to produce non–trivial conformal factors n^2 in the metric (23.8), and thus to design refractive index profiles. But since the conformal transformations in §5 were also coordinate transformations, the generated metrics (23.8) were those of flat space in different coordinate systems; the metrics were flat, not just conformally flat. Any refractive index generated in this fashion was shown to satisfy Eqs. (5.12)–(5.14), since this is the condition that the optical line element be transformable to the Euclidean line element. We now know that this transformation property means the metric has zero curvature, and indeed we see from Eqs. (23.21) and (5.12) that the condition (5.14) states that the curvature scalar is zero, the scalar that completely determines the curvature in two dimensions.

§24. THE HYPERSPHERE

A curved 3–dimensional space implemented by an interesting optical device, Maxwell's fish–eye in three dimensions (§36), is the *hypersphere*. This is the generalization of the sphere obtained by adding one extra dimension. Just as the sphere is the 2–dimensional surface of fixed distance from a point in 3–dimensional Euclidean space, the hypersphere is the 3–dimensional *hypersurface* of fixed distance from a point in 4D. In terms of Cartesian coordinates $\{X, Y, Z, W\}$ in 4–dimensional Euclidean space, the equation of a hypersphere of radius a centered at the origin is

$$X^2 + Y^2 + Z^2 + W^2 = a^2 . \tag{24.1}$$

To compute quantities on the hypersphere we need to place a coordinate grid on it, a set of three independent cordinates x^i. Consider the coordinates $\{\theta, \phi\}$ we have been using on the sphere of radius a; from Eqs. (10.1) these are related as follows to Cartesian coordinates in the ambient 3–dimensional Euclidean space whose origin is the centre of the sphere:

$$x' = a\sin\theta\cos\phi\,, \quad x' = a\sin\theta\sin\phi\,, \quad z' = a\cos\theta\,. \tag{24.2}$$

These equations specify all the points in the 3–dimensional Euclidean space that lie on the sphere $x'^2 + y'^2 + z'^2 = a^2$. There is a generalization of Eqs. (24.2) that connects the 4–dimensional Cartesian coordinates $\{X, Y, Z, W\}$ to coordinates $x^i = \{\chi, \theta, \phi\}$ in the hypersphere (24.1):

$$X = a\sin\chi\sin\theta\cos\phi\,, \quad Y = a\sin\chi\sin\theta\sin\phi\,, \quad Z = a\sin\chi\cos\theta\,, \quad W = a\cos\chi\,,$$
$$0 \le \chi \le \pi\,, \quad 0 \le \theta \le \pi\,, \quad 0 \le \phi \le 2\pi\,. \tag{24.3}$$

It can be easily checked that Eqs. (24.3) give points $\{X, Y, Z, W\}$ that satisfy Eq. (24.1).

Figure 3.25 offers a picture of the coordinates $\{\chi, \theta, \phi\}$ defined by Eqs. (24.3), although it is of course impossible to capture fully the hypersphere in a 2–dimensional drawing (or in a 3–dimensional brain). The origin of the coordinate system $\{\chi, \theta, \phi\}$ is the "North Pole" $\{X = 0, Y = 0, Z = 0, W = a\}$ of the hypersphere. It is reasonably easy to perceive that the coordinates $\{\theta, \phi\}$ cover a sphere in the hypersphere;

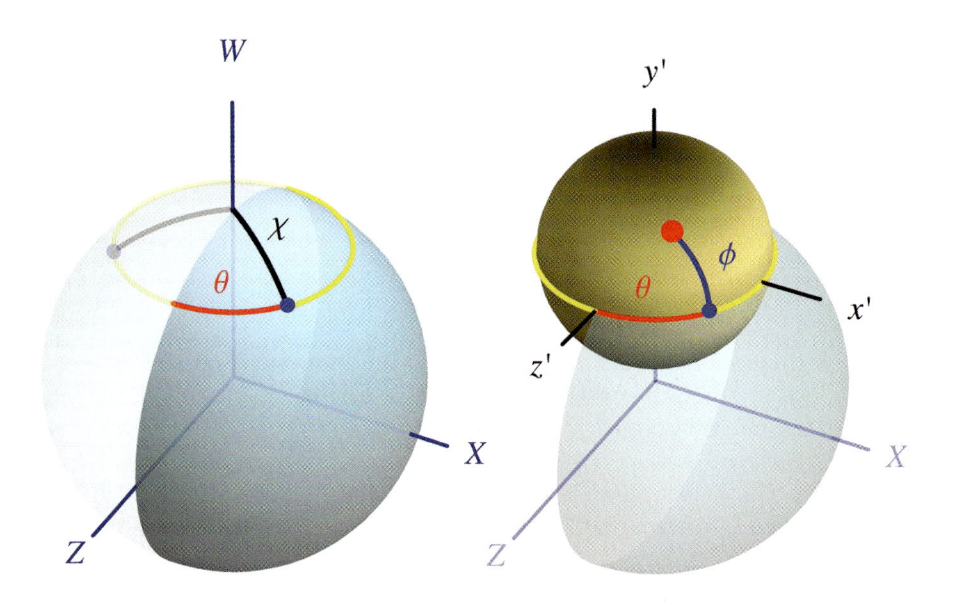

Figure 3.25: The coordinate system $\{\chi, \theta, \phi\}$ in the hypersphere. The left–hand picture shows the hypersphere (24.1) with the Y–dimension suppressed. The coordinate χ of a point in the hypersphere is the angle subtended from the origin $X = Y = Z = W = 0$ in 4–dimensional space by the line in the hypersphere joining the point to the North Pole $W = a$. All points on the yellow and red circle in the left–hand picture, which is in reality the brass sphere in the right–hand picture, have the same χ–coordinate; a fixed value of χ in the hypersphere thus defines a sphere. The remaining coordinates $\{\theta, \phi\}$ in the hypersphere label the points on the sphere given by a fixed value of χ. The brass sphere in the right–hand picture shows all points in the hypersphere with the coordinate value χ shown on the left. Axes parallel to the X–, Y– and Z–axes in 4–dimensional space are drawn through the centre of this sphere; these axes are labelled x', y', z' (they are not the X, Y, Z axes because the points of the brass sphere have $W \neq 0$, as is seen in the left–hand picture). The coordinates $\{\theta, \phi\}$ are related to the x', y', z' Cartesian axes exactly as in the spherical polar system in Fig. 3.1. Notice from the right–hand picture that points with $y' = 0$ and $x' > 0$, and thus with $Y = 0$ and $X > 0$, have $\phi = 0$, whereas points with $y' = 0$ and $x' < 0$, and thus with $Y = 0$ and $X < 0$, have $\phi = \pi$; the darker half of the sphere in the left–hand picture thus has $\phi = 0$ whereas the lighter half has $\phi = \pi$. Notice also that the two blue points in the left–hand picture have the same χ and θ coordinates; they differ in the value of their ϕ coordinate, the blue point in the $X > 0$ region having $\phi = 0$ and the blue point in the $X < 0$ region having $\phi = \pi$.

the centre of this sphere is the origin (North Pole), and its radius (distance in the hypersphere from the North Pole) is $a\chi$. The coordinates $\{\chi, \theta, \phi\}$ are thus quite similar to spherical polar coordinates in flat space, with $\{\theta, \phi\}$ performing the same role in both systems and $a\chi$ playing the role of the radial distance r in spherical polars. In the hypersphere the "radial" coordinate χ only takes values up to π, because of its finite volume.

To quantify the intrinsic geometry of the hypersphere we require its metric tensor. An infinitesimal distance $\mathrm{d}s$ in the ambient 4–dimensional Euclidean space obeys

$$\mathrm{d}s^2 = \mathrm{d}X^2 + \mathrm{d}Y^2 + \mathrm{d}Z^2 + \mathrm{d}W^2 \,, \tag{24.4}$$

and an infinitesimal distance $\mathrm{d}s$ in the hypersphere also obeys Eq. (24.4), but with conditions (24.3) enforced, so that all points lie in the hypersphere. Insertion of Eqs. (24.3) into Eq. (24.4) thus gives the line element of the hypersphere, which is easily simplified to

$$\mathrm{d}s^2 = a^2 \left[\mathrm{d}\chi^2 + \sin^2\chi \left(\mathrm{d}\theta^2 + \sin^2\theta \,\mathrm{d}\phi^2\right)\right] . \tag{24.5}$$

Comparison of this with Eq. (11.18) reveals again the similarity to flat space in spherical polars. Indeed, for small χ the line element (24.5) is exactly of the form (11.18) to lowest order in $\mathrm{d}\chi$, with $a\chi$ corresponding to r. This is an example of the fact noted on p. 135 that coordinates can be found in which the metric tensor at any one point in a curved space is that of flat space; in the case of the coordinates in Eq. (24.5) the metric tensor takes a flat–space form at the North Pole ($\chi = 0$) of the hypersphere.

In §9 we took advantage of the stereographic projection (9.23) for mapping the sphere to the plane. We can do something very similar for the hypersphere: we map it to 3–dimensional space by defining the stereographic coordinates

$$x = \frac{X}{1 - W/a} \,, \quad y = \frac{Y}{1 - W/a} \,, \quad z = \frac{Z}{1 - W/a} \,, \tag{24.6}$$

with the inverse relations

$$X = \frac{2x}{1 - r^2/a^2} \,, \quad Y = \frac{2y}{1 - r^2/a^2} \,, \quad Z = \frac{2z}{1 - r^2/a^2} \,, \quad W = a\,\frac{r^2 - a^2}{r^2 + a^2} \tag{24.7}$$

and $r^2 = x^2 + y^2 + z^2$. In terms of the hyperspherical coordinates (24.2) we have

$$\begin{aligned} x &= a\cot(\chi/2)\sin\theta\cos\phi \,, \\ y &= a\cot(\chi/2)\sin\theta\sin\phi \,, \\ z &= a\cot(\chi/2)\cos\theta \,. \end{aligned} \tag{24.8}$$

Equations (24.8) reveal how the radius of the stereographic coordinates is related to the "radial" coordinate χ in the hypersphere:

$$r = \sqrt{x^2 + y^2 + z^2} = a\cot(\chi/2) \,. \tag{24.9}$$

Since $\cot(\chi/2)$ runs from 0 to $+\infty$ the coordinates x, y, z run all from $-\infty$ to $+\infty$; the stereographic coordinates are infinitely extended, although the hypersphere is a finite space, which is possible if their line element vanishes sufficiently fast at ∞. We obtain the metric in stereographic coordinates from the line element (24.5) using the relationships $2\sin^2(\chi/2)\,\mathrm{d}r = -a\,\mathrm{d}\chi$ and $\sin\chi = 2\sin^2(\chi/2)\cot(\chi/2)$:

$$\mathrm{d}s^2 = n^2\left[\mathrm{d}r^2 + r^2\left(\mathrm{d}\theta^2 + \sin^2\theta\,\mathrm{d}\phi^2\right)\right] = n^2\left(\mathrm{d}x^2 + \mathrm{d}y^2 + \mathrm{d}z^2\right) \tag{24.10}$$

with the scalar function

$$n = \frac{2}{1 + r^2/a^2}\,. \tag{24.11}$$

We see that $\mathrm{d}s$ decreases like $2(a^2/r^2)\mathrm{d}l$ for large r, so the volume element (11.16) falls with the power r^{-6}; thus the volume integral is finite. More importantly, we see that in stereographic coordinates the geometry of the hypersphere (24.10) is conformally flat (§23). The function $n(r)$ is the refractive–index profile (9.13) of Maxwell's fish–eye with a defining the characteristic length of the device. In stereographic projection, the geometry of the hypersphere can be implemented by Maxwell's fish–eye in three dimensions (§36).

In §9 we also found that the Möbius transformations (9.34) represent the rotations on the sphere in stereographic projection. There the Möbius transformations were formulated on the complex plane, by the representation of points with Cartesian coordinates x and y as complex numbers $x + \mathrm{i}y$. Since we can generalize the 2–dimensional stereographic projection (9.23) to 3–dimensional space (24.6), it should also be possible to describe rotations on the hypersphere by 3–dimensional Möbius transformations. The 2–dimensional Möbius transformation (9.36) can be written as

$$\boldsymbol{r}' = \frac{\boldsymbol{r}(1 + a^{-2}r_0^2) - \boldsymbol{r}_0(1 + 2a^{-2}\boldsymbol{r}\cdot\boldsymbol{r}_0 - a^{-2}r^2)}{1 + 2a^{-2}\boldsymbol{r}\cdot\boldsymbol{r}_0 + a^{-4}r^2r_0^2}\,. \tag{24.12}$$

Exercise 24.1
Derive the representation (24.12) with $\boldsymbol{r} = (x, y)$ and $\boldsymbol{r}_0 = (a\tan\gamma, 0)$ from the Möbius transformation (9.36) understanding the complex number z as $x + \mathrm{i}y$ in units of a,

$$z' = a\,\frac{z\cos\gamma - a\sin\gamma}{z\sin\gamma + a\cos\gamma}\,. \tag{24.13}$$

Solution

$$z' = a\,\frac{z\cos\gamma - a\sin\gamma}{z\sin\gamma + a\cos\gamma}\,\frac{z^*\sin\gamma + a\cos\gamma}{z^*\sin\gamma + a\cos\gamma} = a\,\frac{(|z|^2 - a^2)\cos\gamma\sin\gamma + za\cos^2\gamma - z^*a\sin^2\gamma}{|z|^2\sin^2\gamma + (z + z^*)a\cos\gamma\sin\gamma + a^2\cos^2\gamma}\,. \tag{24.14}$$

Writing expression (24.14) in terms of its real and imaginary part gives

$$x' = a\,\frac{ax(1 - \tan^2\gamma) - (a^2 - |z|^2)\tan\gamma}{a^2 + 2xa\tan\gamma + |z|^2\tan^2\gamma}\,,\quad y' = a\,\frac{ay(1 + \tan^2\gamma)}{a^2 + 2xa\tan\gamma + |z|^2\tan^2\gamma}\,, \tag{24.15}$$

which is equivalent to formula (24.12) with $\boldsymbol{r} = (x, y)$, $r = |z|$ and $\boldsymbol{r}_0 = (a\tan\gamma, 0)$.

In the form (24.12) we can easily generalize the Möbius transformations to the 3–dimensional stereographic coordinates of the hypersphere, by regarding $\boldsymbol{r}$ as (x, y, z).

The parameter $\boldsymbol{r}_0 = (a\tan\gamma, 0, 0)$ describes a hypersphere rotation in the (X, W) plane; by rotating the 3–dimensional parameter $(a\tan\gamma, 0, 0)$ to $(0, a\tan\gamma, 0)$ and $(0, 0, a\tan\gamma)$ we capture hypersphere rotations in the (Y, W) and (Z, W) planes. An arbitrary parameter $\boldsymbol{r}_0$ corresponds to an arbitrary hypersphere rotation that involves the W axis. Finally, according to formula (24.6), hypersphere rotations in the remaining (X, Y), (Y, Z) and (Z, X) planes produce ordinary 3–dimensional rotations of the stereographic coordinates. In total we have 6 fundamental hyperspace rotations; 3 correspond to the Möbius transformations (24.12) and the remaining 3 to ordinary rotations in stereographic projection (24.6).

Exercise 24.2
Prove that rotations on the hypersphere preserve the line element (24.10) with conformal factor (24.11):

$$n(|\boldsymbol{r}'|)\,\mathrm{d}l' = n(|\boldsymbol{r}|)\,\mathrm{d}l\,. \tag{24.16}$$

One could perform a direct calculation of the transformed line element $n(|\boldsymbol{r}'|)\mathrm{d}l'$, but it is more economical to consider the 2–dimensional case with Möbius transformation (24.13), prove the Möbius invariance (24.16) using optical conformal mapping (§5) and then simply generalize to 3D.
Solution
The Möbius transformation (24.13) is a conformal mapping that transforms the function $n(r)$ according to the recipe (5.9). We write

$$n' = n\left|\frac{\mathrm{d}z}{\mathrm{d}z'}\right| = \frac{2a^2}{a^2+|z|^2}\left|\frac{\mathrm{d}z}{\mathrm{d}z'}\right| \quad \text{with} \quad z = a\,\frac{a\sin\gamma + z'\cos\gamma}{a\cos\gamma - z'\sin\gamma} \tag{24.17}$$

and thus obtain

$$n' = \frac{2a^2}{|a\cos\gamma - z'\sin\gamma|^2 + |a\sin\gamma + z'\cos\gamma|^2} = \frac{2a^2}{a^2+|z'|^2} = n(|z'|)\,, \tag{24.18}$$

which proves that $n'\mathrm{d}l' = n\,\mathrm{d}l$ in 2D.

It is of course a crucial point for the geometry of the hypersphere that the metric (24.10) is only conformally flat or that Eq. (24.5) is not quite the flat–space line element in spherical polar coordinates. The hypersphere is curved and it is therefore impossible to write its line element in flat–space form. To prove this, we must show that the Riemann tensor is non–zero. Before embarking on a brute–force calculation, however, we can deduce that the Riemann tensor has a very simple form. Recall from Exercise 21.2 that the sphere is a space of constant curvature: its Riemann tensor is given by one (constant) number. We saw in Eq. (21.24) that all 2–dimensional spaces have a curvature determined by one scalar function, so the simplification to one constant in the case of the sphere is not so dramatic. But it is clear that the hypersphere is also a space of constant curvature: it is homogeneous (it looks the same at all of its points) and isotropic (it looks the same in all directions at any of its points) so its local curvature is determined by one number and is the same everywhere. This is a major simplification since the curvature of a general 3–dimensional space is specified by six scalar functions (§21). The constant curvature of the hypersphere is of course determined by its radius a, and its Riemann tensor must be constructed from a in a manner that respects the symmetry properties (21.20)–(21.21). It is clear how to do this from the result (21.28) for the sphere; the change to indices that take three values does not spoil the symmetries of (21.28) so the Riemann tensor of the hypersphere should be of this form, up to an overall numerical factor. In fact the result for the hypersphere is the same, namely

$$R^{ij}{}_{kl} = \frac{1}{a^2}\left(\delta^i_k\delta^j_l - \delta^i_l\delta^j_k\right) , \tag{24.19}$$

as direct calculation shows.

Exercise 24.3
Compute the Riemann tensor of the hypersphere from the line element (24.5) and show that it is given by Eq. (24.19).
Solution
The metric tensor in the coordinates $x^i = \{\chi, \theta, \phi\}$ is shown by Eq. (24.5) to be

$$g_{ij} = a^2 \begin{pmatrix} 1 & 0 & 0 \\ 0 & \sin^2\chi & 0 \\ 0 & 0 & \sin^2\chi\sin^2\theta \end{pmatrix} \quad \Longrightarrow \quad g^{ij} = \frac{1}{a^2}\begin{pmatrix} 1 & 0 & 0 \\ 0 & \csc^2\chi & 0 \\ 0 & 0 & \csc^2\chi\csc^2\theta \end{pmatrix} . \tag{24.20}$$

The non–vanishing Christoffel symbols (19.7) are thereby found to be

$$\begin{gathered} \Gamma^\theta{}_{\chi\theta} = \Gamma^\theta{}_{\theta\chi} = \cot\chi\,, \quad \Gamma^\phi{}_{\chi\phi} = \Gamma^\phi{}_{\phi\chi} = \cot\chi\,, \quad \Gamma^\phi{}_{\theta\phi} = \Gamma^\phi{}_{\phi\theta} = \cot\theta\,, \\ \Gamma^\chi{}_{\theta\theta} = -\sin\chi\cos\chi\,, \quad \Gamma^\chi{}_{\phi\phi} = -\sin\chi\cos\chi\sin^2\theta\,, \quad \Gamma^\theta{}_{\phi\phi} = -\sin\theta\cos\theta\,, \end{gathered} \tag{24.21}$$

and by inserting these into the definition (21.8) we can verify the relation

$$R^i{}_{jkl} = \frac{1}{a^2}\left(\delta^i_k g_{jl} - \delta^i_l g_{jk}\right) . \tag{24.22}$$

Rising an index in Eq. (24.22) gives Eq. (24.19); being tensor equations, both of these hold in any coordinate system on the hypersphere.

We obtain from Eq. (24.19) the Ricci tensor and curvature scalar of the hypersphere:

$$R^i{}_j = R^{ki}{}_{kj} = \frac{2}{a^2}\delta^i_j\,, \quad R = R^i{}_i = \frac{6}{a^2} . \tag{24.23}$$

Comparing with Eqs. (21.29) we see that the curvature scalar of a hypersphere is three times that of a sphere of the same radius.

Exercise 24.4
What is the volume of a sphere in the hypersphere? What is the volume of the hypersphere itself?
Solution
Let the centre of the sphere be the North Pole ($\chi = 0$) and the radius be r_0; the surface of the sphere is then given by $\chi = r_0/a$, as is clear from Fig. 3.25. The volume element (18.3) of the hypersphere is found from Eq. (24.20):

$$\mathrm{d}V = a^3\sin^2\chi\sin\theta\,\mathrm{d}\chi\,\mathrm{d}\theta\,\mathrm{d}\phi\,, \tag{24.24}$$

and so the following integration gives the volume V_S of the sphere:

$$V_S = \int_0^{2\pi}\int_0^{\pi}\int_0^{r_0/a} a^3\sin^2\chi\sin\theta\,\mathrm{d}\chi\,\mathrm{d}\theta\,\mathrm{d}\phi = \pi a^3\left[\frac{2r_0}{a} - \sin\left(\frac{2r_0}{a}\right)\right] . \tag{24.25}$$

Note that as the radius a of the hypersphere goes to infinity so that it becomes a flat space, V_S approaches the flat–space result $4\pi r_0^3/3$, as is seen by using the power expansion for the sine function and then taking the limit $a \to \infty$.

To obtain the volume of the hypersphere we must take the upper limit in the χ integration in Eq. (24.25) to be π; we therefore obtain the hypersphere volume V by setting $r_0 = a\pi$:

$$V = 2\pi^2 a^3 \,. \tag{24.26}$$

Exercise 24.5
Show that the distance between two points P, with coordinates $\{\chi, \theta, \phi\}$, and P_0, with coordinates $\{\chi_0, \theta_0, \phi_0\}$, in the hypersphere is

$$s(P, P_0) = a \arccos\left\{\cos\chi\cos\chi_0 + \sin\chi\sin\chi_0\left[\cos\theta\cos\theta_0 + \sin\theta\sin\theta_0\cos(\phi - \phi_0)\right]\right\} . \tag{24.27}$$

Solution
The easiest way to calculate $s(P, P_0)$ is to embed the hypersphere in a 4–dimensional Euclidean space through Eq. (24.1). It is then clear from Fig. 3.25 that the distance $s(P, P_0)$ is given by

$$s(P, P_0) = a\,\alpha(P, P_0)\,, \tag{24.28}$$

where $\alpha(P, P_0)$ is the angle between the points P and P_0 as measured from the centre of the hypersphere in the 4–dimensional space. The centre of the hypersphere is the origin of the 4–dimensional Cartesian coordinates $\{X, Y, Z, W\}$ in Eq. (24.1), and Eqs. (24.3) show that the position vectors of the points P and P_0 in 4–dimensional space in these Cartesian coordinates are

$$\boldsymbol{R} = \left\{a\sin\chi\sin\theta\cos\phi,\ a\sin\chi\sin\theta\sin\phi,\ a\sin\chi\cos\theta,\ a\cos\chi\right\} \tag{24.29}$$

and

$$\boldsymbol{R}_0 = \left\{a\sin\chi_0\sin\theta_0\cos\phi_0,\ a\sin\chi_0\sin\theta_0\sin\phi_0,\ a\sin\chi_0\cos\theta_0,\ a\cos\chi_0\right\}, \tag{24.30}$$

respectively. Now the angle $\alpha(P, P_0)$ between the points P and P_0 is related to the scalar product of their 4–dimensional position vectors $\boldsymbol{R}$ and $\boldsymbol{R}_0$ through the basic formula of vector algebra

$$\boldsymbol{R}\cdot\boldsymbol{R}_0 = |\boldsymbol{R}|\,|\boldsymbol{R}_0|\cos\left[\alpha(P, P_0)\right] = a^2\cos\left[\alpha(P, P_0)\right]\,, \tag{24.31}$$

where we used the fact that the position vectors $\boldsymbol{R}$ and $\boldsymbol{R}_0$ have length a in the 4–dimensional space. The last equation gives

$$\alpha(P, P_0) = \arccos\left(\boldsymbol{R}\cdot\boldsymbol{R}_0/a^2\right)\,. \tag{24.32}$$

All we need to do now is substitute Eqs (24.29) and (24.30) into Eq. (24.32). After simplification we obtain

$$\alpha(P, P_0) = \arccos\left\{\cos\chi\cos\chi_0 + \sin\chi\sin\chi_0\left[\cos\theta\cos\theta_0 + \sin\theta\sin\theta_0\cos(\phi - \phi_0)\right]\right\}, \tag{24.33}$$

which together with Eq. (24.28) gives Eq. (24.27).

---

§25. SPACE–TIME GEOMETRY

We developed tensor analysis in three–dimensional space, but it is a trivial matter to extend the treatment to four dimensions—the indices just take one more value and the Levi–Civita tensor has one more index. In this manner one can combine space and time into 4–dimensional space–time. Time and distance are measured in different units so in space–time the time coordinate is multiplied by the speed of light c to give a coordinate with the units of distance. It is customary to denote the time coordinate ct by x^0 and the three spatial coordinates by $x^i = \{x^1, x^2, x^3\}$. We use Greek letters for space–time indices that run over four values $\{0, 1, 2, 3\}$, and in particular coordinates in space–time are written $x^\mu = \{x^0, x^1, x^2, x^3\}$.

With the development of the theory of relativity it was discovered that space–time has a non–trivial geometry of its own, characterized by a metric tensor. But the appropriate distance in Cartesian coordinates $x^\mu = \{ct, x, y, z\}$ in flat space–time is not given by the four–dimensional Euclidean line element, but rather by the *Minkowski line element*:

$$\mathrm{d}s^2 = -c^2\mathrm{d}t^2 + \mathrm{d}x^2 + \mathrm{d}y^2 + \mathrm{d}z^2 = \eta_{\mu\nu}\,\mathrm{d}x^\mu \mathrm{d}x^\nu\,, \tag{25.1}$$

where $\eta_{\mu\nu}$ is the *Minkowski metric*

$$\eta_{\mu\nu} = \mathrm{diag}\,(-1,\ 1,\ 1,\ 1)\ . \tag{25.2}$$

The time dimension thus contributes negatively to the line element (25.1) in space–time. In a general space–time coordinate system x^μ the line element is

$$\mathrm{d}s^2 = g_{\mu\nu}\,\mathrm{d}x^\mu \mathrm{d}x^\nu\,, \tag{25.3}$$

where $g_{\mu\nu}$ is the space–time metric tensor. The component g_{00} is usually negative throughout space–time, as in the Minkowski case (25.2), and the determinant g is always negative. We can consider curved space–times with non–zero Riemann tensors determined by the metric $g_{\mu\nu}$ just as in the 3–dimensional case. But flat space–time is not a Euclidean space (it does not have a Euclidean metric) and curved space–time is a distorted version of the flat space–time described by Eq. (25.1); space–time is therefore not strictly speaking a Riemannian manifold, rather it is a *pseudo–Riemannian manifold*.

Tensor equations have the same form in every coordinate system. If an equation is written solely in terms of tensors then it can be evaluated in any coordinate system by using the components of the tensors in that system. This seemingly trivial observation has had a profound impact on physics. Einstein proposed that the laws of physics should have the same form for all observers, and this requires that they be expressed by tensor equations, not tensors in 3–dimensional space but in 4–dimensional space–time. Maxwell's electromagnetism meets this requirement, and his equations in 4–dimensional tensor form will make a brief appearance later in this book (see Appendix).

It turns out that space–time is in fact curved everywhere in the universe, a phenomenon usually called gravity (Misner, Thorne and Wheeler [1973]). The space–time geometry in which the reader resides has a Riemann tensor whose largest components at the surface of the Earth are of the order of $10^{-23}\mathrm{m}^{-2}$. This space–time curvature is the reason the reader does not float off into space.

Further reading

Most textbooks on general relativity provide accessible introductions to differential geometry. We recommend Schutz [2009] followed by Misner, Thorne and Wheeler [1973]. Landau and Lifshitz [1975] explain general relativity with emphasis on the physics, while developing the necessary differential geometry on the way. Van Bladel [1984] describes relativity in engineering. A rigorous development of differential geometry from its foundations requires a knowledge of point–set topology; a superb presentation (in just 45 pages) is given in Chapter 2 of Hawking and Ellis [1973].

Chapter 4

Maxwell's equations

Having developed the mathematical machinery of differential geometry, we are now well–prepared to formulate the foundations of electromagnetism and optics, Maxwell's equations in arbitrary coordinates and arbitrary geometries. We assume that the reader is familiar with the basics of electromagnetism in empty, flat space, as we will not derive Maxwell's equations from other principles, but take electromagnetism in empty space as the starting point of this Chapter. In Cartesian coordinates in flat space, the Maxwell equations for the *electric field strength* $\boldsymbol{E}$ and the *magnetic induction* $\boldsymbol{B}$ are (Jackson [1999])*

$$\nabla \cdot \boldsymbol{E} = \frac{\rho}{\varepsilon_0} \qquad \text{GAUSS'S LAW},$$

$$\nabla \times \boldsymbol{B} = \frac{1}{c^2}\frac{\partial \boldsymbol{E}}{\partial t} + \mu_0 \boldsymbol{j} \qquad \text{AMPÈRE'S LAW WITH MAXWELL'S DISPLACEMENT CURRENT},$$

$$\nabla \times \boldsymbol{E} = -\frac{\partial \boldsymbol{B}}{\partial t} \qquad \text{FARADAY'S LAW OF INDUCTION},$$

$$\nabla \cdot \boldsymbol{B} = 0 \qquad \text{ABSENCE OF MAGNETIC MONOPOLES}.$$

In this Chapter, we show how a geometry appears as a medium and how a medium appears as a geometry. This geometrical perspective will give us a wealth of insights into the properties of visible light and other electromagnetic waves and it lays the foundations for practical applications like invisibility cloaking and perfect imaging. In particular, we develop further the concept of transformation optics where we use the freedom of coordinates to describe transformation media as elegantly as possible. We derive the equation of electromagnetic waves and discuss the regime of geometrical optics. Furthermore, we generalize the geometry of light to space–time. We show that space–time geometries appear as moving media and, in turn, can be made by moving media. We also return to the starting point of this book, Fermat's principle.

*We use SI units with electric permittivity ε_0, magnetic permeability μ_0 and speed of light c in vacuum, $\varepsilon_0\mu_0 = c^{-2}$. Charge and current densities are denoted by ρ and $\boldsymbol{j}$.

§26. Spatial geometries and media

Maxwell's equations in empty, flat space are usually formulated in terms of the electric field $\boldsymbol{E}$ and the magnetic induction $\boldsymbol{B}$, but for what follows it is advantageous to write them as equations for the electric field strength $\boldsymbol{E}$ and the *magnetic field strength* $\boldsymbol{H}$, instead of $\boldsymbol{E}$ and $\boldsymbol{B}$, which is easily done in empty space where the magnetic induction $\boldsymbol{B}$ is simply $\mu_0 \boldsymbol{H}$. Maxwell's equations connect the electric and magnetic fields to each other by curls, and they connect the fields to charges by divergences. Using the expressions (17.16) and (17.19) from differential geometry we can now write Maxwell's equations in arbitrary spatial coordinates:

$$
\begin{aligned}
&\frac{1}{\sqrt{g}}\left(\sqrt{g}\,E^i\right)_{,i} = \frac{\rho}{\varepsilon_0}\,, \quad \frac{1}{\sqrt{g}}\left(\sqrt{g}\,H^i\right)_{,i} = 0\,, \\
&\epsilon^{ijk}E_{k,j} = -\mu_0\frac{\partial H^i}{\partial t}\,, \quad \epsilon^{ijk}H_{k,j} = \varepsilon_0\frac{\partial E^i}{\partial t} + j^i\,.
\end{aligned}
\tag{26.1}
$$

This form of Maxwell's equation is also valid in arbitrary spatial geometries, e.g. in curved space, for the following reason: any geometry, no matter how curved, is *locally flat*—at each spatial point we can always construct an infinitesimal patch of a Cartesian coordinate system (see the discussion on p. 135). In curved space, however, the Cartesian system attached to one point will not agree with the Cartesian system of another point. Only in flat space can all the local Cartesian systems form a global Cartesian frame. Curved space is of course also described by a coordinate frame, but this frame is not globally Cartesian. Nevertheless, since we may regard curved space as a patchwork of locally flat pieces, we may postulate Maxwell's equations on each individual patch. In writing these equations in arbitrary coordinates, we naturally express them in the frame of the curved geometry: Maxwell's equations are not only true locally, but globally.

Let us rewrite the form (26.1) of Maxwell's equations with all the vector indices in the lower position and the Levi–Civita tensor expressed in terms of the permutation symbol according to formula (14.7):

$$
\begin{aligned}
&\left(\sqrt{g}\,g^{ij}E_j\right)_{,i} = \frac{\sqrt{g}\,\rho}{\varepsilon_0}\,, \quad \left(\sqrt{g}\,g^{ij}H_j\right)_{,i} = 0\,, \\
&[ijk]E_{k,j} = -\mu_0\frac{\partial(\pm\sqrt{g}\,g^{ij}H_j)}{\partial t}\,, \quad [ijk]H_{k,j} = \varepsilon_0\frac{\partial(\pm\sqrt{g}\,g^{ij}E_j)}{\partial t} \pm \sqrt{g}\,j^i\,.
\end{aligned}
\tag{26.2}
$$

In this form, Maxwell's equations in empty space, but in curved coordinates or curved geometries, resemble the *macroscopic* Maxwell equations for the electric field strength $\boldsymbol{E}$, the *dielectric displacement* $\boldsymbol{D}$, the magnetic induction $\boldsymbol{B}$ and the magnetic field strength $\boldsymbol{H}$ in dielectric media (Jackson [1999]),

$$
\begin{aligned}
&\nabla\cdot\boldsymbol{D} = \varrho\,, \quad \nabla\cdot\boldsymbol{B} = 0\,, \\
&\nabla\times\boldsymbol{E} = -\frac{\partial\boldsymbol{B}}{\partial t}\,, \quad \nabla\times\boldsymbol{H} = \frac{\partial\boldsymbol{D}}{\partial t} + \boldsymbol{J}\,,
\end{aligned}
\tag{26.3}
$$

written in right–handed Cartesian coordinates:

$$D^i{}_{,i} = \varrho \,, \quad B^i{}_{,i} = 0 \,,$$
$$[ijk]E_{k,j} = -\frac{\partial B^i}{\partial t} \,, \quad [ijk]H_{k,j} = \frac{\partial D^i}{\partial t} + J^i \,. \tag{26.4}$$

In fact, the empty–space equations (26.2) can be expressed exactly in the macroscopic form (26.4) if we rescale the charge and current densities,

$$\varrho = \pm\sqrt{g}\,\rho \,, \quad J^i = \pm\sqrt{g}\,j^i \,, \tag{26.5}$$

and take the *constitutive equations*

$$D^i = \varepsilon_0 \varepsilon^{ij} E_j \,, \quad B^i = \mu_0 \mu^{ij} H_j \tag{26.6}$$
$$\varepsilon^{ij} = \mu^{ij} = \pm\sqrt{g}\, g^{ij} \,, \tag{26.7}$$

where the *electric permittivity* $\boldsymbol{\varepsilon}$ and the *magnetic permeability* $\boldsymbol{\mu}$ are given in terms of the geometry. Consequently, the *empty–space* Maxwell equations in *arbitrary* coordinates and geometries are equivalent to the *macroscopic* Maxwell equations in *right–handed Cartesian* coordinates. Geometries appear as dielectric media.

The electric permittivities $\boldsymbol{\varepsilon}$ and magnetic permeabilities $\boldsymbol{\mu}$ are matrices ε^{ij} and μ^{ij}. These matrices are real and symmetric, because the metric tensor (11.6) is real and symmetric. But, as ε^{ij} is a matrix, the dielectric displacement $\varepsilon_0 \boldsymbol{\varepsilon} \boldsymbol{E}$ will not necessarily point in the direction of the electric field strength $\boldsymbol{E}$ (and the magnetic induction $\boldsymbol{B}$ will deviate from the magnetic field strength $\boldsymbol{H}$). Such materials are called *anisotropic media.* Certain crystals, for example, are anisotropic. But in one important respect geometries differ from most anisotropic materials: the electric permittivities of geometries are equal to their magnetic permeabilities,

$$\boldsymbol{\varepsilon} = \boldsymbol{\mu} \,. \tag{26.8}$$

In electrical engineering, $\sqrt{\mu/\varepsilon}$ is called the *impedance*, usually denoted by Z. In empty flat space, the impedance is obviously 1. Equation (26.8) shows that the impedance of empty space is not changed in curved space nor in curved coordinates: a geometry is *impedance–matched* to the vacuum. Spatial geometries thus appear as anisotropic impedance–matched media.

The converse is also true: anisotropic impedance–matched media appear as geometries. We easily derive this statement from the constitutive equations: calculate the determinant $\det \boldsymbol{\varepsilon}$ of ε^{ij} in definition (26.7) where g^{ij} is the inverse matrix of the metric tensor $\mathbf{g} = g_{ij}$ with determinant g. For a matrix $\mathbf{M}$ in three dimensions $\det(a\mathbf{M}) = a^3(\det \mathbf{M})$ and $\det(\mathbf{M}^{-1}) = (\det \mathbf{M})^{-1}$, hence we get from formula (26.7)

$$\det \boldsymbol{\varepsilon} = \pm\sqrt{\det \mathbf{g}} = \pm\sqrt{g} \,. \tag{26.9}$$

This is the factor in front of the metric in the constitutive equations (26.7). Dividing by this factor we obtain from the constitutive equations of a geometry the metric tensor of a medium

$$g^{ij} = \frac{\varepsilon^{ij}}{\det \boldsymbol{\varepsilon}} \tag{26.10}$$

or, in matrix notation,

$$\mathrm{g}^{-1} = (\det \boldsymbol{\varepsilon})^{-1}\boldsymbol{\varepsilon}\,, \tag{26.11}$$

$$\mathrm{g} = (\det \boldsymbol{\varepsilon})\,\boldsymbol{\varepsilon}^{-1}\,. \tag{26.12}$$

For general impedance–matched media, this geometry is curved, but if and only if the Riemann tensor (21.8) vanishes the spatial geometry is flat. In the flat case, there exists a coordinate transformation of physical space that turns Maxwell's equations into their empty–space Cartesian form. The electromagnetic fields in real, physical space containing the medium are therefore the coordinate–transformed fields of empty space. Media that perform such a feat are called *transformation media.*

Spatial geometries appear as impedance–matched media and impedance–matched media appear as spatial geometries. Note carefully that the electric field strength $\boldsymbol{E}$ in the impedance–matched medium is identical to $\boldsymbol{E}$ in the geometry, and the same is true of the magnetic field strength $\boldsymbol{H}$. The electric displacement $\boldsymbol{D}$ and the magnetic induction $\boldsymbol{B}$, however, are different in the two situations: in the spatial geometry $\boldsymbol{D} = \varepsilon_0 \boldsymbol{H}$ and $\boldsymbol{B} = \mu_0 \boldsymbol{H}$, but in the equivalent medium $\boldsymbol{D}$ and $\boldsymbol{B}$ are related to $\boldsymbol{E}$ and $\boldsymbol{H}$ by the constitutive equations (26.6) and (26.7). Note also that when $\boldsymbol{D}$ and $\boldsymbol{B}$ in the medium are expressed in terms of the geometry, they are different geometrical objects than $\boldsymbol{E}$ and $\boldsymbol{H}$. The field strengths $\boldsymbol{E}$ and $\boldsymbol{H}$ appear as one–forms E_i and H_i, as covariant vectors. What are the $\boldsymbol{D}$ and $\boldsymbol{B}$? They carry upper indices and contain the prefactor $\sqrt{g}$. Recall formula (11.16) for the volume element $\mathrm{d}V$. We see that $\boldsymbol{D}\,\mathrm{d}^3x$ is proportional to the vector $g^{ij}E_j$ multiplied by the volume element $\mathrm{d}V$ with respect to the optical geometry. Since the volume element is a scalar, $\boldsymbol{D}\,\mathrm{d}^3x$ must be a vector. In other words, the dielectric displacement $\boldsymbol{D}$ is a *vector density* and, by the same argument, so is the magnetic induction $\boldsymbol{B}$.

We must also be careful in correctly relating the *electromagnetic potentials* to the fields. Maxwell wrote the electromagnetic fields in terms of electromagnetic potentials, the electric potential U, the voltage, and the magnetic vector potential $\boldsymbol{A}$ (Jackson [1999]). In electrostatics, the electric field is the negative gradient of the voltage, but in electrodynamics the time derivative of the vector potential induces an electric field such that, in total,

$$\boldsymbol{E} = -\boldsymbol{\nabla} U - \frac{\partial \boldsymbol{A}}{\partial t}\,. \tag{26.13}$$

This relationship implies that U is a scalar and $\boldsymbol{A}$ is a one–form in 3–dimensional space.* The magnetic induction is the curl of the vector potential in right–handed Cartesian coordinates,

$$B^i = [ijk]A_{k,j}\,. \tag{26.14}$$

In our interpretation of media as spatial geometries, B^i is not a vector, but a vector density, whereas H_i is a genuine one–form. It is therefore wise to express representation (26.14) in terms of the magnetic field $\boldsymbol{H}$, not the magnetic induction. We

*In 4–dimensional space–time, U and $\boldsymbol{A}$ are the components of a 4–dimensional one–form. These components mix in space–time transformations.

obtain from formula (14.7) for the Levi–Civita tensor and one of the constitutive equations (26.6)

$$H^i = g^{ij} H_j = \frac{1}{\mu_0}\,\epsilon^{ijk} A_{k,j}\,. \tag{26.15}$$

Thus, in the medium representation (26.14), $\boldsymbol{B}$ is the curl of $\boldsymbol{A}$, whereas in the geometry representation (26.15), $\boldsymbol{H}$ is the curl *of the same vector potential* $\boldsymbol{A}$. The electromagnetic potentials bring the advantage that the representations (26.13) and (26.15) naturally satisfy two of Maxwell's equations, Faraday's law of induction and the absence of magnetic monopoles. Their true significance lies in quantum mechanics and quantum field theory, however, which goes beyond the subject of this book.

Exercise 26.1
Show that the representations (26.13)–(26.15) of the fields in terms of the electromagnetic potentials obey the Maxwell equations (in macroscopic and geometrical form)

$$[ijk]E_{k,j} = -\frac{\partial B^i}{\partial t}\,, \quad {B^i}_{,i} = 0 \quad \text{and} \quad \epsilon^{ijk}E_{k;j} = -\mu_0\frac{\partial H^i}{\partial t}\,, \quad {H^i}_{;i} = 0\,. \tag{26.16}$$

Solution
Exploit the antisymmetry of $[ijk]$ and the symmetry of partial derivatives $\partial_i\partial_j = \partial_j\partial_i$, and use formula (14.7) for the Levi–Civita tensor ϵ^{ijk} and expression (17.16) for the divergence.

According to Fermat's principle, the geometry of light in isotropic optical media directly depends on the refractive index. What is the refractive index in anisotropic, impedance–matched materials? How does it depend on the $\boldsymbol{\varepsilon}$ and $\boldsymbol{\mu}$ of the material? We need to relate the refractive index to the geometry of light. At the end of §9 we argued that the refractive index is the ratio between the length element $\mathrm{d}s$ perceived by light and the length element $\mathrm{d}l$ in physical space (and in Cartesian coordinates). In anisotropic media establishing geometries that are not conformally flat, the length element $\mathrm{d}s$ depends on direction. At each spatial point however, we can align a Cartesian system of coordinates such that its axes lie along the eigenvectors of $\boldsymbol{\varepsilon} = \boldsymbol{\mu}$, because $\boldsymbol{\varepsilon}$ is real and symmetric, so its eigenvectors must be orthogonal to each other and hence suitable for establishing a local Cartesian system. Let us denote the local eigenvalues of $\boldsymbol{\varepsilon}$ by $\{\varepsilon_x, \varepsilon_y, \varepsilon_z\}$ such that

$$\boldsymbol{\varepsilon} = \mathrm{diag}\,(\varepsilon_x, \varepsilon_y, \varepsilon_z) \quad \text{and} \quad \det\boldsymbol{\varepsilon} = \varepsilon_x\,\varepsilon_y\,\varepsilon_z\,. \tag{26.17}$$

(The three Cartesian eigenvalues $\{\varepsilon_x, \varepsilon_y, \varepsilon_z\}$ are known as the *principle values*.) Imagine a line element $\mathrm{d}x$ in the x–direction. Formula (26.12) shows that the square of the corresponding optical line element is $\varepsilon_y\,\varepsilon_z\,\mathrm{d}x^2$ and so the square of the refractive index n_x in the x–direction must be $\varepsilon_y\,\varepsilon_z$. Similarly, we deduce the index–components in the other directions and obtain

$$\mathrm{diag}\left(n_x^2,\, n_y^2,\, n_z^2\right) = \mathrm{diag}\left(\varepsilon_y\,\varepsilon_z,\; \varepsilon_z\,\varepsilon_x,\; \varepsilon_x\,\varepsilon_y\right)\,. \tag{26.18}$$

We see that the refractive index in one direction of the local eigensystem of the dielectric matrix does not depend on the ε and μ in that direction, but on the dielectric properties in the orthogonal directions. This is because electromagnetic waves are transversal—their fields point orthogonally to the direction of propagation, as we show in §30. So the wave naturally responds to the electromagnetic properties in these orthogonal directions. Now, regard the left–hand side of relation (26.18) as the square of the matrix of the refractive index $\mathbf{n} = \mathrm{diag}(n_x, n_y, n_z)$, the matrix product of $\mathbf{n}$ with itself. The right–hand side reads in matrix notation

$$\mathbf{n}^2 = (\det \boldsymbol{\varepsilon})\boldsymbol{\varepsilon}^{-1} = \mathbf{g}\,. \tag{26.19}$$

Expression (26.19) is independent of the eigenvector system of $\boldsymbol{\varepsilon}$ (that may vary from point to point in the medium), and hence it is universally true for anisotropic, impedance–matched dielectric materials at rest. Formula (26.19) describes how the matrix of the refractive index depends on the dielectric properties of the material. The formula also generalizes Fermat's principle: the matrix square of the refractive index establishes the metric of light.

Exercise 26.2
Calculate the matrix square of the refractive index for the dielectric tensors

$$\boldsymbol{\varepsilon} = \boldsymbol{\mu} = \begin{pmatrix} \varepsilon_{xx} & \varepsilon_{xy} & \varepsilon_{xz} \\ \varepsilon_{yx} & \varepsilon_{yy} & \varepsilon_{yz} \\ \varepsilon_{zx} & \varepsilon_{zy} & \varepsilon_{zz} \end{pmatrix}. \tag{26.20}$$

Solution
According to Cramer's rule, the inverse matrix $\boldsymbol{\varepsilon}^{-1}$ is the transposed matrix of the cofactors of $\boldsymbol{\varepsilon}$ divided by $\det \boldsymbol{\varepsilon}$. Therefore, $\mathbf{n}^2$ simply is the transposed cofactor matrix, the adjugate matrix $\mathrm{adj}\,\boldsymbol{\varepsilon}$ (also called the adjoint). We obtain

$$\mathbf{n}^2 = \mathrm{adj}\,\boldsymbol{\varepsilon} = \begin{pmatrix} \varepsilon_{yy}\,\varepsilon_{zz} - \varepsilon_{yz}\,\varepsilon_{zy} & \varepsilon_{xz}\,\varepsilon_{zy} - \varepsilon_{xy}\,\varepsilon_{zz} & \varepsilon_{xy}\,\varepsilon_{yz} - \varepsilon_{xz}\,\varepsilon_{yy} \\ \varepsilon_{yz}\,\varepsilon_{zx} - \varepsilon_{yx}\,\varepsilon_{zz} & \varepsilon_{xx}\,\varepsilon_{zz} - \varepsilon_{xz}\,\varepsilon_{zx} & \varepsilon_{xz}\,\varepsilon_{yx} - \varepsilon_{xx}\,\varepsilon_{yz} \\ \varepsilon_{yx}\,\varepsilon_{zy} - \varepsilon_{yy}\,\varepsilon_{zx} & \varepsilon_{xy}\,\varepsilon_{zx} - \varepsilon_{xx}\,\varepsilon_{zy} & \varepsilon_{xx}\,\varepsilon_{yy} - \varepsilon_{xy}\,\varepsilon_{yx} \end{pmatrix}. \tag{26.21}$$

---

§27. Planar media

In §26 an equivalence was established between impedance–matched media and media that create effective spatial geometries for electromagnetism. The correspondence between the (in general anisotropic) medium, with permittivity and permeability $\varepsilon^{ij} = \mu^{ij}$, and the effective geometry, described by the metric g_{ij}, is summarized as

$$\varepsilon^{ij} = \mu^{ij} = \pm\sqrt{g}\,g^{ij}\,, \qquad g^{ij} = \frac{\varepsilon^{ij}}{\det \boldsymbol{\varepsilon}} = \frac{\mu^{ij}}{\det \boldsymbol{\mu}}\,. \tag{27.1}$$

If the metric tensor g_{ij} is flat (if it has zero Riemann tensor) then the medium performs a coordinate transformation, whereas if the metric is curved (non–zero Riemann tensor) the medium creates a curved spatial geometry. These statements hold true for all electromagnetic phenomena (static fields, electromagnetic waves, etc.) because the correspondence (27.1) is exact for Maxwell's equations. Unfortunately

Nature has shown little interest in producing materials with non–trivial magnetic response ($\mu \neq 1$) so the permeability in impedance–matched media must be engineered from scratch. There is a burgeoning expertise in the manufacture of such artificial materials, or metamaterials, but it is still easier to tune the permittivity than the permeability. For this reason it is of considerable practical importance that an electric permittivity alone is sufficient to create effective 2–dimensional geometries, not for all electromagnetism but for electromagnetic waves polarized orthogonal to the effective 2–dimensional space.

Consider a dielectric with $\mu = 1$, whose permittivity ε^{ij} in some Cartesian coordinate system is

$$\varepsilon^{ij} = \begin{pmatrix} \varepsilon^{AB} & 0 \\ 0 & n^2(x,y) \end{pmatrix}, \tag{27.2}$$

where capital indices refer to the 2–dimensional coordinates $\{x, y\}$. We will show that n is the refractive index where, in contrast to the impedance–matched case (26.19), n^2 is given by ε^{zz}. It will be essential for what follows that $\varepsilon^{zz} = n^2(x,y)$ is homogeneous in the z–direction and that the cross components $\varepsilon^{zA} = \varepsilon^{Az}$ vanish. The components ε^{AB} will not appear in the final result.

The source–free Maxwell equations in a medium with $\mu = 1$, in Cartesian coordinates, are

$$\begin{gathered} \left(\varepsilon^{ij} E_j\right)_{,i} = 0 \,, \quad B^i{}_{,i} = 0 \,, \\ [ijk] E_{k,j} = -\frac{\partial B^i}{\partial t} \,, \quad [ijk] B_{k,j} = \frac{1}{c^2} \frac{\partial(\varepsilon^{ij} E_j)}{\partial t} \,. \end{gathered} \tag{27.3}$$

These are not Maxwell's equations in an effective geometry, in contrast to the case of impedance–matching. Nevertheless, we shall show that for the permittivity (27.2) the equation for electromagnetic waves polarized in the z–direction is a 2–dimensional wave equation in a geometry determined by $n(x, y)$.

To obtain the equation of an electromagnetic wave we take the curl of the third equation in the list (27.3) and use the fourth equation to eliminate the magnetic field:

$$[lmi][ijk] E_{k,j,m} = -\frac{1}{c^2} \frac{\partial^2}{\partial t^2} \left(\varepsilon^{lm} E_m\right) \,. \tag{27.4}$$

The left–hand side of Eq. (27.4) can be rewritten using Eq. (14.15), with the result

$$E_{k,k,l} - E_{l,k,k} = -\frac{\varepsilon^{lm}}{c^2} \frac{\partial^2 E_m}{\partial t^2} \,, \tag{27.5}$$

where there is no need to distinguish between upper and lower indices since we are in Cartesian coordinates. The wave equation (27.5) gives coupled equations for the electric field components in a dielectric with general anisotropic permittivity ε^{ij}. In the case of the permittivity (27.2) however, Eq. (27.5) produces an uncoupled equation for a z–component $E_z = E_z(x, y)$ of the electric field, so an electromagnetic wave with this polarization can propagate in the medium (27.2). To see this, try taking $E_x = E_y = 0$ in Eq. (27.5) with permittivity (27.2), and assume that E_z is homogeneous in the z–direction ($E_z = E_z(x,y)$); the only non–trivial content occurs when $\imath = z$, giving

$$E_{z,A,A} = -\frac{n^2(x,y)}{c^2}\frac{\partial^2 E_z}{\partial t^2}\,. \tag{27.6}$$

The behaviour of z–polarized waves is thus determined by the function $n(x,y)$, which is the refractive index experienced by these waves, and propagation is in the 2–dimensional (x,y)–subspace. Consider a 2–dimensional metric that is related to the refractive index $n(x,y)$ by

$$g_{AB} = n^2(x,y)\,\delta_{AB}\,, \quad g^{AB} = \frac{1}{n^2(x,y)}\,\delta_{AB}\,, \quad g = n^4(x,y)\,. \tag{27.7}$$

With the scalar field $E_z(x,y)$ denoted simply by E, it is easy to see that Eq. (27.6) can be written in terms of the metric (27.7) as

$$\frac{1}{\sqrt{g}}\left(\sqrt{g}\,g^{AB}E_{,A}\right)_{,B} = -\frac{1}{c^2}\frac{\partial^2 E}{\partial t^2}\,, \tag{27.8}$$

since $\sqrt{g} = n^2$ and $\sqrt{g}\,g^{AB} = \delta^{AB}$. Now the left–hand side of Eq. (27.8) is precisely the Laplacian of the scalar E in the 2–dimensional geometry defined by the metric (27.7) (recall Eq. (17.23)), so Eq. (27.8) is a 2–dimensional wave equation in an effective geometry.

Dielectrics with permittivity (27.2) thus support effective 2–dimensional propagation of z–polarized light in the (x,y)–plane, a mode of propagation called *transverse electric* (*TE*). As far as TE waves are concerned, such media are essentially planar in physical space, but they create an effective 2–dimensional geometry given by the metric (27.7) that is in general curved. One can create planar transformation media or effective curved 2–dimensional spaces for TE waves using the recipe (27.7), with no need of a magnetic response. Other modes of propagation in the dielectric (27.2) will however not experience an effective geometry. For example, waves with magnetic field in the z–direction, known as *transverse magnetic* (*TM*) waves, satisfy an equation different from Eq. (27.8) because the medium is not impedance–matched.

The results of this Section have been derived for a medium (27.2) homogeneous in one direction, the z–direction in our case, but in practice planar media are of the opposite extreme: they have a very small extension in one direction. In this case electromagnetic waves are confined in the z–direction and one must solve for the non–trivial dependence of the waves on the z–coordinate. But the (x,y)–dependence of TE waves is still described by equation (27.8); the TE waves are confined inside the planar medium and propagate in an effective 2–dimensional geometry (Snyder and Love [1983]).

For monochromatic fields oscillating with frequency ω, the second time derivative $\partial^2\boldsymbol{E}/\partial t^2 = -\omega^2\boldsymbol{E}$ and so the wave equation (27.6) is the Helmholtz equation (4.21) familiar from optical conformal mapping (§5). In §4 we had to resort to Feynman's path integrals to deduce the Helmholtz equation; here we see that it naturally and exactly follows from Maxwell's equations in planar media. Optically conformal mapping is exact in planar media for TE polarized light.

§28. TRANSFORMATION MEDIA

Let us return to media in three dimensions. In §1 we already introduced the idea of transformation optics where an optical material appears to perform a transformation of space. In §26 we made this idea more precise: the optical geometry established by transformation media is flat; it can only differ from Cartesian space by a coordinate transformation. Transformation media thus implement coordinate transformations of Maxwell's equations.

Note carefully how this interpretation of Maxwell's equations (26.1) works: we write the free–space equations in coordinates that are not right–handed Cartesian, but we then interpret these equations as being in a right–handed Cartesian system with an effective medium (26.7). This sounds a bit paradoxical, but the way to think of it is to imagine *two* different spaces as well as two different coordinate systems (look again at Fig. 2.6). In the first space, which we call *virtual space*, we have no medium and we write the empty–space Maxwell equations in right–handed Cartesian coordinates. We then perform a transformation that gives us a non–trivial effective medium (26.7) and we interpret the transformed coordinates as being right–handed Cartesian in a *new* space, physical space, which contains the medium (26.7). The Cartesian grid in virtual space will deform under the transformation and this deformed grid shows ray trajectories in physical space.

There are two aspects of transformation media that make them highly significant. First, we know a good deal about solutions of Maxwell's equation in vacuum (light rays travel in straight lines, etc) and to find the effect of the medium we can just take a vacuum solution in virtual space and transform to physical space using the coordinate transformation that defines the medium: the transformed fields are a solution of the macroscopic Maxwell equations in physical space. Second, since a transformation medium is defined by a coordinate transformation, we can use this as a design tool to find materials with remarkable electromagnetic properties.

Some readers may have nagging doubts about the juxtaposition of the mathematical tools of general relativity with the attribution of a physical significance to coordinate transformations. For a relativist coordinate systems have no physical meaning; the geometry of the space is the important thing, and that is independent of the coordinate grid one chooses to cover the space. But here we wish to consider materials that, as far as electromagnetism is concerned, perform active coordinate transformations. In this theory, the coordinate transformation *is* physically significant, it describes completely the macroscopic electromagnetic properties of the material, and differential geometry is just as useful for these purposes as it is in general relativity.

In our description of transformation media, the starting point of the theory was a right–handed Cartesian system in virtual space; any non–trivial transformation from this system gives an effective medium in physical space. It is, however, often convenient to adjust coordinates to the particular situation under investigation (Fig. 4.1). We should be allowed to use any coordinates we wish in virtual space. In order to implement this freedom of coordinates, we generalize the theory. Suppose that we describe virtual space by a curvilinear system such as cylindrical or spherical polar

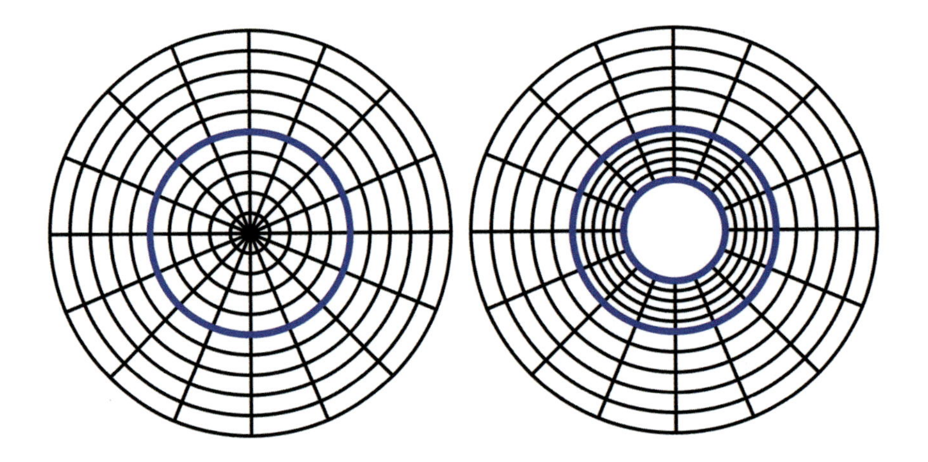

Figure 4.1: Transformation media in cylindrical coordinates. The figure shows the transformation of Fig. 2.6 in cylindrical coordinates; these are the coordinates best adapted to this case.

coordinates; any deformation of this system through a coordinate transformation is to be interpreted as a medium, but to describe electromagnetism in the presence of this medium we employ the original curvilinear grid.

Let x'^i be the curvilinear system in virtual space and we denote its metric tensor by γ_{ij}. Then in virtual space the empty–space Maxwell equations (26.1) are:

$$\frac{1}{\sqrt{\gamma}}(\sqrt{\gamma}\,E^i)_{,i} = \frac{\rho}{\varepsilon_0}\,, \quad \frac{1}{\sqrt{\gamma}}(\sqrt{\gamma}\,H^i)_{,i} = 0\,,$$

$$[ijk]E_{k,j} = -\mu_0\frac{\partial(\sqrt{\gamma}\,H^i)}{\partial t}\,, \quad [ijk]H_{k,j} = \varepsilon_0\frac{\partial(\sqrt{\gamma}\,E^i)}{\partial t} + \sqrt{\gamma}\,j^i\,, \tag{28.1}$$

where we have made the metric dependence of the Levi–Civita tensor (14.7) explicit. Now we perform a coordinate transformation and, as before, we interpret the resulting equations as being macroscopic Maxwell equations written in the *same* (curvilinear) system we started with, but in physical space. The macroscopic equations (26.3) in the curvilinear system with the metric γ_{ij} are:

$$\left(\sqrt{\gamma}\,D^i\right)_{,i} = \sqrt{\gamma}\,\rho\,, \quad \left(\sqrt{\gamma}\,B^i\right)_{,i} = 0\,,$$

$$[ijk]E_{k,j} = -\frac{\partial(\sqrt{\gamma}\,B^i)}{\partial t}\,, \quad [ijk]H_{k,j} = \frac{\partial(\sqrt{\gamma}\,D^i)}{\partial t} + \sqrt{\gamma}\,j^i\,. \tag{28.2}$$

By the same reasoning as before, we can interpret the free–space equations (28.1) as the macroscopic equations (28.2) written in the curvilinear system if we rescale the charge and current densities and take the constitutive equations (26.6) with

$$\varepsilon^{ij} = \mu^{ij} = \pm\frac{\sqrt{g}}{\sqrt{\gamma}}\,g^{ij}\,. \tag{28.3}$$

So far we assumed that the virtual space conjured up by the transformation medium is empty, but it is of course straightforward to include virtual spaces in our theory that contain media. We would not gain any theoretical simplification if these media are arbitrary, but if the virtual media are simple and isotropic we could reduce complicated problems in real space to simple problems in virtual space. We will encounter two examples of this procedure, the transmutation of singularities (Exercise 33.2) and cloaking at a distance (§35). In the case virtual space carries an isotropic medium with permittivity ε' and permeability μ' the dielectric functions in real space are simply

$$\varepsilon^{ij} = \pm\frac{\sqrt{g}}{\sqrt{\gamma}}\, g^{ij}\varepsilon' \,, \quad \mu^{ij} = \pm\frac{\sqrt{g}}{\sqrt{\gamma}}\, g^{ij}\mu' \,. \tag{28.4}$$

If we wish to implement a certain coordinate transformation or the effect of a coordinate transformation on an isotropic medium, formula (28.4) gives a simple and efficient recipe for calculating the required material properties in arbitrary coordinates (Leonhardt and Philbin [2006]).

§29. ELECTROMAGNETIC WAVES

Light is an electromagnetic wave, and so are radio waves, heat radiation and X rays, to name a few examples. In an electromagnetic wave, the field oscillates with a frequency ω that, for visible light, determines the colour. A changing electric field generates a magnetic field that, in turn, induces a further contribution to the electric field and so forth; the electromagnetic oscillation spreads through space as a wave. Let us turn to the mathematical description of electromagnetic waves.

We could always use Maxwell's equations that explicitly show how a changing electric field contributes to the magnetic field and how changing magnetic fields induce electric fields. But it is often advantageous to combine Maxwell's equations in a single wave equation of only one of the fields, the electric or the magnetic field. As this equation includes the mediating influence of the other field, it is a second–order partial differential equation (that contains second derivatives—changes of changes), whereas Maxwell's equations are of first order. We assume that the medium is time–independent and impedance–matched (26.8), generating the spatial geometry (26.12). We also consider a region of space without currents and charges. We obtain from Maxwell's equation (26.1)

$$\epsilon^{ijk}\left(\epsilon_{klm}E^{m;l}\right)_{;j} + \frac{1}{c^2}\frac{\partial^2 E^i}{\partial t^2} = 0 \,. \tag{29.1}$$

We could write down exactly the same wave equation for the magnetic field, because for establishing a perfect geometry the electric and magnetic responses of the medium must be the same (the medium is impedance–matched). Note that we expressed derivatives as covariant derivatives (semicolons instead of commas) in anticipation of simplifications to come. In particular, the covariant derivative of the Levi–Civita tensor (14.6) vanishes (Exercise 20.2). Consequently, we only need to consider differentiating the field, but not ϵ_{klm}. With the notation (15.25) for the covariant derivatives we obtain

$$\begin{aligned}\frac{1}{c^2}\frac{\partial^2 E^i}{\partial t^2} &= -\,\epsilon^{ijk}\epsilon_{klm}\nabla_j\nabla^l E^m \\ &= \left(\delta^i_m\delta^j_l - \delta^i_l\delta^j_m\right)\nabla_j\nabla^l E^m \\ &= \nabla^j\nabla_j E^i - \nabla_j\nabla^i E^j \qquad (29.2)\end{aligned}$$

where we applied formula (14.14). The second term in the last line of the wave equation (29.2) resembles the gradient of the divergence $\nabla_j E^j$. As we have seen in §26, the covariant divergence of the electric field, $\nabla_j E^j$, corresponds to the Cartesian divergence of the dielectric displacement $\boldsymbol{D}$, cf. Eqs. (26.2) and (26.4). In the absence of external charges, the charge density ϱ vanishes and so does $\boldsymbol{\nabla}\cdot\boldsymbol{D}$ and therefore also $\nabla_j E^j$. It would be wise to exploit the vanishing divergence in the wave equation (29.2), but there $\nabla_j\nabla^i E^j$ appears and not $\nabla^i\nabla_j E^j$. We found in §21 that covariant derivatives do not commute in general: the commutator (21.10) of covariant derivatives is given by the Riemann tensor (21.8). Hence we obtain, lowering the index of E^i and using the commutator (21.10)

$$\begin{aligned}\frac{1}{c^2}\frac{\partial^2 E_i}{\partial t^2} &= \nabla^j\nabla_j E_i - \nabla_i\nabla_k E^k - \left(\nabla_k\nabla_i - \nabla_i\nabla_k\right)E^k \\ &= \nabla^j\nabla_j E_i - R^k{}_{jki}E^j\,. \qquad (29.3)\end{aligned}$$

The contracted Riemann tensor, $R^k{}_{jki}$, is the Ricci tensor (21.13) with symmetry (21.22). In this way we arrive at the *electromagnetic wave equation*

$$\nabla^j\nabla_j E_i - R_{ij}E^j - \frac{1}{c^2}\frac{\partial^2 E_i}{\partial t^2} = 0\,. \qquad (29.4)$$

Finally, we supplement the wave equation (29.4) by the Maxwell equation (26.2) in the absence of charges:

$$\nabla_i E^i = 0\,. \qquad (29.5)$$

Equations (29.4) and (29.5) establish a complete description of electromagnetic waves in terms of only the electric field. We would get the same equations for the magnetic field, because we have assumed impedance–matching of the medium (26.8) or, equivalently, an effective spatial geometry.

The wave equation (29.4) describes how electromagnetic waves propagate. Consider, for example, empty space in Cartesian coordinates. A complete set of solutions in this case are the *plane waves* with constant complex amplitudes $\mathcal{E}_i$:

$$E_i = \mathcal{E}_i\,\mathrm{e}^{\mathrm{i}\varphi} \qquad (29.6)$$

and the phase

$$\varphi = \boldsymbol{k}\cdot\boldsymbol{r} - \omega t \qquad (29.7)$$

with wave vector $\boldsymbol{k}$ and frequency ω subject to the dispersion relation

$$\omega = c\,|\boldsymbol{k}|\,. \qquad (29.8)$$

Exercise 29.1
Verify that the plane waves (29.6–29.8) are solutions of the wave equation (29.4) in empty space in Cartesian coordinates.

Equations (29.6)–(29.8) show that a plane wave propagates in the direction of the wave vector $\boldsymbol{k}$ with phase velocity c (remember the discussion at the beginning of §4). One might object that the plane waves (29.6) are complex–valued, but the electric field surely is real, and also that such waves would extend throughout infinite space; plane waves seem rather artificial. But we may represent all solutions of the wave equation (29.4) in empty space, all *wave packets*, as superpositions of plane waves, because the wave equation is a linear partial differential equation with the plane waves representing a complete set of solutions, as is shown in *Fourier analysis* (Riley, Hobson and Bence [2006]):

$$E_i = \int \mathcal{E}_i(\boldsymbol{k}) \exp(\mathrm{i}\boldsymbol{k}\cdot\boldsymbol{r} - \mathrm{i}\omega t)\,\mathrm{d}^3k \quad \text{with} \quad \omega = c\,|\boldsymbol{k}|\,. \tag{29.9}$$

Furthermore, supplementing expression (29.9) by its complex conjugate, we obtain all *real* solutions of the wave equation in empty space with Cartesian coordinates:

$$E_i = \int \left[\mathcal{E}_i(\boldsymbol{k})\mathrm{e}^{\mathrm{i}\varphi} + \mathcal{E}_i^*(\boldsymbol{k})\mathrm{e}^{-\mathrm{i}\varphi}\right]\mathrm{d}^3k\,. \tag{29.10}$$

For example, to describe the real electric field of a plane wave we should add to the complex wave (29.6) with amplitude $\mathcal{E}_i$ and wave number $\boldsymbol{k}$ the complex conjugate wave $\mathcal{E}_i^*\exp(-\mathrm{i}\boldsymbol{k}\cdot\boldsymbol{r} + \mathrm{i}\omega t)$, i.e. a plane wave with complex–conjugate amplitude, wave vector $-\boldsymbol{k}$ and frequency $-\omega$. Note that $-\omega$ is a *negative frequency*, which is the defining characteristics of the required supplement to the complex field (29.9): we must add the negative–frequency part to obtain a real field. For the plane wave (29.6) we get

$$E_i = \mathcal{E}_i\,\mathrm{e}^{\mathrm{i}\varphi} + \mathcal{E}_i^*\,\mathrm{e}^{-\mathrm{i}\varphi} = 2\,|\mathcal{E}_i|\cos(\varphi + \varphi_i)\,, \tag{29.11}$$

where the φ_i denote the arguments of the complex constants $\mathcal{E}_i = |\mathcal{E}_i|\exp(\mathrm{i}\varphi_i)$. Expression (29.11) describes the actual electric field of an electromagnetic oscillation that is propagating through space in time, an electromagnetic wave. Since in the superposition (29.10) the negative–frequency components are always "shadowing" the positive–frequency components, we may as well just focus on the positive–frequency part of the field in our theory, but bearing in mind that the negative–frequency counterpart must be there in reality.

The wave equation (29.4) determines how electromagnetic waves propagate. The condition (29.5), on the other hand, constrains the amplitude vector of the electric field. For plane waves (29.6) in empty space, the amplitude must be orthogonal to the wave vector,

$$k_i\,\mathcal{E}^i = 0\,, \tag{29.12}$$

as we easily see by inserting expression (29.6) in condition (29.5). The electric field is said to be *transversal*, and so is the magnetic field (since, because of impedance–matching (26.8), they obey the same equations). Suppose we rotate our system of coordinates such that $\boldsymbol{k}$ points in the z–direction. Because of the orthogonality condition (29.12) the amplitude vector must lie in the (x, y) plane, but recall that $\mathcal{E}_i$ is a complex vector. Let us describe the components of $\mathcal{E}_i$ by the complex unit vector (ψ_x, ψ_y), the *polarization*, and the real overall amplitude $\mathcal{E}$ such that

$$\mathcal{E}_x = \mathcal{E}\psi_x\,, \quad \mathcal{E}_y = \mathcal{E}\psi_y \quad \text{with} \quad |\psi_x|^2 + |\psi_y|^2 = 1\,. \tag{29.13}$$

We can express any complex unit vector (ψ_x, ψ_y) as

$$\psi_x = \mathrm{e}^{\mathrm{i}\phi_0 + \mathrm{i}\delta} \cos\gamma\,, \quad \psi_y = \mathrm{e}^{\mathrm{i}\phi_0 - \mathrm{i}\delta} \sin\gamma \tag{29.14}$$

in terms of the three real constants ϕ_0, δ and γ and therefore obtain, from representation (29.13), for the actual electric field strengths (29.11):

$$E_x = 2\,\mathcal{E}\cos\gamma\,\cos(\varphi + \varphi_0 + \delta)\,, \quad E_y = 2\,\mathcal{E}\sin\gamma\,\cos(\varphi + \varphi_0 - \delta)\,. \tag{29.15}$$

Figure (4.2) shows the characteristic curves the field vectors trace out when the phase φ progresses in space or time. For $\delta = 0$ the components E_x and E_y are proportional to each other; so they lie on a straight line: the wave is *linearly polarized.* For $\delta = \pm\pi/4$ and $\gamma = \pi/4$, the curve is a circle, going clockwise (as the wave approaches) for $\delta = +\pi/4$ and counter–clockwise for $\delta = -\pi/4$. The wave is *circularly polarized, right–circularly* when the field vector is going clockwise and *left–circularly* when it is going counter–clockwise (for historic reasons). Otherwise the field vector traces out an ellipse and so the light is *elliptically polarized.*

Exercise 29.2
Argue why the field vector (29.15) draws an ellipse when the phase φ progresses. For example, describe (29.15) in the complex plane as $z = \mathcal{E}^{-1}(E_x + \mathrm{i}E_y)$ and compare $z(\varphi)$ with Hooke's ellipse $z(\xi)$ given by Eq. (3.17).
Solution
Comparison with Eq. (3.17) and representing ξ as $\varphi + \xi_0$ with some constant ξ_0 gives

$$\mathrm{e}^{\mathrm{i}\delta}\cos\gamma + \mathrm{i}\,\mathrm{e}^{-\mathrm{i}\delta}\sin\gamma = \mathrm{e}^{\mathrm{i}\alpha}(a+b)\mathrm{e}^{\mathrm{i}\xi_0}\,, \quad -\mathrm{e}^{\mathrm{i}\delta}\sin\gamma + \mathrm{i}\,\mathrm{e}^{-\mathrm{i}\delta}\cos\gamma = \mathrm{e}^{\mathrm{i}\alpha}(a-b)\mathrm{e}^{-\mathrm{i}\xi_0}\,. \tag{29.16}$$

We may thus understand $\alpha + \xi_0$ as the argument and $a + b$ as the modulus of the complex number $\mathrm{e}^{\mathrm{i}\delta}\cos\gamma + \mathrm{i}\,\mathrm{e}^{-\mathrm{i}\delta}\sin\gamma$, while $\alpha - \xi_0$ is the argument and $a - b$ the modulus of $-\mathrm{e}^{\mathrm{i}\delta}\sin\gamma + \mathrm{i}\,\mathrm{e}^{-\mathrm{i}\delta}\cos\gamma$. As these identifications are always possible, the curve of the polarization vector is identical to the ellipse (3.17).

We discussed Eq. (29.5) for plane waves in empty space. For transformation media or optical materials that establish curved spatial geometries, condition (29.5) defines the *transversality* of electromagnetic waves in general. Moreover, we show in §30 that, within the validity of geometrical optics, the field–amplitude vector is orthogonal to the direction of propagation and parallel–transported along ray trajectories.

The wave equation (29.4) and the transversality condition (29.5), containing covariant derivatives, are compact notations for rather complex explicit expressions, involving the metric tensor or, alternatively, the dielectric tensors (26.7). To appreciate the degree of conciseness achieved, it is instructive to compare this geometric notation with the explicit form of the wave equation in the special case of purely

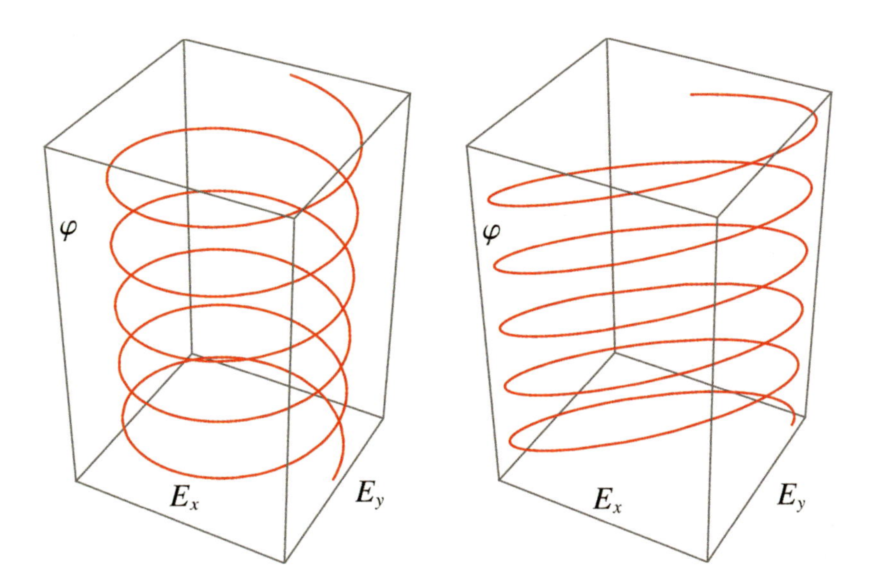

Figure 4.2: The path traced out by the end–point of the electric field vector as the phase increases. In the left picture the path is a circular spiral (circular polarization); in the right picture the path is an elliptical spiral (elliptical polarization).

isotropic, impedance–matched media. Such media establish Fermat's geometry (1.1) in 3–dimensional space:

$$g_{ij} = n^2\,\delta_{ij}\,, \quad g^{ij} = \frac{1}{n^2}\,\delta_{ij}\,, \quad g = n^6\,. \tag{29.17}$$

We are in Cartesian coordinates in physical space, but with the conformally flat geometry (29.17) for electromagnetic waves. The upper–index field E^i thus is $n^{-2}E_i$. We get from the transversality condition (29.5) with expression (17.16) for the divergence:

$$0 = \partial_i \sqrt{g}\, g^{ij} E_j = \partial_j n E_j\,, \tag{29.18}$$

where we sum over repeated indices, although they are not always upper and lower. From relationship (29.18) follows

$$0 = E_j \partial_j n + n \partial_j E_j \quad \text{and so} \quad \partial_j E_j = -E_j \frac{\partial_j n}{n} = -(\partial_j \ln n) E_j\,. \tag{29.19}$$

We obtain from Maxwell's equations (26.1) along the same lines as in the derivation (29.2), and using $\epsilon^{ijk} = n^{-3}[ijk]$ and formula (14.15):

$$\frac{1}{c^2}\frac{\partial^2 E_i}{\partial t^2} = -\frac{1}{n}\,[ijk]\,\partial_j \frac{1}{n}\,[klm]\,\partial_l E_m = \frac{1}{n}\left(\partial_j \frac{1}{n} \partial_j E_i - \partial_j \frac{1}{n} \partial_i E_j\right). \tag{29.20}$$

We re–write the right–hand side:

$$\frac{1}{c^2}\frac{\partial^2 E_i}{\partial t^2} = \frac{1}{n}\left(\partial_j \frac{1}{n} \partial_j E_i + \frac{1}{n^2}(\partial_j n)\,\partial_i E_j - \frac{1}{n}\partial_i \partial_j E_j\right) \tag{29.21}$$

and apply the transversality condition (29.19) in the last term:

$$\partial_i \partial_j E_j = -\partial_i (\partial_j \ln n) E_j = -(\partial_i \partial_j \ln n) E_j - (\partial_j \ln n)\, \partial_i E_j \,. \tag{29.22}$$

We write the other two terms in Eq. (29.21) as

$$\frac{1}{n^2}(\partial_j n)\, \partial_i E_j = \frac{1}{n}(\partial_j \ln n)\, \partial_i E_j \,, \tag{29.23}$$

$$\partial_j \frac{1}{n} \partial_j E_i = \frac{1}{n}\left[\partial_j \partial_j E_i - (\partial_j \ln n) \partial_j E_i\right] , \tag{29.24}$$

and, combining the terms (29.22)–(29.24) and multiplying by n^2, arrive at the wave equation

$$\frac{n^2}{c^2}\frac{\partial^2 E_i}{\partial t^2} = \partial_j \partial_j E_i - (\partial_j \ln n)\, \partial_j E_i + 2(\partial_j \ln n)\, \partial_i E_j + (\partial_i \partial_j \ln n) E_j \,. \tag{29.25}$$

This fairly complicated equation is encoded in the concise geometric form (29.4). Note that for anisotropic media, the explicit wave equation is significantly more convoluted. The first derivative in the explicit wave equation (29.25) stems from the covariant differentiation in the Laplacian $\nabla^j \nabla_j E_i$ in the geometric formulation (29.4). The $(\partial_i \partial_j \ln n) E_j$ term originates from the curvature contribution; further terms from the Laplacian cancel the other terms in the Ricci tensor (23.6). We see that the curvature term in the wave equation (29.25) and also in the general wave equation (29.4) directly couples the field components. As the components of the electric field define the polarization of the electromagnetic wave, a non–zero Ricci tensor in the wave equation describes polarization mixing; the polarization state of light may not be maintained in curved geometries.

For non–Euclidean geometries, plane waves (29.6) are no longer exact solutions of the wave equation (29.4). An incident plane wave, coming from regions where the medium is uniform, is thus scattered into several outgoing plane waves. It is often stated that impedance–matching eliminates scattering, but here we see that even impedance–matched media may scatter light, when they establish non–Euclidean geometries with non–vanishing curvature. The Riemann tensor (21.8) quantifies the degree of curvature, irrespective of the coordinates. When the Riemann tensor vanishes the geometry is flat—the apparent curvature of light rays in physical space is an illusion; physical space can be transformed into empty virtual space with Euclidean geometry in Cartesian coordinates: the material makes a transformation medium. As we discussed in §21, the Riemann tensor in three dimensions vanishes if and only if the Ricci tensor vanishes,

$$R_{ij} = 0 \,. \tag{29.26}$$

In this case, the wave equation (29.4) reduces to the *vector Helmholtz equation*

$$\nabla^j \nabla_j E_i - \frac{1}{c^2}\frac{\partial^2 E_i}{\partial t^2} = 0 \,. \tag{29.27}$$

In virtual space, we replace the covariant derivatives by partial derivatives (in Cartesian coordinates), because there the metric is Euclidean. Consequently, in virtual space

$$(\partial^{j'}\partial_{j'})E_{i'} - \frac{1}{c^2}\frac{\partial^2 E_{i'}}{\partial t^2} = 0 \tag{29.28}$$

with the plane–wave solutions

$$E_{i'} = \mathcal{E}_{i'} \exp(\mathrm{i}\boldsymbol{k}\cdot\boldsymbol{r}' - \mathrm{i}\omega t)\,. \tag{29.29}$$

In physical space, we obtain the electric field by transforming the plane waves (29.29) with the coordinate transformation that defines the medium, replacing $\boldsymbol{r}'$ by the function $\boldsymbol{r}'(\boldsymbol{r})$. But note that we must also transform the amplitude $\mathcal{E}'_i$, because the electric field is a one–form. Hence in physical space, the electromagnetic waves (29.29) appear as the modulated plane waves

$$E_i = \Lambda^{i'}_{\ i}\,\mathcal{E}_{i'} \exp(\mathrm{i}\boldsymbol{k}\cdot\boldsymbol{r}'(\boldsymbol{r}) - \mathrm{i}\omega t)\,, \quad \Lambda^{i'}_{\ i} = \frac{\partial x^{i'}}{\partial x^i}\,. \tag{29.30}$$

As we can describe all real solutions of the wave equation in empty virtual space as superpositions (29.10) of plane waves, all electromagnetic waves in spatial transformation media are superpositions of the modulated plane waves (29.30) in physical space. It is thus sufficient to consider only plane waves in transformation optics. Furthermore, an incident plane wave may be modulated while it propagates through the transformation medium, but when it leaves the optical material the modulation ceases and the wave turns into an ordinary plane wave, as if the medium were not there in the first place. The medium is invisible and does not scatter light. Some space–time transformation media, however, cause scattering in situations when the topologies of virtual and physical space differ from each other. In §38 and §39 we discuss two examples of such topological scattering, the optical Aharonov–Bohm effect and Hawking radiation at horizons.

Exercise 29.3

In this Exercise we solve one of the basic problems of electromagnetic theory: the behaviour of an electromagnetic plane wave at the boundary between two isotropic, homogeneous dielectrics. Let the plane wave propagate from medium 1, with permittivity ε_1 and permeability μ_1, into medium 2, with permittivity ε_2 and permeability μ_2, across a sharp boundary between the media (Fig. 4.3). We take the boundary to be located at $y = 0$, with medium 1 in the region $y < 0$ and medium 2 in the region $y > 0$. Consider first an incident plane wave in medium 1. The electric displacement and magnetic field strength in medium 1 are $\boldsymbol{D}_1 = \varepsilon_0\varepsilon_1\boldsymbol{E}_1$ and $\boldsymbol{H}_1 = \boldsymbol{B}_1/(\mu_0\mu_1)$, respectively. By a similar procedure used to derive the wave equation (29.4), show that the macroscopic Maxwell equations (26.4) without charges or currents give the following wave equation for the electric field in medium 1, in Cartesian coordinates:

$$\partial^i\partial_i\boldsymbol{E}_1 - \frac{\varepsilon_1\mu_1}{c^2}\frac{\partial^2\boldsymbol{E}_1}{\partial t^2} = 0\,. \tag{29.31}$$

This equation has the plane–wave solution

$$\boldsymbol{E}_1 = \boldsymbol{\mathcal{E}}_1 \exp\left(\mathrm{i}\boldsymbol{k}_1\cdot\boldsymbol{r} - \mathrm{i}\omega t\right)\,, \quad k_1 \equiv |\boldsymbol{k}_1| = \frac{\sqrt{\varepsilon_1\mu_1}\,\omega}{c}\,. \tag{29.32}$$

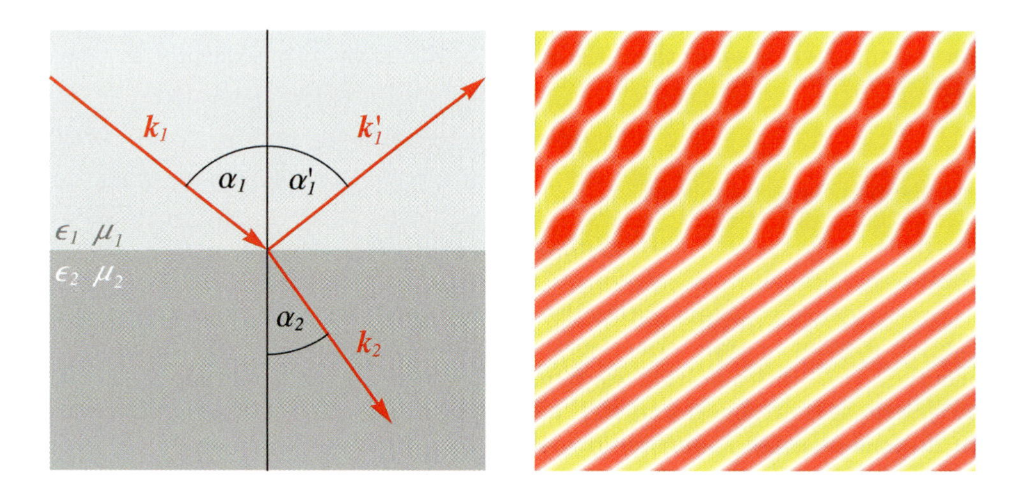

Figure 4.3: A plane wave at the boundary between two isotropic, homogeneous media. The left figure shows the wave vectors of the incident ($\boldsymbol{k}_1$), reflected ($\boldsymbol{k}_1'$) and transmitted ($\boldsymbol{k}_2$) waves. The right figure shows the wave amplitude, with the colours yellow, red and white representing a positive, negative and zero amplitude, respectively; the interference between the incident and reflected waves is visible in the wavy pattern in the upper region of the picture.

By substituting the plane wave (29.32) into the macroscopic Maxwell equations (26.4) without charges or currents, show that

$$\boldsymbol{k}_1 \cdot \boldsymbol{\mathcal{E}}_1 = 0\,, \quad \boldsymbol{B}_1 = \boldsymbol{\mathcal{B}}_1 \exp\left(\mathrm{i}\boldsymbol{k}_1 \cdot \boldsymbol{r} - \mathrm{i}\omega t\right)\,, \quad \boldsymbol{\mathcal{B}}_1 = \frac{\sqrt{\varepsilon_1 \mu_1}}{ck}\, \boldsymbol{k}_1 \times \boldsymbol{\mathcal{E}}_1\,, \quad \boldsymbol{k}_1 \cdot \boldsymbol{\mathcal{B}}_1 = 0\,. \tag{29.33}$$

A plane wave in medium 2 is obviously given by Eqs. (29.32) and (29.33) with $\varepsilon_1 \to \varepsilon_2$, $\mu_1 \to \mu_2$. The plane–wave solutions in each medium must be matched together at the boundary. We have assumed an idealized situation where the boundary between the two media is infinitely sharp, with a discontinuous permittivity and/or permeability at the boundary $y = 0$. Maxwell's equations contain derivatives and we must be careful in regard to derivatives with respect to y, since the dielectric functions are discontinuous in this direction. For clarity let us replace the infinitely sharp boundary by a very narrow slab centred at $y = 0$, inside which the permittivity and permeability change continuously from ε_1, μ_1 to ε_2, μ_2. In this case there is no problem taking derivatives in the y–direction and we demand that the macroscopic Maxwell equations (26.4) (with no external charges or currents) are satisfied everywhere. Consider Gauss's law $\boldsymbol{\nabla} \cdot \boldsymbol{D} = 0$; by integrating this with respect to y from y_1 to y_2 we obtain

$$D_y\big|_{y=y_2} - D_y\big|_{y=y_1} = -\int_{y_1}^{y_2} \left(\partial_x D_x + \partial_z D_z\right) \mathrm{d}y\,. \tag{29.34}$$

Now let y_1 lie on one edge of the narrow slab discussed above, the edge where the dielectric functions take the values $\varepsilon = \varepsilon_1$, $\mu = \mu_1$, and let y_2 lie on the other edge of the narrow slab, where $\varepsilon = \varepsilon_2$, $\mu = \mu_2$. Consider the limit in which the narrow slab becomes infinitesimally thin, and the boundary becomes infinitely sharp; this limit implies $y_1 \to y_2$ so that the right–hand side of (29.34) goes to zero. But from the left–hand side of (29.34) we see that this in turn implies that D_y on each side of the sharp boundary must be the same, i.e. D_y is continuous across the

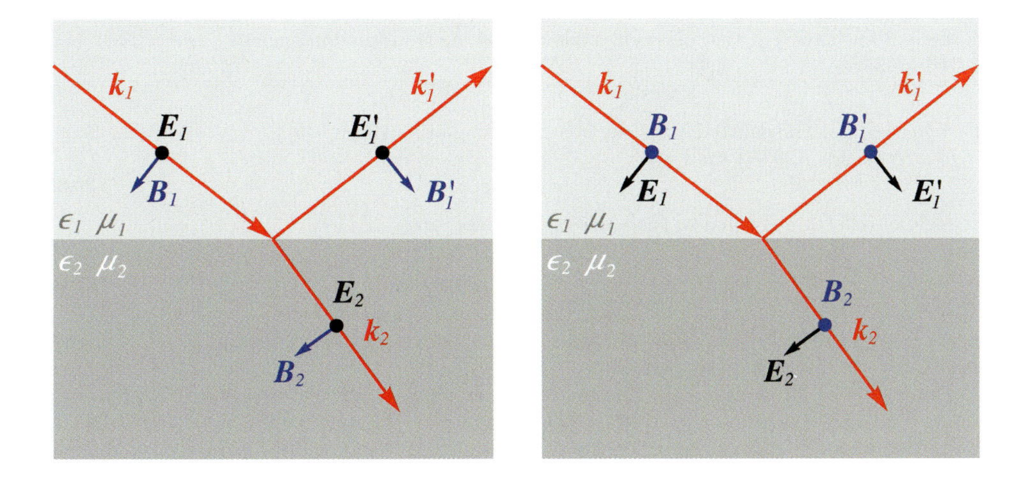

Figure 4.4: Electric and magnetic fields of the incident, reflected and transmitted waves in Fig. 4.3. The wave vectors $\boldsymbol{k}_1$, $\boldsymbol{k}_1'$ and $\boldsymbol{k}_2$ lie in a plane, the plane of incidence. In the left figure the waves have electric field orthogonal to the plane of incidence ($\boldsymbol{E}$–field points out of the page); in the right figure the magnetic field is orthogonal to the plane of incidence ($\boldsymbol{B}$–field points into the page).

sharp boundary. A similar analysis can be applied to derivatives of field components with respect to y in the remaining Maxwell equations. In this manner one finds that the field components in Maxwell's equations that are differentiated with respect to y must be continuous across the sharp boundary (Stratton [1941]). By inspecting the macroscopic Maxwell equations (26.4) in Cartesian coordinates, verify that the conditions on the field components at the boundary $y = 0$ are thus

$$\text{continuity of} \quad D_y,\ E_x,\ E_z,\ B_y,\ H_x,\ H_z\,. \tag{29.35}$$

Consider the plane wave (29.32) in medium 1, incident on the boundary $y = 0$. This wave can be reflected at the boundary, giving another wave $\boldsymbol{E}_1' = \boldsymbol{\mathcal{E}}_1' \exp(\mathrm{i}\boldsymbol{k}_1' \cdot \boldsymbol{r} - \mathrm{i}\omega t)$ in medium 1, and it can also be transmitted into medium 2, giving a plane wave $\boldsymbol{E}_2 = \boldsymbol{\mathcal{E}}_2 \exp(\mathrm{i}\boldsymbol{k}_2 \cdot \boldsymbol{r} - \mathrm{i}\omega t)$ in the region $y > 0$ (in general both reflection and transmission will occur). The frequencies of the reflected and transmitted waves must be the same as that of the incident wave (29.32), because the conditions (29.35) hold at the boundary at all times, which is only possible if the waves on either side are oscillating at the same frequency. In addition, the conditions (29.35) hold at all points on the boundary plane $y = 0$, so the changing phases of the waves on either side must agree as one moves along the boundary, i.e.

$$\boldsymbol{k}_1 \cdot \boldsymbol{r}|_{y=0} = \boldsymbol{k}_1' \cdot \boldsymbol{r}|_{y=0} = \boldsymbol{k}_2 \cdot \boldsymbol{r}|_{y=0}\,. \tag{29.36}$$

As $\boldsymbol{r} = \{x, y, z\}$, and Eqs. (29.36) hold for all values of x and z, it follows that the 2–dimensional vectors obtained by projecting the wave vectors $\boldsymbol{k}_1$, $\boldsymbol{k}_1'$ and $\boldsymbol{k}_2$ to the xz–plane at $y = 0$ are equal; this means that the 3–dimensional vectors $\boldsymbol{k}_1$, $\boldsymbol{k}_1'$ and $\boldsymbol{k}_2$ lie in the same plane. Light rays follow the wave vectors, so we have obtained one of the laws of reflection and transmission for isotropic, homogeneous media: the incident, reflected and transmitted rays lie in the same plane. This result is obvious from symmetry considerations, but here we have derived it from Maxwell's equations. But there is much more information in Eqs. (29.36). The components (29.36) of the three wave vectors $\boldsymbol{k}_1$, $\boldsymbol{k}_1'$ and $\boldsymbol{k}_2$ in the boundary plane $y = 0$ are in each case equal to the magnitude of the vector times the sine of the angle between the vector and a normal to the boundary; these angles are the incident angle α_1, the reflection angle α_1' and the transmission angle α_2, respectively (Fig. 4.3), so we have

$$k_1 \sin\alpha_1 = k_1' \sin\alpha_1' = k_2 \sin\alpha_2 \,. \tag{29.37}$$

The reflected and transmitted wave vectors have magnitudes $k_1' \equiv |\boldsymbol{k}_1'| = k_1$ and $k_2 \equiv |\boldsymbol{k}_2| = \sqrt{\varepsilon_2\mu_2}\,\omega/c$, so Eqs. (29.37) show that

$$\alpha_1 = \alpha_1' \,, \tag{29.38}$$

the angle of reflection is equal to the angle of incidence, and

$$\frac{\sin\alpha_1}{\sin\alpha_2} = \frac{\sqrt{\varepsilon_2\mu_2}}{\sqrt{\varepsilon_1\mu_1}} = \frac{n_2}{n_1} \,, \tag{29.39}$$

where $n_1 = \sqrt{\varepsilon_1\mu_1}$ and $n_2 = \sqrt{\varepsilon_2\mu_2}$ are the refractive indices of the two media. We have derived Snell's law (1.3) from Maxwell's equations.

The results so far can be stated in terms of light rays. A complete solution of the problem requires use of the boundary conditions (29.35) to relate the amplitudes of the reflected and transmitted waves to that of the incident wave. It is simplest to consider separately two different polarizations of the incident wave (Fig. 4.4); because of the linearity of Maxwell's equations (and the boundary conditions), an arbitrary linear sum of these two polarizations is the solution for a general plane wave. The first polarization we consider is one where the electric field is orthogonal to the plane in which all three wave vectors $\boldsymbol{k}_1$, $\boldsymbol{k}_1'$ and $\boldsymbol{k}_2$ lie, the plane of incidence (Fig. 4.4). At the outset we have not proved that a solution exists for this polarization (perhaps the interaction with the boundary changes the polarization of the incident wave); we will find that we can satisfy the boundary conditions (29.35) with all waves having this polarization, which will prove that this is a solution of Maxwell's equations. Without loss of generality we can rotate the x– and z–axes of our Cartesian system so that the plane of incidence is the xy–plane and then the electric field has an z–component only, orthogonal to the plane of incidence. From the first of Eqs. (29.33) follows that the three wave vectors have zero z–component; making use also of the result $\alpha_1 = \alpha_1'$ above, and the equality (29.36) of the wave–vector components in the boundary plane, we obtain for the wave vectors

$$\begin{aligned} \boldsymbol{k}_1 = (k_{1x}, k_{1y}, 0)\,, \quad \boldsymbol{k}_1' = (k_{1x}, -k_{1y}, 0)\,, \\ \boldsymbol{k}_2 = (k_{1x}, k_{2y}, 0)\,, \quad k_{2y} = \sqrt{\frac{\varepsilon_2\mu_2\omega^2}{c^2} - k_{1x}^2}\,. \end{aligned} \tag{29.40}$$

Calculate the magnetic induction $\boldsymbol{B}$ of the three plane waves using Eqs. (29.33) and show that the boundary conditions (29.35) reduce to

$$\mathcal{E}_1 + \mathcal{E}_1' = \mathcal{E}_2 \,, \tag{29.41}$$

$$\sqrt{\varepsilon_1\mu_1}\,(\mathcal{E}_1 + \mathcal{E}_1')\sin\alpha_1 = \sqrt{\varepsilon_2\mu_2}\,\mathcal{E}_2 \sin\alpha_2 \,, \tag{29.42}$$

$$\sqrt{\frac{\varepsilon_1}{\mu_1}}\,(\mathcal{E}_1 - \mathcal{E}_1')\cos\alpha_1 = \sqrt{\frac{\varepsilon_2}{\mu_2}}\,\mathcal{E}_2 \cos\alpha_2 \,. \tag{29.43}$$

Use of Snell's law (29.39) shows that the second boundary condition (29.42) is the same as the first (29.41). Equations (29.41) and (29.43) provide the two conditions that allows us to find the amplitudes $\mathcal{E}_1' =$ and $\mathcal{E}_2$ of the reflected and transmitted waves in terms of that of the incident wave ($\mathcal{E}_1$); show that, with use of Snell's law (29.39), the result is

$$\frac{\mathcal{E}_1'}{\mathcal{E}_1} = \frac{\mu_2 k_{1z} - \mu_1 k_{2z}}{\mu_2 k_{1z} + \mu_1 k_{2z}} = \frac{\mu_2 n_1 \cos\alpha_1 - \mu_1\sqrt{n_2^2 - n_1^2\sin^2\alpha_1}}{\mu_2 n_1 \cos\alpha_1 + \mu_1\sqrt{n_2^2 - n_1^2\sin^2\alpha_1}} \,, \tag{29.44}$$

$$\frac{\mathcal{E}_2}{\mathcal{E}_1} = \frac{2\mu_2 k_{1z}}{\mu_2 k_{1z} + \mu_1 k_{2z}} = \frac{2\mu_2 n_1 \cos\alpha_1}{\mu_2 n_1 \cos\alpha_1 + \mu_1\sqrt{n_2^2 - n_1^2\sin^2\alpha_1}} \,. \tag{29.45}$$

Equations (29.44) and (29.45) are the *Fresnel reflection and transmission coefficients* for polarization orthogonal to the plane of incidence.

The second polarization we consider is one where the electric field lies in the plane of incidence (Fig. 4.4) and we prove that this is a solution to the problem by successfully imposing the boundary conditions. We again take the plane of incidence to be the yz–plane so that the wave vectors are still described by Eqs. (29.40). In this case the first of Eqs. (29.33) implies the incident electric–field amplitude is given by $\boldsymbol{\mathcal{E}}_1 = \mathcal{E}_1(-k_{1y}/k_1, k_{1x}/k_1, 0)$. The magnetic induction $\boldsymbol{B}_1$ now has only an x–component, as is seen from Eqs. (29.33); we assume that the amplitude $\boldsymbol{\mathcal{B}}_1$ is in the same direction for the reflected and transmitted waves. Show that in this case the conditions (29.35) give

$$\varepsilon_1(\mathcal{E}_1 + \mathcal{E}_1') \sin\alpha_1 = \varepsilon_2 \mathcal{E}_2 \sin\alpha_2\,, \tag{29.46}$$

$$(\mathcal{E}_1 - \mathcal{E}_1') \cos\alpha_1 = \mathcal{E}_2 \cos\alpha_2\,, \tag{29.47}$$

$$\sqrt{\frac{\varepsilon_1}{\mu_1}}\,(\mathcal{E}_1 + \mathcal{E}_1') = \sqrt{\frac{\varepsilon_2}{\mu_2}}\,\mathcal{E}_2\,. \tag{29.48}$$

Equation (29.46) combined with Snell's law (29.39) gives Eq. (29.48). Solve Eqs. (29.47) and (29.48) for the reflected and transmitted amplitudes and show that

$$\frac{\mathcal{E}_1'}{\mathcal{E}_1} = \frac{\varepsilon_2 k_{1z} - \varepsilon_1 k_{2z}}{\varepsilon_2 k_{1z} + \varepsilon_1 k_{2z}} = \frac{\varepsilon_2 n_1 \cos\alpha_1 - \varepsilon_1\sqrt{n_2^2 - n_1^2\sin^2\alpha_1}}{\varepsilon_2 n_1 \cos\alpha_1 + \varepsilon_1\sqrt{n_2^2 - n_1^2\sin^2\alpha_1}}\,, \tag{29.49}$$

$$\frac{\mathcal{E}_2}{\mathcal{E}_1} = \frac{2\sqrt{\dfrac{\varepsilon_1}{\mu_1}}\,n_2 k_{1z}}{\varepsilon_2 k_{1z} + \varepsilon_1 k_{2z}} = \frac{2\varepsilon_1 n_2 \cos\alpha_1}{\varepsilon_2 n_1 \cos\alpha_1 + \varepsilon_1\sqrt{n_2^2 - n_1^2\sin^2\alpha_1}}\,. \tag{29.50}$$

Equations (29.44) and (29.45) are the Fresnel reflection and transmission coefficients for polarization in the plane of incidence.

A general linearly polarized wave is a superposition of waves with the two polarizations we have considered. As the reflection and transmission coefficients are different for the two basis polarizations we see that the polarization of a plane wave in general changes when it is reflected and transmitted at a sharp boundary. Note that even in the case where the media are impedance–matched, $(\varepsilon_1/\mu_1) = (\varepsilon_2/\mu_2)$, there is still reflection of the wave at the boundary for general incidence angle, and also a change of polarization (unless the wave has one of the two polarizations treated above). Show that for perpendicular incidence $\alpha_1 = 0$ there is no reflected wave if the media are impedance–matched. Show also that that for perpendicular incidence the amplitude of the electric field changes sign on reflection if $\sqrt{\varepsilon_2/\mu_2} > \sqrt{\varepsilon_1/\mu_1}$, corresponding to a phase change of π (since $-1 = \mathrm{e}^{\mathrm{i}\pi}$). If $\mu_2 = \mu_1 = 1$, the condition for phase reversal at perpendicular incidence becomes $n_2 > n_1$.

A consequence of Snell's law (1.3) familiar from ray optics is *total internal reflection*: if $n_2 < n_1$ there exists an angle of incidence α_1 such that the refracted angle α_2 is $\pi/2$. In this case light cannot be transmitted into medium 2 and this critical angle is the threshold for total internal reflection. Show from relation (29.40) that $k_{2y} = (\omega/c)(n_2^2 - n_1^2\sin^2\alpha_1)^{1/2}$. At the critical angle, Snell's law implies that $k_{2y} = 0$, and for greater angles of incidence k_{2y} is imaginary. An imaginary value of k_{2y} gives an exponentially decaying transmitted wave, $\boldsymbol{E}_2 \propto \exp(-|k_{2y}|y)$; no energy is therefore transmitted into medium 2, all of the incident energy is reflected at the boundary—we have total reflection.

Exercise 29.4

In §26 we argued that Maxwell's equations (26.1) hold also in curved space, although we only postulated them in flat space, because we can imagine curved space as a patchwork of locally flat pieces and require Maxwell's equations on each piece. Transforming to the global frame of the curved space gives us the equations (26.1). What happens if we follow the same procedure for the wave equation? Suppose we postulate the wave equation of flat space in Cartesian coordinates, the vector Helmholtz equation (29.27), on each locally flat patch. What would we get for the wave equation? Why does the actual wave equation (29.4) differ from this result?

Solution

As the vector Helmholtz equation (29.27) is invariant under coordinate transformations, we obtain this as the wave equation in curved space by postulating it in every locally flat patch. The vector Helmholtz equation (29.27) differs from the true wave equation (29.4) by the curvature term $R_{ij}E^j$. The reason for the missing term is the following: the wave equation is of second order, as it contains the Laplacian and the second time derivative. In curved space, the second derivative of the metric cannot be made to vanish at a general point, for otherwise the Riemann tensor (21.8) would vanish (see p. 140). Our procedure thus amounts to neglecting local curvature (but it may still represent a good approximation if the curvature is small).

---

§30. Geometrical optics

Impedance–matched media establish geometries and geometries appear as impedance–matched media. In general, these media are optically anisotropic, but we still expect that light rays follow a form of Fermat's principle of the extremal optical path. We anticipate that the metric g_{ij} measures the optical path length and that light rays follow geodesics. How do we deduce Fermat's principle from Maxwell's equations? How is the amplitude transported? And when are light rays a useful concept at all, given that light is a wave?

Ray optics is called *geometrical optics*. Rays are used to illustrate the propagation of light, but geometrical optics is far more than ray tracing: it is an approximative wave theory, making approximate statements about full electromagnetic waves. The waves are regarded as modulated plane waves with a slowly varying amplitude and a rapidly oscillating phase. In some cases, geometrical optics is exact; in particular, the modulated plane waves (29.30) in transformation optics are exactly the waves in the description of geometrical optics. As one can construct any other electromagnetic waves, however complicated, as superpositions of these modulated plane waves, it is completely sufficient to consider such transformation media within geometrical optics. In transformation optics, geometrical optics is all we need.

Let us quantify the conditions for geometrical optics. Throughout this Section we use Cartesian coordinates in physical space, but consider the medium as establishing the optical geometry (26.12). For mathematical convenience, we regard the electromagnetic field strengths as complex quantities oscillating at positive frequencies, similar to the superposition (29.9) of plane waves (29.6)—we can always add the complex conjugate part of the field that is oscillating with negative frequencies. We assume that both the electric and the magnetic fields oscillate with a common phase φ such that

$$E_i = \mathcal{E}_i\,\mathrm{e}^{\mathrm{i}\varphi}\,, \quad H_i = \mathcal{H}_i\,\mathrm{e}^{\mathrm{i}\varphi}\,. \tag{30.1}$$

The field amplitudes $\mathcal{E}_i$ and $\mathcal{H}_i$ are complex. Furthermore, we assume that the variations of the phase, describing the oscillations of the field, are much more rapid than any changes in the amplitudes and the medium. We also assume that the overall phase φ should progress in time with the constant angular frequency

$$\omega = -\frac{\partial\varphi}{\partial t} = \text{const}\,. \tag{30.2}$$

In space, the phase propagates with wave vector

$$k_i = \nabla_i\varphi \quad \text{or, in index–free notation,} \quad \boldsymbol{k} = \boldsymbol{\nabla}\varphi\,. \tag{30.3}$$

Note that k_i is not a vector, but a one–form, but the term wave vector is normally used. From Eqs. (30.2) and (30.3) follows the structure of the phase:

$$\varphi = \varphi_0(\boldsymbol{r}) - \omega t\,, \quad \varphi_0 = \int \boldsymbol{k}\cdot \mathrm{d}\boldsymbol{r}\,. \tag{30.4}$$

The wave number k is the modulus of the wave vector $\boldsymbol{k}$, and the wavelength λ and period T are given by relations (4.1). In isotropic media, the wavenumber depends on the refractive index according to the dispersion relation (4.2); in general k will depend on the optical properties of the medium and k must vary when they vary. For geometrical optics to be a good approximation, we assume that the wavelength does not vary much over short distances,

$$|\nabla\lambda| \ll 1\,. \tag{30.5}$$

For having approximate plane waves, we also need to require that the effective curvature of the medium is small on the scale of the wavelength. As the curvature tensor is constructed from second spatial derivatives of the metric, R_{ij} carries the units of an inverse square length in Cartesian coordinates. We thus require

$$|R_{ij}|\lambda^2 \ll 1\,. \tag{30.6}$$

The conditions (30.5) and (30.6) roughly quantify the validity range of geometrical optics. Note that both conditions (30.5) and (30.6) are invariant under coordinate transformation: in the simplest case, if we locally shrink space by a factor n, the wavelength shrinks by n, derivatives increase by n and the curvature increases by n^2, so the gradient $\nabla\lambda$ and the product $|R_{ij}|\lambda^2$ remain the same. If geometrical optics is valid in virtual space (that may not be empty), geometrical optics must also be valid in physical space.

Assume that the electromagnetic wave has the structure (30.1) and that the conditions (30.5) and (30.6) for geometrical optics are met. We calculate the Laplacian and the time derivatives that occur in the wave equation (29.4):

$$\begin{aligned}
\nabla^j\nabla_j E_i &= \nabla^j \mathrm{e}^{\mathrm{i}\varphi}\left(\nabla_j\mathcal{E}_i + \mathrm{i}k_j\mathcal{E}_i\right) \\
&= \mathrm{e}^{\mathrm{i}\varphi}\left(\nabla^j\nabla_j\mathcal{E}_i + 2\mathrm{i}k^j\nabla_j\mathcal{E}_i + \mathrm{i}\mathcal{E}_i\nabla_j k^j - k^j k_j\mathcal{E}_i\right)\,, \qquad (30.7)\\
\frac{\partial^2 E_i}{\partial t^2} &= \mathrm{e}^{\mathrm{i}\varphi}\left(\frac{\partial^2\mathcal{E}_i}{\partial t^2} - 2\mathrm{i}\omega\frac{\partial\mathcal{E}_i}{\partial t} - \omega^2\mathcal{E}_i\right)\,. \qquad (30.8)
\end{aligned}$$

In the regime of geometrical optics, the dominant contribution to the wave equation (29.4) comes from the highest powers of the derivatives of the phase, $k^j k_j$ and ω^2, because the phase is by far the most rapidly evolving physical feature of the wave. Hence in the wave equation (29.4) we ignore the Ricci tensor and approximate the Laplacian (30.7) by $-\mathrm{e}^{\mathrm{i}\varphi}k^j k_j \mathcal{E}_i$ and the second time derivative (30.8) by $-\mathrm{e}^{\mathrm{i}\varphi}\omega^2\mathcal{E}_i$. Requiring the wave equation (29.4) to be satisfied for all amplitudes $\mathcal{E}_i$ we thereby obtain the dispersion relation:

$$k^j k_j = \frac{\omega^2}{c^2}\,. \tag{30.9}$$

It is instructive to write the dispersion relation (30.9) explicitly in terms of the metric tensor (26.10) of the optical geometry,

$$g^{ij}k_i k_j = \frac{\omega^2}{c^2}\,, \tag{30.10}$$

and to express relation (30.10) in terms of the electric permittivity tensor $\boldsymbol{\varepsilon}$ and the refractive index $\mathbf{n}$ in matrix notation. We obtain from Eqs. (26.11) and (26.19):

$$\omega^2 = c^2(\det\boldsymbol{\varepsilon})^{-1}\boldsymbol{k}\cdot\boldsymbol{\varepsilon}\boldsymbol{k} = c^2\boldsymbol{k}\cdot\mathbf{n}^{-2}\boldsymbol{k}\,. \tag{30.11}$$

Formula (30.11) is the expression of the refractive index in an anisotropic medium with $\boldsymbol{\varepsilon} = \boldsymbol{\mu}$ (Born and Wolf [1999]). Hence it confirms the geometrical definition of the refractive index we adopted in §26. Equation (30.11) also shows that $\boldsymbol{k}$ must vary in space if the medium with dielectric tensor $\boldsymbol{\varepsilon} = \boldsymbol{\mu}$ varies. The form (30.10) of the dispersion relation reveals how the wave vector is related to the frequency in geometrical terms; the norm of k_i with respect to the optical geometry established by the medium is ω/c. The original dispersion relation (30.9) implies that light rays follow geodesics of the optical geometry, as we now show. We obtain by covariant differentiation

$$0 = \nabla_i(k^j k_j) = 2k^j k_{j;i} = 2k^j\nabla_i\nabla_j\varphi\,. \tag{30.12}$$

For scalars φ covariant derivatives commute (see Eq. (21.31)), so

$$\nabla_i\nabla_j\varphi = \nabla_j\nabla_i\varphi = \nabla_j k_i\,, \tag{30.13}$$

and therefore we obtain from Eq. (30.12)

$$k^j k_{i;j} = 0 \quad \text{or, equivalently,} \quad k^j k^i{}_{;j} = 0\,. \tag{30.14}$$

We understand a light ray as a curve in space that always follows the wave vector; we thus define a ray as a curve $x^i(\zeta)$ with parameter ζ to which the wave vector is tangent, i.e.

$$\frac{\mathrm{d}x^i}{\mathrm{d}\zeta} = k^i\,. \tag{30.15}$$

Adopting definition (30.15) for light rays $x^i(\zeta)$, we see that Eq. (30.14) agrees with the geodesic equation (19.16) with k^i as the tangent vector U^i: light rays are the geodesics of the optical geometry (26.12). Geodesics are defined as the paths with extremal length in a given geometry; so light rays follow the shortest (or the longest) optical path: light obeys Fermat's principle. We have deduced Fermat's principle from Maxwell's equations for impedance–matched media.

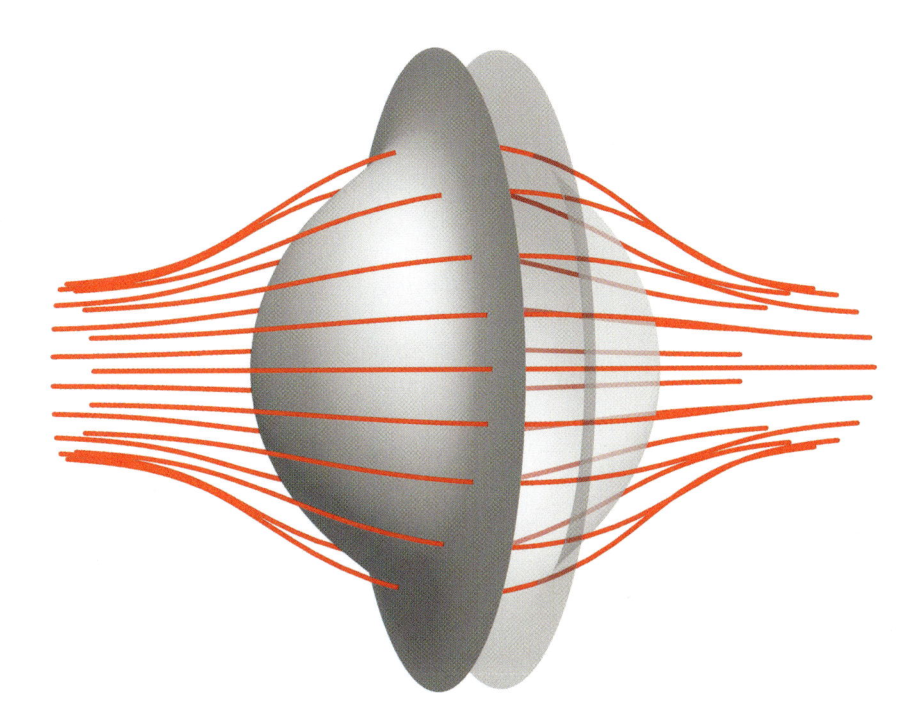

Figure 4.5: Light rays (red lines) and surfaces of constant phase (grey) in an anisotropic medium. The wave vector k^i is tangent to the rays, but it is not equal to the one–form k_i, which is the gradient of the phase. This difference between the vector and the one–form in the effective geometry means that the rays are not orthogonal to the phase surfaces in the medium. The anisotropic medium shown is a cloaking device (Fig. 5.3) to be described in §33.

Note that the wave vector must appear with upper index in the definition (30.15) of light rays, because $\mathrm{d}x^i$ is a spatial vector. On the other hand, k_i is a one–form, the gradient (30.3) of the phase φ. Thus the wave vector appears as two distinct geometrical objects: (1) as the phase gradient k_i, a one–form, and (2) as the ray direction k^i, a vector, the *propagation vector*. They are not the same, but they are related by

$$k^i = g^{ij} k_j \,. \tag{30.16}$$

In physical space, k_i and k^i do not point in the same direction, unless the optical geometry is conformally flat, which is the case for isotropic media. We conclude that in anisotropic optical materials light rays do not directly follow the phase gradients, but rather phase gradients turned from one–forms into vectors (Fig. 4.5).

We previously considered light rays as solutions of Hamilton's equation (4.8). Let us establish how our new definition (30.15) is consistent with Hamilton's ray optics. The variables describing a light ray are the position $\boldsymbol{r}$ and the wave vector $\boldsymbol{k}$ or, to be geometrically precise, the position x^i and the wave one–form k_i. Hamilton's equations (4.8) read in our notation

$$\frac{\mathrm{d}x^i}{\mathrm{d}t} = \frac{\partial\omega}{\partial k_i}\,, \quad \frac{\mathrm{d}k_i}{\mathrm{d}t} = -\frac{\partial\omega}{\partial x^i}\,. \tag{30.17}$$

We prove that Hamilton's equations (30.17) follow from the geodesic equation (30.14) of light rays (30.15) with the frequency ω given by the dispersion relation (30.9):

$$\omega = c\sqrt{g^{ij}(\boldsymbol{r})k_i k_j}\,. \tag{30.18}$$

Hamilton's equations (30.17) contain partial derivatives of the frequency with respect to x^i and k_i. We differentiate ω with respect to k_i and obtain

$$\frac{\mathrm{d}x^i}{\mathrm{d}t} = \frac{\partial\omega}{\partial k_i} = \frac{ck^i}{\sqrt{g^{lj}k_l k_j}} = \frac{c^2}{\omega}\,k^i\,. \tag{30.19}$$

Equation (30.19) reveals that we should require for ζ in the definition (30.15) of light rays:

$$t = \frac{\omega}{c^2}\,\zeta\,. \tag{30.20}$$

We have thus deduced how the parameter ζ of the ray trajectory is related to the time parameter t in Hamilton's ray optics. Now consider the derivative of the frequency with respect to x^i. We obtain along similar lines as before

$$\frac{\partial\omega}{\partial x^i} = \frac{c^2 k_l k_m}{2\omega}\frac{\partial g^{lm}}{\partial x^i} = \frac{c^2}{2\omega}\,g_{ql}g_{mp}\,k^q k^p \frac{\partial g^{lm}}{\partial x^i}\,. \tag{30.21}$$

Then we express the derivative of the inverse metric tensor g^{lm} by the derivative of the metric tensor g_{qp} using the rules of differentiating inverse matrices: we get from

$$0 = \partial(\mathbf{g}^{-1}\mathbf{g}) = (\partial\mathbf{g}^{-1})\mathbf{g} + \mathbf{g}^{-1}(\partial\mathbf{g}) \tag{30.22}$$

the relationship

$$\partial\mathbf{g} = -\mathbf{g}(\partial\mathbf{g}^{-1})\mathbf{g} \tag{30.23}$$

that implies in index notation

$$g_{qp,j} = -g_{ql}\,\frac{\partial g^{lm}}{\partial x^j}\,g_{mp}\,. \tag{30.24}$$

We thus obtain from Eq. (30.21) and definition (30.15):

$$-\frac{\partial\omega}{\partial x^i} = \frac{c^2}{2\omega}\,g_{qp,i}\,k^q k^p = \frac{c^2}{2\omega}\,g_{qp,i}\,\frac{\mathrm{d}x^q}{\mathrm{d}\zeta}\frac{\mathrm{d}x^p}{\mathrm{d}\zeta}\,. \tag{30.25}$$

Now compare this expression with the time derivative of k_i that occurs in Hamilton's equations (30.17). As k_i carries a lower index but expression (30.25) contains upper–index vectors, we should relate k_i to k^i. We obtain

$$\frac{\mathrm{d}k_i}{\mathrm{d}t} = \frac{\mathrm{d}g_{ij}(\boldsymbol{r})k^j}{\mathrm{d}t} = g_{ij}\frac{\mathrm{d}k^j}{\mathrm{d}t} + g_{ij,l}\frac{\mathrm{d}x^l}{\mathrm{d}t}k^j = \frac{c^2}{\omega}\left(g_{ij}\,\frac{\mathrm{d}^2x^j}{\mathrm{d}\zeta^2} + g_{iq,p}\,\frac{\mathrm{d}x^q}{\mathrm{d}\zeta}\frac{\mathrm{d}x^p}{\mathrm{d}\zeta}\right). \tag{30.26}$$

In the last step we applied definitions (30.15) and (30.20) and relabeled indices. The two expressions (30.25) and (30.26) agree if

$$g_{ij}\,\frac{\mathrm{d}^2x^j}{\mathrm{d}\zeta^2}+g_{iq,p}\,\frac{\mathrm{d}x^q}{\mathrm{d}\zeta}\frac{\mathrm{d}x^p}{\mathrm{d}\zeta}=\frac{1}{2}\,g_{qp,i}\,\frac{\mathrm{d}x^q}{\mathrm{d}\zeta}\frac{\mathrm{d}x^p}{\mathrm{d}\zeta}\,, \tag{30.27}$$

but this is equivalent to the equation (19.6) of a geodesic (with use of the identity (19.5)). Therefore, if light rays follow geodesics with respect to the optical geometry they also obey Hamilton's equations (30.17). Conversely, Hamilton's equations imply that the geodesic equation (30.27) is satisfied. Fermat's principle follows from Hamilton's optics and vice versa.

We can draw another important conclusion from the geodesic equation (30.27). We have seen in §19 that the geodesic equation appears in the form (30.27) only when the parameter increment $\mathrm{d}t$ is proportional to the line element $\mathrm{d}s$, the optical line element in our case. We just need to deduce the proportionality factor: imagine the simplest case, light rays in empty flat space. There the optical path length s agrees with the physical path length l. From Hamilton's equations (30.17) in free space we obtain for the speed of light $c=|\mathrm{d}\boldsymbol{r}/\mathrm{d}t|=\mathrm{d}l/\mathrm{d}t=\mathrm{d}s/\mathrm{d}t$; so the proportionality factor between s and t is c, the speed of light in vacuum. Let us return to an arbitrary optical geometry. There the travel time of light is the optical path length divided by c. We thus conclude that the Hamiltonian time t is the travel time of light.

Hamilton's equations (30.17) are practically useful for computing the trajectories of light rays, because they tend to be numerically stable. But geometrical optics goes beyond ray tracing; it approximately describes full electromagnetic waves. Consider the amplitudes $\mathcal{E}_i$ and $\mathcal{H}_i$ of the electric and magnetic fields. Instead of discussing the wave equation, we directly approximate Maxwell's equations in the regime of geometrical optics. We substitute the ansatz (30.1) in Maxwell's equations (26.1) without charges and currents, and ignore derivatives of the metric and the field amplitudes $\mathcal{E}$ and $\mathcal{H}$ in comparison with the derivatives of the phase φ:

$$k_i\mathcal{E}^i=0\,,\quad k_i\mathcal{H}^i=0\,, \tag{30.28}$$

$$\epsilon^{ijk}k_j\mathcal{E}_k=\omega\mu_0\mathcal{H}^i\,,\quad \epsilon^{ijk}k_j\mathcal{H}_k=-\omega\varepsilon_0\,\mathcal{E}^i\,. \tag{30.29}$$

Equations (30.28) state that the field amplitudes are orthogonal to the wave vector k_i, the gradient of the phase: electromagnetic waves are *transversal*, but note that $\mathcal{E}^i$ and $\mathcal{H}^i$ are upper–index amplitudes that, according to the constitutive equations (26.6) and (26.7), correspond to the dielectric displacement $\boldsymbol{D}$ and the magnetic induction $\boldsymbol{B}$. The lower–index field amplitudes $\mathcal{E}_i$ and $\mathcal{H}_i$ that describe the electric and magnetic field strengths $\boldsymbol{E}$ and $\boldsymbol{H}$ are orthogonal to the propagation vector k^i, as we see by moving the index in the Maxwell equations (30.28),

$$k^i\mathcal{E}_i=0\,,\quad k^i\mathcal{H}_i=0\,. \tag{30.30}$$

Equations (30.29) imply that $\boldsymbol{B}$ is orthogonal to $\boldsymbol{k}$ and $\boldsymbol{E}$, and $\boldsymbol{D}$ is orthogonal to $\boldsymbol{k}$ and $\boldsymbol{H}$ (where $\boldsymbol{k}$ is the wave vector k_i). Only in isotropic media does it follow that

$\boldsymbol{E}$ is orthogonal to $\boldsymbol{H}$, because only in that case is the geometry of light conformally flat so that $\boldsymbol{D}$ points in the same direction as $\boldsymbol{E}$, and $\boldsymbol{B}$ in the direction of $\boldsymbol{H}$.

We define the moduli $\mathcal{E}$ and $\mathcal{H}$ of the field amplitudes with respect to the effective geometry of the medium:

$$\mathcal{E}^2 = g^{ij}\mathcal{E}_i^*\mathcal{E}_j\,, \quad \mathcal{H}^2 = g^{ij}\mathcal{H}_i^*\mathcal{H}_j\,, \tag{30.31}$$

where the symmetry of the tensor g^{ij} implies that $\mathcal{E}^2$ and $\mathcal{H}^2$ are real. From Maxwell's equations (30.28) and (30.29) follows that $\mathcal{E}$ and $\mathcal{H}$ are identical within the regime of geometrical optics, apart from a factor of $\varepsilon_0 c$ that accounts for the different physical dimensions of $\boldsymbol{E}$ and $\boldsymbol{H}$ in SI units, as we see from the calculation below:

$$\begin{aligned}\varepsilon_0^2 c^2\,\mathcal{E}^2 &= \frac{c^2}{\omega^2}\,\epsilon^{ijk}k_j\mathcal{H}_k^*\,\epsilon_{ilm}k^l\mathcal{H}^m = \frac{c^2}{\omega^2}\left(\delta_l^j\delta_m^k - \delta_m^j\delta_l^k\right)k_jk^l\,\mathcal{H}_k^*\mathcal{H}^m \\ &= \frac{c^2}{\omega^2}\left(k_jk^j\mathcal{H}_m^*\mathcal{H}^m - k_m\mathcal{H}^m k^l\mathcal{H}_l^*\right) \\ &= \mathcal{H}^2\,,\end{aligned} \tag{30.32}$$

where we used the dispersion relation (30.9) and Maxwell's equation (30.28) in the last step. Furthermore, from $\varepsilon_0\mu_0 = c^{-2}$ follows

$$\varepsilon_0\mathcal{E}^2 = \mu_0\mathcal{H}^2\,. \tag{30.33}$$

The moduli (30.31) of the field amplitudes have an important physical interpretation that we deduce next.

Consider the next order of accuracy of the ansatz (30.1) with rapidly oscillating phase φ. In the wave equation (29.4) we use expressions (30.7) and (30.8) for the derivatives together with the dispersion relation (30.9) we already derived. We now retain the terms linear in the derivatives of the rapidly varying phase and ignore the much smaller curvature term and the Laplacian $\nabla^j\nabla_j\mathcal{E}_i$ of the field amplitude. In this way we obtain from the wave equation (29.4) the *amplitude–transport equation*

$$2k^j\nabla_j\mathcal{E}_i + \mathcal{E}_i\nabla_jk^j + 2\frac{\omega}{c^2}\frac{\partial\mathcal{E}_i}{\partial t} = 0\,. \tag{30.34}$$

Consider the modulus squared of the amplitude given by definition (30.31). We have

$$\nabla_j\mathcal{E}^2 = g^{il}\left(\mathcal{E}_i^*\nabla_j\mathcal{E}_l + \mathcal{E}_l\nabla_j\mathcal{E}_i^*\right) \tag{30.35}$$

(remember that the covariant derivative of the metric vanishes). We contract $\nabla_j\mathcal{E}^2$ with k^j, apply the amplitude–transport equation (30.34) and its complex conjugate, and obtain

$$k^j\nabla_j\mathcal{E}^2 + \frac{\omega}{c^2}\frac{\partial\mathcal{E}^2}{\partial t} = -g^{il}\mathcal{E}_i^*\mathcal{E}_l\nabla_jk^j = -\mathcal{E}^2\nabla_jk^j\,. \tag{30.36}$$

Collecting the covariant derivatives in one term and using the Leibnitz rule we arrive at:

$$\frac{\partial\mathcal{E}^2}{\partial t} + \nabla_i\left(\frac{c^2}{\omega}\,\mathcal{E}^2k^i\right) = 0\,. \tag{30.37}$$

For stationary electromagnetic waves the modulus of the amplitude $\mathcal{E}$ does not change in time; we obtain in this case the *continuity equation*

$$\nabla_i \, \mathcal{E}^2 k^i = 0 \, . \tag{30.38}$$

We see that, for maintaining a stationary flow of light, the amplitude must adjust to the wave vector when k^i varies in non–uniform media.

Let us return to the general case. Equation (30.37) describes in differential form a conservation law, because we obtain from Gauss' theorem (17.17) for an arbitrary spatial volume Ω with boundary surface $\partial\Omega$:

$$\frac{\partial}{\partial t} \int_\Omega \sqrt{g}\, \mathcal{E}^2 \, \mathrm{d}^3 x + \oint_{\partial\Omega} \frac{c^2}{\omega} \sqrt{g}\, \mathcal{E}^2 \, k^i \, n_i \mathrm{d}^2 x = 0 \, . \tag{30.39}$$

Formula (30.39) states that changes in the total amount of the quantity $\sqrt{g}\,\mathcal{E}^2$ must come from fluxes across the boundary; $\sqrt{g}\,\mathcal{E}^2$ is neither created nor destroyed, just redistributed (Fig. 3.14). What is this quantity? We express the inverse metric tensor g^{ij} in the definition (30.31) of $\mathcal{E}^2$ and $\mathcal{H}^2$ in terms (26.7) of the dielectric tensors $\varepsilon^{ij} = \mu^{ij}$, and get

$$\sqrt{g}\,\mathcal{E}^2 = \pm\,\varepsilon^{ij} \mathcal{E}_i^* \mathcal{E}_j \, , \quad \sqrt{g}\,\mathcal{H}^2 = \pm\,\mu^{ij} \mathcal{H}_i^* \mathcal{H}_j \, . \tag{30.40}$$

Now, the *instantaneous energy densities* of the electric and magnetic field are given by (Jackson [1999])

$$I_E^{\text{inst}} = \pm\,\frac{\varepsilon_0}{2}\,\varepsilon^{ij} E_i E_j \, , \quad I_M^{\text{inst}} = \pm\,\frac{\mu_0}{2}\,\mu^{ij} H_i H_j \, . \tag{30.41}$$

The $\pm$ sign ensures that the field energy is always positive, even when ε^{ij} and μ^{ij} are negative.* In geometrical optics, we have focused on the positive–frequency part of the field with phase (30.4), where it is understood that the field is always "shadowed" by the negative–frequency part:

$$E_i^{\text{inst}} = \mathcal{E}_i \exp\left[\mathrm{i}\varphi_0(\boldsymbol{r}) - \mathrm{i}\omega t\right] + \mathcal{E}_i^* \exp\left[-\mathrm{i}\varphi_0(\boldsymbol{r}) + \mathrm{i}\omega t\right] \, . \tag{30.42}$$

Hence for the real electromagnetic wave we obtain the instantaneous electric–field energy density

$$I_E^{\text{inst}} = \pm\,\varepsilon_0\,\varepsilon^{ij} \left[\mathcal{E}_i^* \mathcal{E}_j + |\mathcal{E}_i \mathcal{E}_j| \cos\left(2\varphi_0 - 2\omega t + \varphi_j - \varphi_i\right)\right] , \tag{30.43}$$

where φ_i denotes the argument of $\mathcal{E}_i$ and we have used the symmetry of the dielectric tensors. As the field of the electromagnetic wave is rapidly oscillating with frequency ω, it is wise to average over several oscillations in time. After averaging, the contribution to the energy (30.43) oscillating with frequency 2ω vanishes, only the slowly

*Note that these expressions for the energy density are only valid for dispersionless media (when ε^{ij} and μ^{ij} do not depend on frequency for the relevant spectral region); they need to be modified in dispersive dielectrics (Landau, Lifshitz and Pitaevskii [1984]).

varying term $\mathcal{E}_i^*\mathcal{E}_j$ remains in the time–averaged electric field energy density that we simply call the *field energy density*:

$$I_E = \pm\,\varepsilon_0\,\varepsilon^{ij}\mathcal{E}_i^*\mathcal{E}_j\,. \tag{30.44}$$

We can of course run the same argument for the *magnetic–field energy* I_M. We see from expression (30.40) that $\sqrt{g}\,\mathcal{E}^2$ and $\sqrt{g}\,\mathcal{H}^2$ correspond to the electric and magnetic field energy densities:

$$I_E = \varepsilon_0\sqrt{g}\,\mathcal{E}^2\,, \quad I_M = \mu_0\sqrt{g}\,\mathcal{H}^2\,. \tag{30.45}$$

The equality (30.33) of $\varepsilon_0\mathcal{E}^2$ and $\mu_0\mathcal{H}^2$ thus means that, within the regime of geometrical optics, the electric and magnetic fields have the same energy densities:

$$I_E = I_M\,. \tag{30.46}$$

The total energy density is thus

$$I = 2\varepsilon_0\sqrt{g}\,\mathcal{E}^2 = 2\varepsilon_0\sqrt{g}\,g^{ij}\mathcal{E}_i^*\mathcal{E}_j\,. \tag{30.47}$$

Note that the geometry established by the medium ensures that the field energy is always positive, even when ε^{ij} and μ^{ij} are negative.

We have seen that the modulus square of the field amplitude with respect to the metric established by the medium corresponds to the field–energy density (30.47). We obtain from formula (30.39) the law of *energy conservation*:

$$\frac{\partial}{\partial t}\int_\Omega I\,\mathrm{d}^3x + \oint_{\partial\Omega} S^i\,n_i\mathrm{d}^2x = 0\,, \tag{30.48}$$

where the *energy flux* S^i is given by the expression

$$S^i = \frac{2\varepsilon_0 c^2}{\omega}\sqrt{g}\,\mathcal{E}^2k^i = \frac{c^2}{\omega}Ik^i \tag{30.49}$$

having the physical units of energy per time and area. In the following we show that S^i is the *Poynting vector* of the complex fields,*

$$\boldsymbol{S} = 2\,\boldsymbol{E}^*\times\boldsymbol{H}\,, \tag{30.50}$$

where the vector product is taken in physical space, not with respect to the metric of the medium. Note that $2\,\boldsymbol{E}^*\times\boldsymbol{H}$ must be real–valued if Eqs. (30.49) and (30.50) agree, because S^i of Eq. (30.49) is real. To prove the equivalence of the two expressions, we write the vector product $\boldsymbol{E}^*\times\boldsymbol{H}$ in Cartesian coordinates as $[ijk]E_j^*H_k$ and then put the antisymmetric symbol $[ijk]$ in terms of the Levi–Civita tensor (14.7)

*Similarly to the field energy density, the vector $2\,\boldsymbol{E}^*\times\boldsymbol{H}$ is the time average of the instantaneous Poynting vector $\boldsymbol{E}\times\boldsymbol{H}$ for the real field (30.42) of the electromagnetic wave.

with respect to the geometry of the medium. We obtain from Maxwell's equations (30.28) and (30.29) for the amplitudes within the regime of geometrical optics:

$$\begin{aligned} S^i &= 2\sqrt{g}\,\epsilon^{ijk}\mathcal{E}_j^*\mathcal{H}_k \\ &= \frac{2\varepsilon_0 c^2}{\omega}\sqrt{g}\,\epsilon^{ijk}\mathcal{E}_j^*\epsilon_{klm}k^l\mathcal{E}^m = \frac{2\varepsilon_0 c^2}{\omega}\sqrt{g}\left(k^i\mathcal{E}_j^*\mathcal{E}^j - \mathcal{E}_j^*k^j\mathcal{E}^i\right) \\ &= \frac{2\varepsilon_0 c^2}{\omega}\sqrt{g}\,\mathcal{E}^2 k^i\,, \end{aligned} \tag{30.51}$$

which shows that Eqs. (30.49) and (30.50) are identical; the Poynting vector $2\,\boldsymbol{E}^*\times\boldsymbol{H}$ thus describes the energy flux of electromagnetic waves.

Although we are dealing with classical electromagnetic fields, it is instructive to cast the conservation of field energy (30.48) in terms of light quanta, *photons*. Imagine the energy density I is brought about by photons with density N and energy $\hbar\omega$ where $\hbar$ is Planck's constant divided by 2π. Adopting this—rather simplistic—notion of photons, the energy density (30.47) is

$$I = \hbar\omega N\,, \quad N = \frac{2\varepsilon_0\sqrt{g}\,\mathcal{E}^2}{\hbar\omega} \tag{30.52}$$

and the Poynting vector (30.49) appears as

$$S^i = c^2 p^i N \tag{30.53}$$

with the *photon momentum*

$$p^i = \hbar k^i\,. \tag{30.54}$$

The momentum p^i is known as the *Abraham momentum* and it carries an upper index; p^i is a vector. On the other hand, in Hamilton's equations (30.17) the momentum corresponds to the wave vector with lower index; the corresponding momentum is the wave momentum or *Minkowski momentum*

$$p_i = \hbar k_i = \hbar\nabla\varphi\,. \tag{30.55}$$

The two momenta are connected by relationship (30.16), but they are clearly not the same, because k^i and k_i are not the same geometrical objects. In isotropic media that generate the conformally flat geometries (29.17) the Minkowski momentum is n^2 times the Abraham momentum. Which momentum is the right one? One may argue that the Abraham momentum appears when the particle aspects of light are relevant and the Minkowski momentum when the wave aspects are probed (Leonhardt [2006c]). But note that the momentum of light in media is still subject of considerable debate lasting more than a century (Barnett and Loudon [2010]).

Exercise 30.1
We showed that the momentum density of photons (30.52) with the Abraham momentum (30.54) is the Poynting vector $\boldsymbol{E}\times\boldsymbol{H}$. Use a similar calculation to prove that the momentum density associated with the Minkowski momentum (30.55) is $\boldsymbol{D}\times\boldsymbol{B}$.

We realized that light rays are geodesics in the geometry created by the medium and that the intensity of the amplitude (30.31) is the energy density (30.47). How is the field–strength vector transported along the ray trajectory? Similar to the polarization (29.13) of plane waves, we represent the amplitude vector $\mathcal{E}_i$ in terms of the modulus $\mathcal{E}$ and the complex polarization vector ψ_i such that

$$\mathcal{E}_i = \mathcal{E}\psi_i\,, \tag{30.56}$$

which, according to the definition (30.31) of $\mathcal{E}$, implies the normalization

$$\psi_i^*\psi^i = g^{ij}\psi_i^*\psi_j = 1\,. \tag{30.57}$$

We considered a time–dependent amplitude $\mathcal{E}$ (in order to deduce the conservation (30.37) of energy). Suppose now that the polarization does not vary in time. We obtain from the transversality (30.30) of the electric–field amplitude

$$k^i\psi_i = 0\,, \tag{30.58}$$

so ψ_i must be orthogonal to the propagation vector k^i, but k^i may vary along the ray trajectory; in general ψ_i will not remain in the same direction in physical space. We obtain from the amplitude–transport equation (30.34) and relationship (30.36):

$$0 = \mathcal{E}k^j\nabla_j\psi_i + \psi_i\left(k^j\nabla_j\mathcal{E} + \frac{\mathcal{E}}{2}\nabla_j k^j + \frac{\omega}{c^2}\frac{\partial\mathcal{E}}{\partial t}\right) = \mathcal{E}k^j\nabla_j\psi_i\,. \tag{30.59}$$

Hence, dropping the amplitude $\mathcal{E}$, we get from the definition (30.15) of light rays:

$$\frac{\mathrm{d}x^j}{\mathrm{d}\zeta}\nabla_j\psi_i = 0\,. \tag{30.60}$$

This is the defining equation (20.1) of parallel transport. Consequently, the polarization is *parallel–transported* along the ray trajectory. Inhomogeneous media may bend light rays and create a curved geometry for light, but the polarization stays parallel to the ray, in as much as staying parallel makes sense in curved space.

In geometrical optics, we thus arrive at the following picture (Fig. 4.5) for electromagnetic waves in impedance–matched media.* Light rays are solutions of Hamilton's equations (30.17) with dispersion relation (30.18). The Hamiltonian time is the physical travel time of light. The rays are geodesics with respect to the geometry of light (26.12) created by the medium. The light rays act as fictitious lines the electric and magnetic fields $\boldsymbol{E}$ and $\boldsymbol{H}$ move along, pointing orthogonal to the propagation vector—the fields are transversal. The propagation vector k^i is the vector (30.16) derived from the gradient (30.3) of the phase, the one–form k_i given by Hamilton's equations (30.17). The modulus of the field amplitude with respect to the effective

*Without impedance matching, anisotropic media are birefringent (Born and Wolf [1999]): the dispersion relation depends on the polarization state and so the ray trajectories are different for different polarizations. Note that in isotropic media all the results we obtained in geometrical optics are valid even without impedance matching (Born and Wolf [1999]).

metric corresponds to the energy density (30.47) and, together with the propagation vector k^i, gives the energy flux, the Poynting vector (30.50). Electric and magnetic fields carry the same energy (30.46). The amplitude of stationary waves adjusts to changes in the propagation vector for maintaining a stationary light flux (30.38). The polarization is parallel–transported along the ray trajectories. Thus, although light rays are fictitious lines, they guide and govern the entire electromagnetic field within the validity range (30.5) and (30.6) of geometrical optics.

Exercise 30.2

Imagine a point dipole, an infinitely small antenna, that is continuously emitting electromagnetic waves in free space (Fig. 4.6). Model the dipole as an infinitely localised current density, pointing in the z–direction, that is oscillating at frequency ω. The electromagnetic wave will oscillate at the same frequency. As the dipole is assumed to be infinitely small, the phase φ of the electromagnetic wave should only depend on the distance r from the dipole. We also expect from Ampère's law that the magnetic field of the wave curls around the direction of the dipole such that $\boldsymbol{H}$ points in the ϕ direction in spherical polars $\{r, \theta, \phi\}$; it has only the component $\mathcal{H}_\phi$. The axial symmetry of the dipole implies that $\mathcal{H}$, the amplitude of $\boldsymbol{H}$, cannot depend on ϕ. But $\mathcal{H}$ depends on θ, because we expect that the magnetic field sees only the projection of the dipole in the direction of the observation point; this projection is given by $\sin\theta$. Otherwise $\mathcal{H}$ should only depend on r and not on ϕ. We thus arrive at the following mathematical model of dipole radiation:

$$\varphi = \varphi_0(r) - \omega t\,, \quad \mathcal{H}_i = (0, 0, \mathcal{H}_\phi)\,, \quad \mathcal{H} = \mathcal{H}_0(r)\sin\theta\,. \tag{30.61}$$

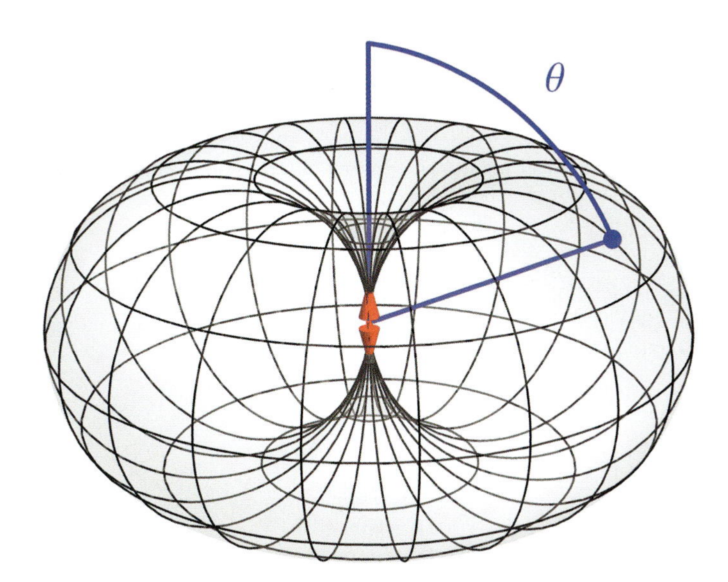

Figure 4.6: A point dipole is an infinitesimal current density (red) that oscillates along an axis. By Ampère's law this gives rise to a magnetic field that circulates around the axis of the dipole. The geometrical–optics solution for the magnetic and electric fields is given in Exercise 30.2; it shows that the moduli $\mathcal{H}$ and $\mathcal{E}$ of the field amplitudes are constant on the surface shown in the picture. (The exact solution of this problem is given in Exercise 34.2.)

Show that this model of dipole radiation is completely consistent with geometrical optics. In particular, deduce from the dispersion relation the wave vector k_i and the propagation vector k^i. Show that

$$\mathcal{H}_\phi = \mathcal{H}_0 r \sin^2\theta\,,\quad \mathcal{E}_i = (0, \mathcal{E}_\theta, 0)\,,\quad \mathcal{E}_\theta = \mu_0 c \mathcal{H}_0 r \sin\theta\,, \tag{30.62}$$

and that from the continuity equation (30.38) follows

$$\mathcal{H}_0(r) = H_0/r \quad \text{with} \quad H_0 = \text{const}\,. \tag{30.63}$$

Finally, verify that the polarization is parallel–transported.
Solution
Calculation of the wave vector: as the gradient of φ points in the r–direction $k_i = (k_r, 0, 0)$. We obtain from the dispersion relation (30.9) in empty space $k_r = \omega/c$ and so

$$k_i = (\omega/c, 0, 0) \quad \text{and from the inverse metric (13.9)} \quad k^i = (\omega/c, 0, 0)\,. \tag{30.64}$$

In this way we also obtain $\varphi = \omega(r/c - t)$, the phase (4.5) of spherical waves.
Calculation of the magnetic and electric field: we obtain from the inverse metric tensor (13.9) in spherical polars $\mathcal{H}^2 = \mathcal{H}_\phi^2\,(r\sin\theta)^{-2}$ and hence from our requirements (30.61) $\mathcal{H}_\phi = \mathcal{H}_0 r \sin^2\theta$. From Eq. (30.29) and our result for the wave vector follows $\varepsilon_0 c\,\mathcal{E}^i = -\epsilon^{ir\phi}\mathcal{H}_\phi$, which, according to the antisymmetry of the Levi–Civita tensor, implies that $\mathcal{E}^i = (0, \mathcal{E}^\theta, 0)$ and from the metric (11.17) $\mathcal{E}_i = (0, \mathcal{E}_\theta, 0)$. Furthermore, from Eqs. (14.7) and (11.19) follows $\varepsilon_0 c\,\mathcal{E}^\theta = \mathcal{H}_\phi/(r^2\sin\theta)$ and so, lowering the index with metric (11.17) and using $\varepsilon_0\mu_0 = c^{-2}$, we get $\mathcal{E}_\theta = \mu_0 c \mathcal{H}_\phi/\sin\theta = \mu_0 c \mathcal{H}_0 r \sin\theta$.
Continuity: as $\varepsilon_0\mathcal{E}^2 = \mu_0\mathcal{H}^2$ we obtain from the continuity equation (30.38)

$$0 = \nabla_i \mathcal{H}^2 k^i = \partial_i\left(\sqrt{g}\,g^{ij}\mathcal{H}^2 k_j\right) \quad \text{with determinant (11.19) and inverse metric (13.9)}\,,$$

and from Eqs. (30.64) and (30.62) follows the result (30.63).
Parallel transport of polarization: according to definition (30.56) the polarization ψ_i is given by $\mathcal{E}_i$ divided by $\mathcal{E}$. As the electric field points in the θ–direction, $\psi_i = (0, \psi_\theta, 0)$, and from the normalization (30.57) in spherical polars with inverse metric (13.9) follows $\psi_\theta = r$. We see from the Christoffel symbols (15.26) in spherical polars that

$$\psi_{\theta,r} - \Gamma^\theta_{\ \theta r}\psi_\theta = 0\,, \tag{30.65}$$

which gives the equation (30.60) of parallel transport in the r–direction, the direction of the propagation vector (30.64). The polarization is indeed parallel–transported along the light rays.

Exercise 30.3
Imagine light incident on the surface of a glass of water, as described in Exercise 29.3 with

$$\varepsilon_1 = 1\,,\quad \varepsilon_2 = n\,,\quad \mu_1 = \mu_2 = 1\,. \tag{30.66}$$

Suppose the light is incident directly from above and illuminates the surface uniformly. Which total momentum does a photon transfer to the water surface on average? Consider the momentum balance: the influx of momentum is given by the momentum of the incident light plus the recoil, and the transmitted light carries away momentum. Let us denote the reflection coefficient of the amplitude by ϱ. Photons are reflected with probability ϱ^2 and transmitted with probability $1-\varrho^2$, so the recoil is given by ϱ^2 times the incident momentum, and the transmitted momentum is $(1-\varrho^2)$ times the momentum in the water. Suppose the latter is the Abraham or the Minkowski momentum. Calculate the momentum balance in the two cases and interpret the results.
Solution
For photons with frequency ω the Minkowski momentum p_i and the Abraham momentum p^i are

$$p_i = \hbar k = n\frac{\hbar\omega}{c}\,,\quad p^i = \frac{\hbar k}{n^2} = \frac{\hbar\omega}{nc}\,. \tag{30.67}$$

From Eqs. (29.44) and (29.49) follows that the reflection coefficients are identical for the two polarizations (for vertical incidence) and given by

$$\varrho = \frac{\mathcal{E}_1'}{\mathcal{E}_1} = \frac{1-n}{1+n}\,. \tag{30.68}$$

We thus obtain for the Abraham/Minkowski momentum balance Δp^i and Δp_i, respectively:

$$\Delta p^i = (1+\varrho^2)\frac{\hbar\omega}{c} - (1-\varrho^2)\frac{\hbar\omega}{nc} = +2\left(\frac{n-1}{n+1}\right)\frac{\hbar\omega}{c}\,, \tag{30.69}$$

$$\Delta p_i = (1+\varrho^2)\frac{\hbar\omega}{c} - (1-\varrho^2)n\frac{\hbar\omega}{c} = -2\left(\frac{n-1}{n+1}\right)\frac{\hbar\omega}{c}\,. \tag{30.70}$$

In both cases, the magnitude of the momentum transfer is the same, but assuming the Abraham momentum the transfer is positive, the surface is pushed inwards, whereas in the case of the Minkowski momentum the surface is pulled by the light. Which one is correct, the push or the pull of light? An early experiment on a water surface seemed to confirm the pull of light (Ashkin and Dziedzic [1972]) but in a later experiment on a glass interface a push force was observed (She, Yu and Feng [2008]).

---

§31. SPACE–TIME GEOMETRIES AND MEDIA

So far we have considered spatial geometries or coordinate transformations in space, but there are also important examples of media that establish space–time geometries or space–time transformations. In §37 we show that moving media alter the geometry of space *and* time for light, and in §38 and §39 we discuss two cases of space–time transformation media in detail, the optical Aharonov–Bohm effect and optical analogues of the event horizon. In this Section we write down the foundations of electromagnetism in space–time geometries, Maxwell's equations with appropriate constitutive equations.

Space–time coordinates were introduced in §25. Let us briefly recap the essentials: we use four coordinates x^α with Greek indices running from 0 to 3 where x^0 denotes time t multiplied by c; latin indices refer to spatial coordinates. In flat space–time and Cartesian coordinates the space–time geometry is described by the Minkowski metric (25.2), otherwise by the general metric $g_{\alpha\beta}$. The space–time volume element is characterized by $\sqrt{-g}$ where the minus sign comes from the different signature of space and time; the determinant g is negative. It turns out that the theory of transformation media and materials that mimic curvature also works in 4–dimensional space–time. This is proved in the Appendix, which provides a more challenging example of tensor algebra than the 3–dimensional case. There we show that the free–space Maxwell equations in arbitrary right–handed space–time coordinates can be written as the macroscopic Maxwell equations (26.4) in right–handed Cartesian coordinates $\{ct, x, y, z\}$ with Plebanski's constitutive equations (Plebanski [1960]):

$$D^i = \varepsilon_0\varepsilon^{ij}E_j + \frac{1}{c}[ijk]w_jH_k\,, \quad B^i = \mu_0\mu^{ij}H_j - \frac{1}{c}[ijk]w_jE_k\,, \tag{31.1}$$

$$\varepsilon^{ij} = \mu^{ij} = -\frac{\sqrt{-g}}{g_{00}}\,g^{ij}\,, \quad w_i = \frac{g_{0i}}{g_{00}}\,. \tag{31.2}$$

We see that space–time geometries appear as media similar to spatial geometries (26.6) and (26.7). We can generalize the constitutive equations (31.1) and (31.2) to allow for a handedness change in the spatial part of the space–time coordinate transformation, and also to allow for a curvilinear spatial coordinate system in virtual space. Our previous result (28.3) shows how to incorporate these possibilities in the permittivity and permeability in the constitutive equations (31.2). In addition, we can express the constitutive equations (31.1) and (31.2) in index–free form if we denote the permittivity and permeability matrices by $\boldsymbol{\varepsilon}$ and $\boldsymbol{\mu}$, respectively, and understand $\boldsymbol{\varepsilon E}$, etc., as a matrix product. Our final constitutive relations are then (Leonhardt and Philbin [2006])

$$\boldsymbol{D} = \varepsilon_0 \boldsymbol{\varepsilon E} + \frac{\boldsymbol{w}}{c} \times \boldsymbol{H}\,, \quad \boldsymbol{B} = \mu_0 \boldsymbol{\mu H} - \frac{\boldsymbol{w}}{c} \times \boldsymbol{E}\,, \tag{31.3}$$

$$\varepsilon^{ij} = \mu^{ij} = \mp \frac{\sqrt{-g}}{\sqrt{\gamma}\, g_{00}}\, g^{ij}\,, \quad w_i = \frac{g_{0i}}{g_{00}}\,. \tag{31.4}$$

In addition to the familiar impedance–matched electric permittivity $\boldsymbol{\varepsilon}$ and magnetic permeability $\boldsymbol{\mu}$, a geometry that mixes space and time ($g_{0i} \neq 0$) mixes electric and magnetic fields. A space–time geometry thus appears as a *magneto–electric medium*, also called a *bi–anisotropic medium* (Sihvola et al. [1994], Serdyukov et al. [2001]). The mixing of electric and magnetic fields is brought about by the *bi–anisotropy vector* $\boldsymbol{w}$ that, in Cartesian coordinates, has the physical dimension of a velocity. In §37 we show that $\boldsymbol{w}$ is closely related to the velocity of the medium (for slow media, $\boldsymbol{w}$ turns out to be proportional to the velocity). Moving media are naturally magneto–electric (Landau, Lifshitz and Pitaevskii [1984])—a moving dielectric responds to the electromagnetic field in its local frame, but this frame is moving and motion mixes electric and magnetic fields by Lorentz transformations (Jackson [1999]). Such phenomena have been observed before special relativity was discovered, for example in the Röntgen effect (Röntgen [1888]) or, indirectly, in Fizeau's 1851 demonstration (Fizeau [1851]) of the Fresnel drag (Fresnel [1818]). More recently, moving optical media have been shown to generate analogues of the event horizon (Philbin et al. [2008]).

The constitutive equations (31.2) turn out to reveal an important hidden property of electromagnetism: electromagnetic fields are *conformally invariant* in space–time. Suppose we compare two geometries that are related by a conformal transformation, one with the metric $g_{\alpha\beta}$ and one with the metric $g'_{\alpha\beta}$ where we rescale equally space and time at each point, but the scaling factor Ω may vary over space–time:

$$g'_{\alpha\beta} = \Omega^2(x^\nu)\, g_{\alpha\beta}\,. \tag{31.5}$$

This is not a coordinate transformation in general: we compare two different geometries, as we discussed in §23 where we considered conformal transformations in 3–dimensional space, but here we are in space–time. The two geometries $g_{\alpha\beta}$ and $\Omega^2(x^\nu)\, g_{\alpha\beta}$ measure space–time distances differently. Since the metric tensors differ by a common factor, the "angles" between space–time trajectories remain the same. Let us work out why electromagnetism is invariant under the transformation (31.5).

If $g_{\alpha\beta}$ is multiplied by Ω^2, the inverse metric tensor $g^{\alpha\beta}$ scales with Ω^{-2} and the determinant g with Ω^8. The g_{00} element is multiplied by Ω^2, the elements g^{ij} by Ω^{-2}, the space–time volume $\sqrt{-g}$ by Ω^4 and g_{0i} by Ω^2. Inspecting the constitutive equations (31.2) and counting the powers of Ω, we see that ε^{ij}, μ^{ij} and w_i do not change; but these are the only quantities that depend on the geometry. Therefore the electromagnetic field is not affected by a conformal space–time transformation; electromagnetism is invariant if space and time are rescaled equally. However, if we alter only the measure of space in the Minkowski metric (25.2) by a spatially dependent factor, the refractive index $n(\boldsymbol{r})$, with the result

$$g_{\alpha\beta} = \mathrm{diag}\left(-1, n^2, n^2, n^2\right) , \tag{31.6}$$

the new geometry behaves like a dielectric medium with refractive index profile n, as we know from Fermat's principle (1.1). On the other hand, if we alter the measure of time in the Minkowski metric (25.2) by n^{-2}, we get the space–time metric

$$g'_{\alpha\beta} = \mathrm{diag}\left(-n^{-2}, 1, 1, 1\right) . \tag{31.7}$$

The metric tensors $g_{\alpha\beta}$ and $g'_{\alpha\beta}$ differ only by a common factor $\Omega^2 = n^{-4}$, but the factor $n(\boldsymbol{r})$ may vary. From the conformal invariance of electromagnetism follows that both space–time geometries (31.6) and (31.7) are exactly the same as far as electromagnetic fields are concerned. It does not matter whether we put the refractive index n in space or its inverse n^{-1} in time, the speed of light is the same and so is the optical geometry. In §37 we apply this conformal invariance of electromagnetism to discuss the space–time geometry generated by moving media.

Conformal invariance is the basis of *Penrose diagrams* where the entire causal structure of infinitely extended space–time is condensed, by a conformal factor, into a finite map one can draw and discuss. Penrose diagrams are perhaps best introduced by an elementary example in 1–dimensional space, i.e. (1+1)–dimensional space–time. We assume space–time to be flat with Cartesian coordinates z and ct. Let us introduce the new coordinates $t_\pm$, called *null coordinates*, defined as

$$t_\pm = t \mp \frac{z}{c} . \tag{31.8}$$

The $t_\pm$ are convenient for describing light rays in a space–time diagram (Fig. 4.7); rays propagating in positive spatial direction draw 45° lines in space–time $\{ct, z\}$ characterized by the equation $t_+ = \mathrm{const}$, because in this case we have $ct = z + z_0$, the equation of a straight line with slope 1. Similarly, rays going in negative spatial direction draw 135° lines with equation $t_- = \mathrm{const}$. For the right–moving rays $\mathrm{d}t_+ = 0$, so they are the t_- coordinate lines, and, by the same token, the left–moving rays are the t_+ coordinate lines with $\mathrm{d}t_- = 0$. For the null coordinates (31.8) we get

$$t = \frac{1}{2}\left(t_- + t_+\right) , \quad z = \frac{c}{2}\left(t_- - t_+\right) , \tag{31.9}$$

$$\mathrm{d}t = \frac{1}{2}\left(\mathrm{d}t_- + \mathrm{d}t_+\right) , \quad \mathrm{d}z = \frac{c}{2}\left(\mathrm{d}t_- - \mathrm{d}t_+\right) , \tag{31.10}$$

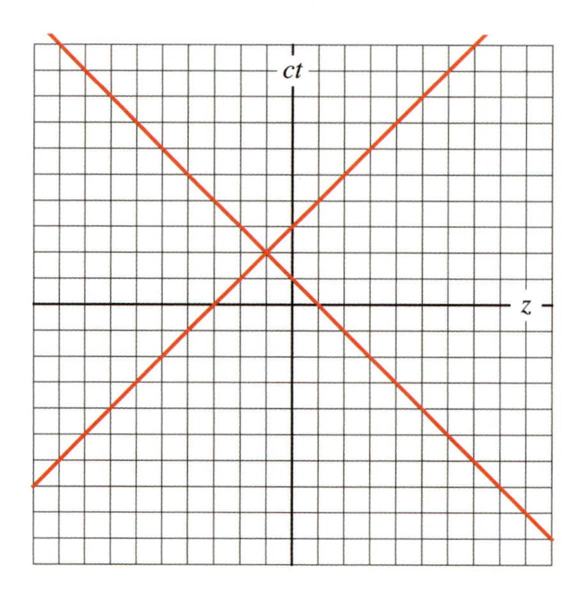

Figure 4.7: Light rays (red) in flat (1+1)–dimensional space–time with Cartesian axes ct and z. The ray with slope 1 (−1) has a constant value of the null coordinate t_+ (t_-), defined in Eq. (31.8).

and so we obtain for the space–time line element with Minkowski metric (25.2) in null coordinates

$$\mathrm{d}s^2 = \mathrm{d}z^2 - c^2\mathrm{d}t^2 = -c^2\,\mathrm{d}t_+\mathrm{d}t_-\,. \tag{31.11}$$

For the light rays in the space–time diagram (Fig. 4.7) the line element (31.11) vanishes: this is the reason why their coordinate lines are called null coordinates. Imagine we transform $t_\pm$ by functions that only depend on either one of the null coordinates:

$$t'_\pm = f_\pm(t_\pm)\,. \tag{31.12}$$

As in this case $\mathrm{d}t'_\pm = (\mathrm{d}t'_\pm/\mathrm{d}t_\pm)\mathrm{d}t_\pm$, we obtain for the line element (31.11) in space–time

$$\mathrm{d}s^2 = -c^2\,\frac{\mathrm{d}t_+}{\mathrm{d}t'_+}\frac{\mathrm{d}t_-}{\mathrm{d}t'_-}\,\mathrm{d}t'_+\mathrm{d}t'_- \tag{31.13}$$

and hence we obtain for the line element in transformed space–time

$$\mathrm{d}z'^2 - c^2\mathrm{d}t'^2 = \mathrm{d}s'^2 = \Omega^2\,\mathrm{d}s^2\,, \quad \Omega^2 = \frac{\mathrm{d}t'_+}{\mathrm{d}t_+}\frac{\mathrm{d}t'_-}{\mathrm{d}t_-}\,. \tag{31.14}$$

The coordinate transformation (31.12) is thus a conformal transformation (31.5). In §5 we used a similar trick for optical conformal mapping in 2–dimensional space where we employed functions $w(z)$ and $w^*(z^*)$ for transformations of the complex z plane that depend on either z or z^*. Consider a simple, instructive example of a conformal transformation in (1+1)–dimensional space–time:

$$t'_\pm = \arctan(t_\pm/t_0)\,, \quad t_0 = \mathrm{const}\,. \tag{31.15}$$

The new null coordinates range from $-\pi/2$ to $+\pi/2$, reaching $\pm\pi/2$ for $t_\pm$ going to $\pm\infty$: infinite space–time has been condensed into a finite region, the *Penrose diagram* (Fig. 4.8). According to relation (31.14), we obtain by differentiating the transforming function (31.15) the conformal factor

$$\Omega = t_0^{-2} \cos^2(t'_+) \cos^2(t'_-) \,. \tag{31.16}$$

As the conformal factor (31.16) vanishes when the transformed coordinates $t'_\pm$ of the Penrose diagram reach the boundaries of their world, increasingly large instances in physical space–time are condensed into small increments in $t'_\pm$. Similar to conformal mapping in 2–dimensional space or the stereographic and Mercator projection we encountered in §9, the Penrose diagram does not faithfully represent space–time distances. However, the 90° angles between light rays are preserved. In (3+1)–dimensional space–time these lines are extended to form 3–dimensional light cones. In general, we cannot simply perform space–time transformations to obtain Penrose diagrams, but we could alter the geometry by a conformal factor (31.5) that vanishes at the boundaries of the diagram. Electromagnetism would not be affected, because it is conformally invariant. According to relativity, physical effects cannot travel

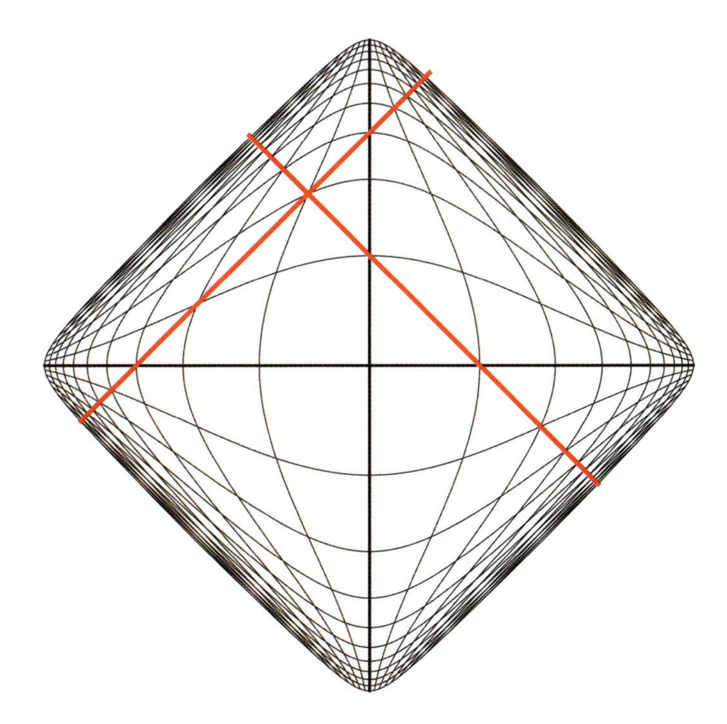

Figure 4.8: The space–time of Fig. 4.7 conformally transformed by Eq. (31.15). The transformation (31.15) compresses the infinite space–time sheet in Fig. 4.7 into a finite region, a Penrose diagram. Light rays in the Penrose diagram are straight lines with slope ±1, just as in the original space–time.

faster than the speed of light in vacuum; so the effects of a cause must lie within the light cone originating from that cause. As Penrose diagrams preserve light cones, they allow us to see the entire causal structure of a given space–time geometry in one glance. In §39 we apply Penrose diagrams to visualise the Hawking radiation of black holes.

Further reading

The classic book on electromagnetism is Jackson [1999]. We also recommend Feynman, Leighton and Sands [1964] and Landau and Lifshitz [1975]. An important book on electromagnetism in media is Landau, Lifshitz and Pitaevskii [1984].

Misner, Thorne and Wheeler [1973] briefly describe geometrical optics from the perspective of space–time geometry; Born and Wolf [1999] contains a chapter that focuses on the physical aspects of geometrical optics. Luneburg [1964] has a special perspective on geometrical optics. We also recommend Stavroudis [2006].

Optics on curved surfaces, where effects of extrinsic curvature can play a role, is treated by Batz and Peschel [2008].

Chapter 5

Geometries and media

This final Chapter combines and applies the physical insights and mathematical techniques developed so far. To recapitulate, in Chapter 2 several interesting optical effects and devices were elucidated using the relatively modest mathematical machinery of Fermat's principle and geometrical optics. A detailed discussion of geometrical optics and its accuracy was given in §30; this reveals the degree to which the geometry of Fermat's principle can be applied to arbitrary media. But in Chapter 4 we learned that the idea of media constituting effective geometries for electromagnetism becomes exact in certain circumstances. This followed from an appreciation that the fundamental equations of electromagnetism, Maxwell's equations, are tensor equations. These geometrical insights therefore required a knowledge of tensor analysis and the powerful mathematics of differential geometry presented in Chapter 3. We saw that media can perform coordinate transformations, or create curved spaces. In this Chapter we discuss specific examples of the various possibilities. Although in most cases the manufacture of devices that exhibit the required electromagnetic properties is a considerable engineering challenge, progress in this regard has been remarkable in recent years. The quest to achieve the degree of control over electromagnetic fields required for implementing the ideas of transformation optics and effective geometries now constitutes a rapidly growing area of material science: metamaterials.

The concept of transformation media has been the key to designing invisibility devices (§33), an idea that has been put into practice using metamaterials. Moreover, the idea that inspired the original surge of interest in metamaterials, the negatively refracting perfect lens, is also an example of transformation optics (§34). Negatively refracting transformation media lead to further possibilities for fascinating invisibility effects such as cloaking at a distance (§35). Maxwell's fish–eye, discussed from the point of view of ray optics in §9, is a positively refracting medium that creates a curved 3–dimensional space for light—a hypersphere. An analysis of wave propagation in the fish–eye shows that it is a perfect imaging device for light waves as well as for rays (§36). Moving media create space–time geometries (§37), and the optical Aharonov–Bohm effect and optical analogues of the event horizon can be understood as cases of media that perform space–time transformations (§38, §39).

§32. Spatial transformation media

Let us focus on some general properties of spatial transformation media, first for the case where virtual space is empty. These are media that perform purely spatial coordinate transformations of electromagnetic fields. We then generalize the theory to isotropic media in virtual space. We develop an economic form of the formalism that allows quick calculations and rapid judgment on the physical properties of the medium.

Recall from §28 that we considered a virtual space, whose coordinates we here denote by $x^{i'}$, and physical space, with coordinates x^i. These were two different flat spaces but the metric–tensor components in both spaces in these respective coordinate systems were the same. We denoted the metric of physical space in the coordinates x^i by γ_{ij}. The metric in virtual space in the coordinates $x^{i'}$ is to be transformed, so we use the usual notation $g_{i'j'}$ for the metric tensor. To restate, in the recipe for transformation media the two matrices γ_{ij} and $g_{i'j'}$ of metric–tensor components are the same with the replacements $x^i \leftrightarrow x^{i'}$. The rather subtle idea of transformation media was to use non–trivial equations $x^{i'} = x^{i'}(x^j)$, relating the coordinates $x^{i'}$ to x^i, to perform a coordinate transformation of Maxwell's equations in virtual space and then to recognize the result of the transformation as Maxwell's equations with non–trivial $\boldsymbol{\varepsilon}$ and $\boldsymbol{\mu}$ in physical space in the coordinates x^i with metric γ_{ij}. This meant that the electromagnetic fields in the x^i coordinates in physical space, containing the medium characterized by $\boldsymbol{\varepsilon}$ and $\boldsymbol{\mu}$, were obtained from those in virtual space in the $x^{i'}$ coordinates by the transformation $x^{i'} = x^{i'}(x^j)$. The coordinate transformation applied to the metric tensor $g_{i'j'}$ in virtual space produces a metric g_{ij} that is not the metric of physical space; rather, the transformed metric g_{ij} defines the effective medium in physical space through Eq. (28.3). As well as the determinant g of g_{ij}, and the inverse metric g^{ij}, the recipe (28.3) contains the determinant γ of the metric γ_{ij} of physical space. Applying the transformation rule (13.8), where the Λ–matrices are for the transformation $x^{i'} = x^{i'}(x^j)$ described above, we can write the inverse metric g^{ij} in Eq. (28.3) in terms of the inverse metric $g^{i'j'}$ of virtual space, obtaining

$$\varepsilon^{ij} = \mu^{ij} = \pm \frac{\sqrt{g}}{\sqrt{\gamma}} \, g^{i'j'} \Lambda^i{}_{i'} \Lambda^j{}_{j'} \,. \tag{32.1}$$

The matrix version (11.9) of the transformation of the metric is

$$\mathbf{g}' = \boldsymbol{\Lambda}^T \mathbf{g} \, \boldsymbol{\Lambda} \qquad \text{(matrix equation)}, \tag{32.2}$$

where $\mathbf{g}'$ is the matrix of components $g_{i'j'}$, $\mathbf{g}$ is the matrix of components g_{ij}, and $\boldsymbol{\Lambda}$ is the transformation matrix $\Lambda^i{}_{i'}$. Taking the determinant of both sides of Eq. (32.2) yields

$$g' = g \, (\det \boldsymbol{\Lambda})^2 \,, \tag{32.3}$$

where the usual notation for the determinant of a metric is employed. Since the determinant of the spatial metric is always positive the square root of Eq. (32.3) is

$$\sqrt{g'} = \sqrt{g} \, | \det \boldsymbol{\Lambda} | \,. \tag{32.4}$$

We use Eq. (32.4) to eliminate $\sqrt{g}$ in Eq. (32.1). Recall from §26 that the $\pm$ in Eq. (32.1) refers to whether $(-)$ or not $(+)$ the transformation changes the handedness of the coordinate system. This sign is equal to the sign of the determinant $\det \mathbf{\Lambda}$ of the transformation matrix $\mathbf{\Lambda}$, so when Eq. (32.4) is substituted into Eq. (32.1) the absolute value sign in $|\det \mathbf{\Lambda}|$ and the $\pm$ can be removed without changing the result. We write the result in matrix form, using the fact that the inverse metric $g^{i'j'}$ is the inverse of the matrix $\mathbf{g}'$:

$$\boldsymbol{\varepsilon} = \boldsymbol{\mu} = \frac{\sqrt{g'}}{\sqrt{\gamma}} \frac{\mathbf{\Lambda}(\mathbf{g}')^{-1}\mathbf{\Lambda}^T}{\det \mathbf{\Lambda}} \quad \text{(matrix equation).} \tag{32.5}$$

This useful formula gives the dielectric tensors $\boldsymbol{\varepsilon}$ and $\boldsymbol{\mu}$ of the transformation medium in terms of the metric tensors in virtual and physical space and the matrix $\mathbf{\Lambda} = (\Lambda^i{}_{i'}) = (\partial x^i / \partial x^{i'})$ of the transformation defining the medium. We can easily extend formula (32.5) to the case when virtual space is not empty, but contains an isotropic medium with permittivity ε' and permeability μ'. There we obtain from Eq. (28.4) and our previous transformation procedure

$$\boldsymbol{\varepsilon} = \frac{\sqrt{g'}}{\sqrt{\gamma}} \frac{\mathbf{\Lambda}(\mathbf{g}')^{-1}\mathbf{\Lambda}^T}{\det \mathbf{\Lambda}} \varepsilon' , \quad \boldsymbol{\mu} = \frac{\sqrt{g'}}{\sqrt{\gamma}} \frac{\mathbf{\Lambda}(\mathbf{g}')^{-1}\mathbf{\Lambda}^T}{\det \mathbf{\Lambda}} \mu' . \tag{32.6}$$

Transformation media are in general made of anisotropic materials characterized by the tensors $\boldsymbol{\varepsilon}$ and $\boldsymbol{\mu}$. When virtual space is empty they are impedance matched to the vacuum ($\boldsymbol{\varepsilon} = \boldsymbol{\mu}$). The matrices of components ε^{ij} and μ^{ij} are symmetric, because the inverse metric g^{ij} in Eq. (28.4) is symmetric, so they have three real eigenvalues corresponding to three orthogonal eigenvectors. In the case virtual space contains an isotropic virtual medium, the ε^{ij} and μ^{ij} tensors are proportional to each other according to Eq. (28.4), so they have the same eigenvectors. Since there are six independent components in the symmetric ε^{ij} (or μ^{ij}) it is simpler to characterize the medium by the three eigenvalues of the matrix ε^{ij} and, furthermore, the eigenvalues of μ^{ij} differ only by the factor μ'/ε' from the eigenvalues of ε^{ij}. But the components ε^{ij} change upon changing the coordinate system in physical space, and in general so do the eigenvalues of the matrix ε^{ij}. It is conventional and convenient to characterize the dielectric properties by the eigenvalues associated with a Cartesian coordinate system (the eigenvalues are the same in all Cartesian systems, as we will shortly prove). The three Cartesian eigenvalues ε_i are called the *principle values* of the dielectric tensors and the resulting orthogonal triad of eigenvectors define the *principle axes*. The principal axes would seem to provide a preferred Cartesian basis in which ε^{ij} is diagonal with components equal to the three eigenvalues, but since the eigenvectors in general rotate as one moves through the medium there is in general no Cartesian coordinate system that is adapted to the medium in this way. It is often the case in transformation optics that the suitable coordinates in physical space are not Cartesian and it would be tedious to always have to transform to Cartesian coordinates to compute the principle values of the dielectric tensors; we therefore show how the principle values can be found from the components ε^{ij} given by Eq. (32.6), which are in general non–Cartesian.

A familiar result from linear algebra is that the eigenvalues of a matrix $\mathbf{M}$ are unchanged after a similarity transformation $\mathbf{M} \to \mathbf{S}^{-1}\mathbf{M}\mathbf{S}$. If $\mathbf{M}$ has an eigenvalue M corresponding to an eigenvector $\mathbf{v}$, i.e.

$$\mathbf{M}\mathbf{v} = M\mathbf{v} \qquad \text{(matrix equation)}, \tag{32.7}$$

then the transformed matrix $\mathbf{S}^{-1}\mathbf{M}\mathbf{S}$ possesses the same eigenvalue M but with an associated eigenvector $\mathbf{S}^{-1}\mathbf{v}$, as easily follows from Eq. (32.7):

$$\left(\mathbf{S}^{-1}\mathbf{M}\mathbf{S}\right)\mathbf{S}^{-1}\mathbf{v} = \mathbf{S}^{-1}\mathbf{M}\mathbf{v} = M\mathbf{S}^{-1}\mathbf{v} \qquad \text{(matrix equation)}. \tag{32.8}$$

Now, the transformation rule for the components $T^i{}_j$ of a 2–index tensor with one upper and one lower index is a similarity transformation when written as a matrix equation: denoting the matrices of components $T^i{}_j$ and $T^{i'}{}_{j'}$ by $\mathbf{T}$ and $\mathbf{T}'$, respectively, the transformation rule (13.11) gives

$$\mathbf{T}' = \mathbf{\Lambda}^{-1}\mathbf{T}\mathbf{\Lambda} \qquad \text{(matrix equation)}, \tag{32.9}$$

where $\mathbf{\Lambda}$ is the transformation matrix $\Lambda^i{}_{i'}$. This means that the eigenvalues of the matrix of tensor components $T^i{}_j$ are the same in every coordinate system. In particular the eigenvalues are those in every Cartesian system, in which raising and lowering the indices does not change the tensor components. Hence the eigenvalues of the matrix $T^i{}_j$, when computed in an arbitrary system, are the eigenvalues of the matrix $T^{ij} = g^{jk}T^i{}_k$ of components of the raised–index tensor when the latter is expressed in a Cartesian system. Applying this to the permittivity tensor ε^{ij} we see that the eigenvalues of the matrix $\varepsilon^i{}_j$ in every coordinate system are equal to the eigenvalues of the matrix ε^{ij} in Cartesian coordinates. Thus, to find the principle values of the dielectric tensors in our preferred coordinates in physical space, we need only lower one of the indices and compute the eigenvalues of the resulting matrix. We have denoted the metric in physical space by γ_{ij} so the index is lowered by

$$\varepsilon^i{}_j = \varepsilon^{ik}\gamma_{kj}\,. \tag{32.10}$$

Note that the matrix $\varepsilon^i{}_j$ is not symmetric in general, and its eigenvectors are not orthogonal; nevertheless, the eigenvalues of $\varepsilon^i{}_j$ directly give the principle values of the dielectric tensors for the transformation medium.

To illustrate how the use of appropriate coordinate systems offers considerable economy in the calculations of transformation optics, we discuss the example of a cylindrical transformation. Suppose that the medium transforms the radial coordinate r of cylindrical polars by an arbitrary function $r(r')$, but preserves the angle ϕ and the vertical coordinate z:

$$\begin{gathered} r = r(r')\,, \qquad \phi = \phi'\,, \qquad z = z'\,, \\ x = \frac{r\,x'}{\sqrt{x'^2 + y'^2}}\,, \qquad y = \frac{r\,y'}{\sqrt{x'^2 + y'^2}}\,, \qquad z = z'\,. \end{gathered} \tag{32.11}$$

As before, the primed coordinates refer to virtual space and the unprimed coordinates to physical space. The first line of Eqs. (32.11) writes the transformation

in cylindrical polars while the second line is the same transformation in Cartesian coordinates according to relation (10.2).

Exercise 32.1
Calculate the dielectric properties required to implement the transformation (32.11) using Cartesian coordinates. Find the principal values ε_i of the dielectric tensor.
Solution
We calculate the transformation matrix ${\Lambda^i}_{i'}$ from $x^i = \{x, y, z\}$ to $x^{i'} = \{x', y', z'\}$ and find

$$ {\Lambda^i}_{i'} = \begin{pmatrix} R\cos^2\phi + \frac{r}{r'}\sin^2\phi & \left(R - \frac{r}{r'}\right)\cos\phi\sin\phi & 0 \\ \left(R - \frac{r}{r'}\right)\cos\phi\sin\phi & \frac{r}{r'}\cos^2\phi + R\sin^2\phi & 0 \\ 0 & 0 & 1 \end{pmatrix}, \tag{32.12} $$

with the abbreviation

$$ R = \frac{\mathrm{d}r}{\mathrm{d}r'} \tag{32.13} $$

and use of the relationship (10.2) between Cartesian and cylindrical coordinates. Now we apply the recipe (32.5). The transformation matrix $\mathbf{\Lambda}$ is given by Eq. (32.12), which has the determinant

$$ \det \mathbf{\Lambda} = \frac{r}{r'} R. \tag{32.14} $$

The use of Cartesian coordinates in virtual and physical space means the matrix $\mathbf{g}'$ is the identity and $g' = \gamma = 1$, so formula (32.5) simplifies to $\boldsymbol{\varepsilon} = \boldsymbol{\mu} = \mathbf{\Lambda}\mathbf{\Lambda}^T / \det \mathbf{\Lambda}$ and we obtain

$$ \varepsilon^{ij} = \mu^{ij} = \frac{r'}{rR} \begin{pmatrix} R^2\cos^2\phi + \frac{r^2}{r'^2}\sin^2\phi & \left(R^2 - \frac{r^2}{r'^2}\right)\cos\phi\sin\phi & 0 \\ \left(R^2 - \frac{r^2}{r'^2}\right)\cos\phi\sin\phi & \frac{r^2}{r'^2}\cos^2\phi + R^2\sin^2\phi & 0 \\ 0 & 0 & 1 \end{pmatrix}. \tag{32.15} $$

The principle values of the dielectric tensors are the eigenvalues of the matrix (32.15); after some algebra one finds the three eigenvalues

$$ \varepsilon_i = \left\{ R\frac{r'}{r},\ \frac{r}{r'R},\ \frac{r'}{rR} \right\}. \tag{32.16} $$

Alternatively, we can calculate $\boldsymbol{\varepsilon} = \boldsymbol{\mu}$ in the coordinate system that is best adapted to cylindrical transformations. With cylindrical polar coordinates in virtual and physical space the transformation (32.11) from $x^i = \{r, \phi, z\}$ to $x^{i'} = \{r', \phi', z'\}$ gives the matrix

$$ \mathbf{\Lambda} = \mathrm{diag}\,(R, 1, 1)\,, \quad \det \mathbf{\Lambda} = R\,, \tag{32.17} $$

with the abbreviation (32.13). From the cylindrical–polar metric (11.21) we see that $\mathbf{g}' = \mathrm{diag}\,(1, r'^2, 1)$, $g' = r'^2$ and $\gamma = r^2$; hence the dielectric tensors (32.5) in cylindrical polars are easily found to be

$$ \varepsilon^{ij} = \mu^{ij} = \mathrm{diag}\left(R\frac{r'}{r}, \frac{1}{rr'R}, \frac{r'}{rR} \right). \tag{32.18} $$

Lowering an index with the metric $\gamma_{ij} = \text{diag}\,(1, r^2, 1)$ of physical space we obtain

$$\varepsilon^i{}_j = \mu^i{}_j = \text{diag}\left(R\frac{r'}{r}, \frac{r}{r'R}, \frac{r'}{rR}\right), \tag{32.19}$$

which obviously has the eigenvalues (32.16), the principle values of the dielectric tensors. The calculation is elementary in cylindrical polar coordinates compared to some tedious algebra if Cartesian coordinates are used.

We deduced the dielectric properties required for implementing the transformation (32.11) of the radial coordinate. Now, if the transformation (32.11) opens a hole in physical space (Fig. 4.1), anything inside this hole is decoupled from the electromagnetic field: it has become invisible (Fig. 5.1).

§33. Perfect invisibility devices

Transformation media that have holes in their coordinate grids in physical space act as perfect invisibility devices. Consider the following simple example in spherical polar coordinates (Pendry, Schurig and Smith [2006]). Suppose the device performs the transformation

$$r = r(r'), \quad \theta = \theta', \quad \phi = \phi', \tag{33.1}$$

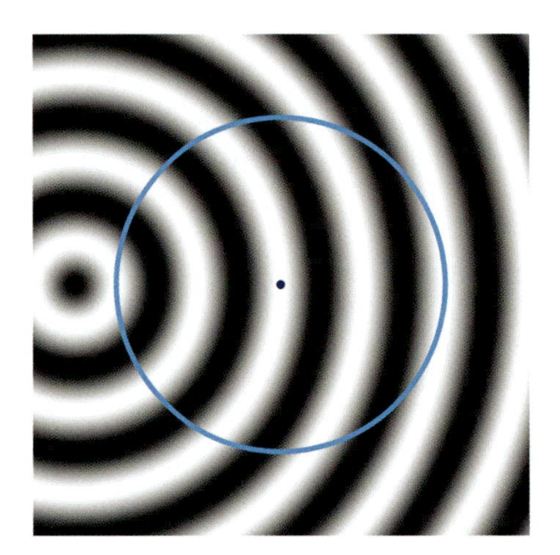

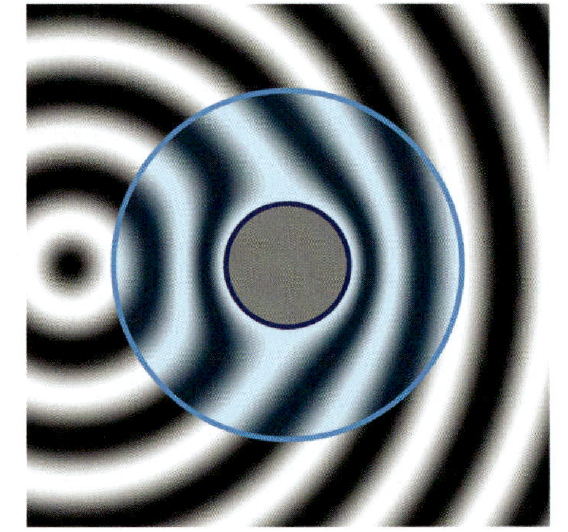

Figure 5.1: Waves in a perfect cloaking device. The left picture shows virtual space, where monochromatic waves from a point source propagate in vacuum. The right picture shows physical space containing the device which performs the transformation of Figs. 2.6 and 4.1. The light blue circle is the outer radius of the device, beyond which virtual and physical space agree; the origin in virtual space (dark blue dot) is blown up to the inner lining of the device (dark blue circle) by the transformation. The cloaking shell in physical space (blue shaded region) guides the wave around the invisible inner core.

so that the coordinates $x^{i'} = \{r', \theta', \phi'\}$ in virtual space differ from the coordinates $x^i = \{r, \theta, \phi\}$ in physical space by a transformation of the radial coordinate. This is the spherical–polar version of the cylindrical transformation (32.11) and it has the transformation matrix

$$\mathbf{\Lambda} = \mathrm{diag}\,(R,\ 1,\ 1)\ , \quad R = \frac{\mathrm{d}r}{\mathrm{d}r'}\,, \quad \det \mathbf{\Lambda} = R\,. \tag{33.2}$$

The metric in virtual and physical space is the spherical polar metric (11.17), so $\mathbf{g}' = \mathrm{diag}\,(1,\ r'^2,\ r'^2 \sin^2\theta')$, $g' = r'^4 \sin^2\theta'$ and $\gamma = r^4 \sin^2\theta$. The transformation medium (32.5) is therefore given by

$$\varepsilon^{ij} = \mu^{ij} = \frac{r'^2}{r^2 R}\,\mathrm{diag}\left(R^2,\ \frac{1}{r'^2},\ \frac{1}{r'^2 \sin^2\theta}\right). \tag{33.3}$$

Lowering an index using the physical–space metric $\gamma_{ij} = \mathrm{diag}\,(1,\ r^2, r^2\sin^2\theta)$ we obtain

$$\varepsilon^i{}_j = \mu^i{}_j = \mathrm{diag}\left(R\frac{r'^2}{r^2},\ \frac{1}{R},\ \frac{1}{R}\right). \tag{33.4}$$

The three diagonal elements in Eq. (33.4) are the principle values of the dielectric tensors, as discussed in §32.

Figure 5.2 shows an example of a radial transformation where $r(r')$ reaches a finite value R_1 at the origin $r' = 0$ in virtual space. This means that in physical space the entire spherical surface of radius R_1 corresponds to a single point in virtual space. Anything lurking inside the sphere of radius R_1 in physical space has become excluded from the electromagnetic field outside this sphere: it has become invisible. Figure 5.2 also shows that beyond the radius R_2 virtual and physical coordinates agree, so in this region of physical space electromagnetic waves are indistinguishable from waves that have propagated through empty space. The device (33.4) has an outer radius R_2 and an inner radius R_1 and it guides electromagnetic waves around the inner region without causing disturbances (Figs. 5.3 and 5.1). Hence any object inside the inner radius R_1 has not only disappeared from sight, but the act of hiding has become undetectable; the scenery behind the device would show no sign of the object nor of the device. Light emitted inside the cloaked inner region cannot escape, since this is the time–reverse of light entering this region, something that has just been ruled out. In short, this transformation makes a perfect cloaking device (Pendry, Schurig and Smith [2006]).

Fortunately perhaps, perfect invisibility is severely limited. Figure 5.3 shows the trajectories of light rays passing through the cloaking shell. The rays make a detour around the hidden core of the device, but they must arrive at the same time as if they were propagating through empty space. There is only one option: inside the cloaking device the phase velocity of light must exceed c. This is possible in principle, although only in regions of the spectrum with narrow bandwidth that correspond to resonances in the material (Milonni [2004]). But, to make matters worse, light rays straddling the inner lining of the cloak must go around the inner core in precisely the same time it would take them to traverse a single point in

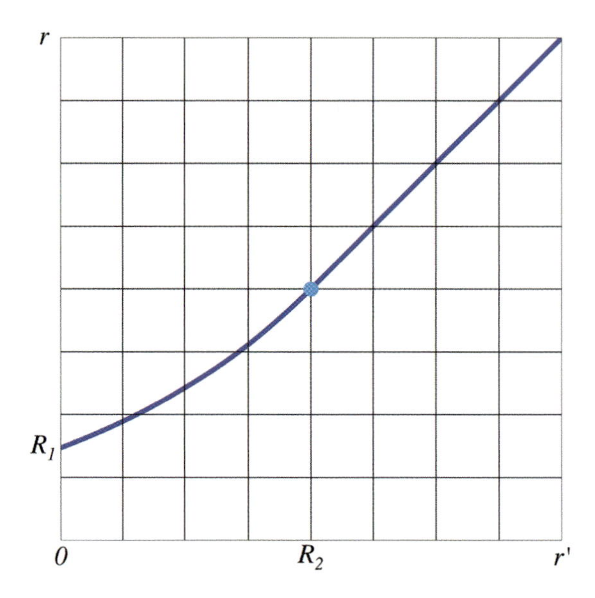

Figure 5.2: Radial transformation performed by a spherical cloaking device. The radius r' of spherical polar coordinates in virtual space is transformed to the radius r in physical space. R_1 denotes the inner and R_2 the outer radius of the cloaking shell: physical regions with radius $r < R_1$ are not reached by electromagnetic waves and for $r > R_2$ the virtual coordinates agree with the coordinates of physical space.

virtual space, i.e. zero time. Light must propagate at infinite phase velocity! One can put these thoughts into precise mathematical terms by considering the product $\varepsilon_1\varepsilon_2\varepsilon_3$ of the three principle values of the dielectric tensors of the transformation medium. The product $\varepsilon_1\varepsilon_2\varepsilon_3$ is the product of the eigenvalues of the matrix ${\varepsilon^i}_j$, which is equal to the determinant of the matrix. Equation (32.5) gives a matrix equation for ε^{ij} and Eq. (32.10) shows that this equation must be right–multiplied by the matrix γ_{ij} to obtain ${\varepsilon^i}_j$. We therefore obtain from Eqs. (32.5) and (32.10)

$$\varepsilon_1\,\varepsilon_2\,\varepsilon_3 = \det\left({\varepsilon^i}_j\right) = \left(\frac{\sqrt{g'}}{\sqrt{\gamma}}\right)^3 \frac{(\det\boldsymbol{\Lambda})^2\gamma}{(\det\boldsymbol{\Lambda})^3 g'} = \frac{\sqrt{g'}}{\sqrt{\gamma}}\frac{1}{\det\boldsymbol{\Lambda}}\,. \tag{33.5}$$

We showed in §11 that $\sqrt{g'}$ determines the volume element, here the volume element in virtual space. The volume of the single point in virtual space that is mapped to the hidden interior of the perfect cloaking device is zero. Hence Eq. (33.5) proves that on the inner surface of the cloak at least one of the principle values of the dielectric tensor is zero—the speed of light reaches infinity at the inner lining of the cloak. This inevitable feature restricts the performance of perfect invisibility devices to a single frequency set by the particular metamaterial used. A simplified version of the invisibility recipe has been implemented at a microwave frequency by Schurig et al. [2006]. This device hides objects from microwave radiation of one frequency and polarization.

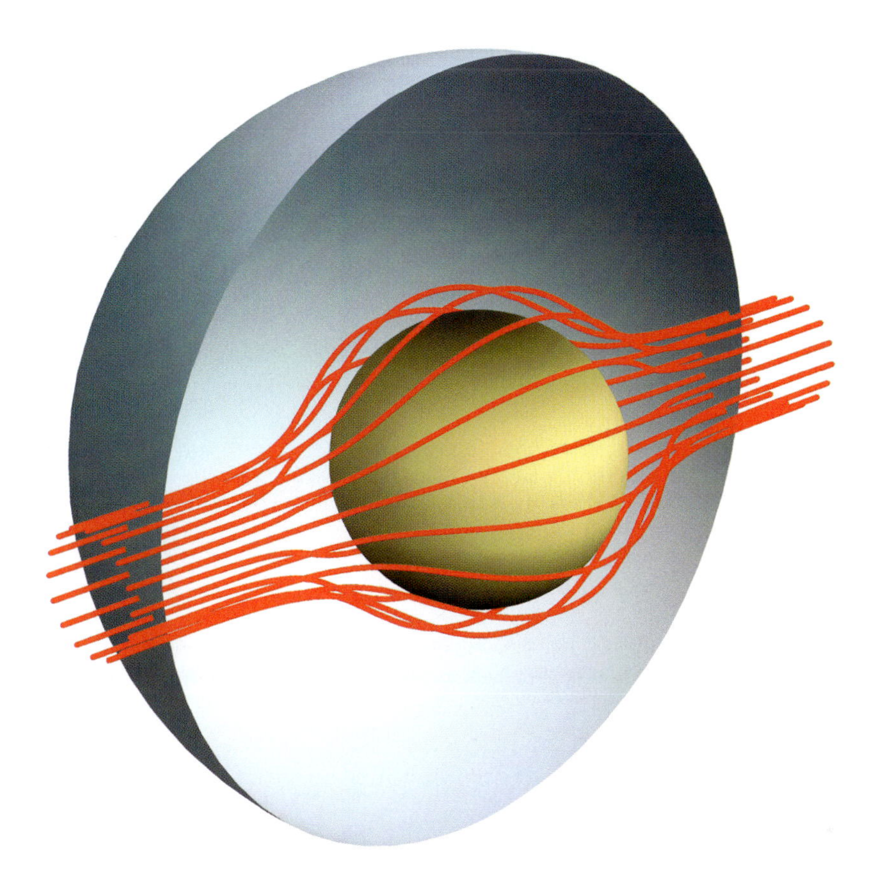

Figure 5.3: Light rays (red) in a perfect cloaking device. The inner and outer surfaces show the inner and outer linings of the cloak. Light rays are guided around the cloaked region inside the inner lining and leave in the direction they entered. (Picture inspired by Pendry, Schurig and Smith [2006]).

Perfect invisibility devices are impractical, but imperfect devices inspired by these ideas may be possible. A theoretical scheme for cloaking that does not require infinite phase velocities has been given by Leonhardt and Tyc [2009]; this device is not a transformation medium but rather creates an effective curved space for light and achieves cloaking within the accuracy of geometrical optics.

Exercise 33.1

Consider a plane wave incident on a spherical cloaking device (with arbitrary radial scaling function). Calculate the $\boldsymbol{E}$–, $\boldsymbol{D}$–, $\boldsymbol{H}$– and $\boldsymbol{B}$–fields in the device and discuss their behavior near the inner boundary of the cloak. Use these expressions to calculate the energy density and the Poynting vector. Calculate also the phase gradient k_i and the corresponding wave vector k^i with index raised according to the geometry of light implemented by the cloaking device. Discuss the results.

Solution

Because of the symmetry of the spherical cloaking device, the modification inside the device of the electromagnetic fields of the incoming plane wave is the same regardless of the polarization and direction of the wave. For simplicity, and without loss of generality, we take the plane wave outside the device to propagate in the z–direction and have electric field along the x–axis. In Cartesian coordinates in virtual space, which agrees with physical space outside the device, the electric field strength of the plane wave is therefore

$$E_{x'} = \mathcal{E}\,\mathrm{e}^{\mathrm{i}(kz'-\omega t)}\,, \quad E_{y'} = E_{z'} = 0\,, \quad k = \frac{\omega}{c}\,, \tag{33.6}$$

where the amplitude of the electric field is denoted by $\mathcal{E}$ and the plane wave has wave number k and frequency ω. It then follows from Maxwell's equations (30.29) that the magnetic field in (empty) virtual space is

$$H_{y'} = \mathcal{H}\mathrm{e}^{\mathrm{i}(kz'-\omega t)}\,, \quad H_{x'} = H_{z'} = 0\,, \quad \mathcal{H} = \varepsilon_0 c\,\mathcal{E}\,. \tag{33.7}$$

We express these fields in spherical polar coordinates in virtual space; the transformation matrix for the one–forms (33.6) and (33.7) is given by Eq. (10.9) and in spherical polars the fields are

$$\begin{aligned} (E_{r'},\ E_{\theta'},\ E_{\phi'}) &= (\sin\theta'\cos\phi',\ r'\cos\theta'\cos\phi',\ -r'\sin\theta'\sin\phi')\mathcal{E}\,\mathrm{e}^{\mathrm{i}(kr'\cos\theta'-\omega t)}\,, \\ (H_{r'},\ H_{\theta'},\ H_{\phi'}) &= (\sin\theta'\sin\phi',\ r'\cos\theta'\sin\phi',\ r'\sin\theta'\cos\phi')\mathcal{H}\,\mathrm{e}^{\mathrm{i}(kr'\cos\theta'-\omega t)}\,. \end{aligned} \tag{33.8}$$

The fields E_i and H_i in spherical polar coordinates in physical space are given by applying the transformation (33.1) to the one–forms (33.8). The transformation matrix for the one–forms is $\Lambda^{i'}_{\ i} = \partial x^{i'}/\partial x^i$, which is the inverse of $\boldsymbol{\Lambda}$ given in Eqs. (33.2); we therefore obtain

$$\begin{aligned} (E_r,\ E_\theta,\ E_\phi) &= \left(R^{-1}\sin\theta\cos\phi,\ r'\cos\theta\cos\phi,\ -r'\sin\theta\sin\phi\right)\mathcal{E}\,\mathrm{e}^{\mathrm{i}(kr'\cos\theta-\omega t)}\,, \\ (H_r,\ H_\theta,\ H_\phi) &= \left(R^{-1}\sin\theta\sin\phi,\ r'\cos\theta\sin\phi,\ r'\sin\theta\cos\phi\right)\mathcal{H}\,\mathrm{e}^{\mathrm{i}(kr'\cos\theta-\omega t)}\,. \end{aligned} \tag{33.9}$$

Now whatever the quantitative details of the function $r = r(r')$ in the transformation (33.1), in order to be a perfect invisibility recipe the point $r' = 0$ must correspond to a non–zero value of r, the inner surface of the cloaking device (Fig. 5.2). We therefore see from our result (33.9) that at the inner lining of the cloak the angular components of the $\boldsymbol{E}$– and $\boldsymbol{H}$–fields vanish; at the inner boundary these fields are perpendicular to the boundary.

The $\boldsymbol{D}$– and $\boldsymbol{B}$–fields in physical space are obtained from the fields (33.9) using the dielectric tensors (33.3). The result is

$$\begin{aligned} (D^r,\ D^\theta,\ D^\phi) &= \left(\frac{r'^2}{r^2}\sin\theta\cos\phi,\ \frac{r'}{r^2R}\cos\theta\cos\phi,\ -\frac{r'}{r^2R\sin\theta}\sin\phi\right)\varepsilon_0\mathcal{E}\,\mathrm{e}^{\mathrm{i}(kr'\cos\theta-\omega t)}\,, \\ (B^r,\ B^\theta,\ B^\phi) &= \left(\frac{r'^2}{r^2}\sin\theta\sin\phi,\ \frac{r'}{r^2R}\cos\theta\sin\phi,\ \frac{r'}{r^2R\sin\theta}\cos\phi\right)\mu_0\mathcal{H}\,\mathrm{e}^{\mathrm{i}(kr'\cos\theta-\omega t)}\,. \end{aligned} \tag{33.10}$$

At the inner surface of the cloak ($r' = 0$) we see that the $\boldsymbol{D}$– and $\boldsymbol{B}$–fields vanish completely.

For the plane wave transformed by the cloaking device geometrical optics is exact (§30). In this case the electric and magnetic fields carry the same energy (30.46) with the electric energy density (30.44). We obtain from the permittivity tensor (33.3) and our result (33.9) for the total energy density

$$I = 2\varepsilon_0|\mathcal{E}|^2\frac{r'^2}{r^2R}\,. \tag{33.11}$$

In order to calculate the Poynting vector (30.50) we express the vector product $\boldsymbol{E}^* \times \boldsymbol{H}$ with Levi–Civita tensor (14.7) in spherical polars with determinant of the metric $\gamma = r^4\sin^2\theta$. We obtain

$$S^i = 2\,\mathcal{E}\mathcal{H}\left(\frac{r'^2}{r^2}\cos\theta,\ -\frac{r'}{r^2R}\sin\theta,\ 0\right). \tag{33.12}$$

The wave vector in physical space is the gradient of the phase of the wave (33.8); more accurately, the gradient is the one–form k_i. In spherical polar coordinates we find

$$(k_r,\ k_\theta,\ k_\phi) = (\partial_r,\ \partial_\theta,\ \partial_\phi)(kr'\cos\theta - \omega t) = k(R^{-1}\cos\theta,\ -r'\sin\theta,\ 0)\,. \tag{33.13}$$

For the optical geometry implemented by the cloaking device the line element is

$$\mathrm{d}s^2 = \mathrm{d}r'^2 + r'^2(\mathrm{d}\theta^2 + \sin^2\theta\,\mathrm{d}\phi^2) = R^{-2}\mathrm{d}r^2 + r'^2(\mathrm{d}\theta^2 + \sin^2\theta\,\mathrm{d}\phi^2)\,, \tag{33.14}$$

so the inverse metric tensor is

$$g^{ij} = \mathrm{diag}\left(R^2,\ \frac{1}{r'^2},\ \frac{1}{r'^2\sin^2\theta}\right). \tag{33.15}$$

Raising the index on k_i using the inverse metric (33.15) gives the wave vector components k^i:

$$(k^r,\ k^\theta,\ k^\phi) = k\left(R\cos\theta,\ -r'^{-1}\sin\theta,\ 0\right), \tag{33.16}$$

which is consistent with relation (30.49) for the Poynting vector (33.12) and the energy density (33.11). We see that at the inner boundary of the cloak the angular components of k_i vanish (the fact that the ϕ–components are zero everywhere is a consequence of the wave impinging along the z–axis). The wave encircles the inner lining of the cloak with uniform phase (Fig. 5.1), so the phase must not change in angular directions; the angular components of the phase gradient are zero. On the other hand, the θ–component of the wave vector k^i goes to infinity. From formula (30.19) follows that the speed of light approaches infinity at the inner lining of the cloak.

Exercise 33.2

In §9 we discussed a class of optical instruments—Eaton lenses, invisible spheres etc.—with spherically–symmetric index profiles that diverge near the origin, which makes them impossible to build in practice. This exercise shows how transformation optics can be used to implement them. Consider a non–empty virtual space containing a spherically symmetric device with $\varepsilon' = \mu' = n(r')$. Applying a general radial transformation (33.1), the dielectric tensors in physical space are given by Eq. (33.4), apart from a factor of $n(r')$ according to Eq. (32.6):

$$\varepsilon^i{}_j = \mu^i{}_j = n(r')\,\mathrm{diag}\left(R\frac{r'^2}{r^2},\ \frac{1}{R},\ \frac{1}{R}\right), \qquad R = \frac{\mathrm{d}r}{\mathrm{d}r'}\,. \tag{33.17}$$

Suppose the refractive–index profile $n(r')$ goes as $n \sim r'^p$ for $r' \to 0$. Note that if p is not equal to 0 then the refractive index in virtual space vanishes or diverges as $r' \to 0$, giving an infinite or zero phase velocity of light at the centre of the device. Choose the transformation to physical space to be $r = r'^{p+1}$ and show that the dielectric tensors (33.17) in physical space are finite. Discuss the result.

Solution

We obtain by differentiation:

$$\varepsilon^i{}_j = \mu^i{}_j \sim \mathrm{diag}\left(p+1,\ \frac{1}{p+1},\ \frac{1}{p+1}\right) \quad \text{for} \quad r \to 0\,. \tag{33.18}$$

In contrast to the original index profile, the principle values of the dielectric tensors (33.18) in physical space do not vanish or diverge as $r \to 0$, provided $p > -1$. This shows how refractive–index profiles with singularities in the material properties, for example the Eaton lens and the invisible sphere of §9, can be implemented with an anisotropic medium free of such singularities (Tyc and Leonhardt [2008]). Only one harmless trace is left of the index singularity: the principal axes of the anisotropic material are not defined at the origin; the eigenvectors of the dielectric tensors point in all directions there, a feature known as a topological defect. One can never completely remove singularities, but one can transmute them into harmless topological defects. Note that this procedure works because the singular behaviour of $n(r')$ occurs at the single point $r' = 0$. If the device in virtual space is the perfect invisibility device of this Section then the transformation used in this Exercise will not remove the singular properties of the inner cloak of the device; this is because the singular behaviour occurs on a sphere rather than at a point.

§34. Negative refraction and perfect lenses

The development of metamaterials has been initiated and inspired by the idea (Pendry [2000]) that a flat lens with negative refractive index (Veselago [1968]) makes a perfect lens. A medium with a negative index, also called a left–handed medium, is one in which both the permittivity and permeability are negative, a circumstance that does not occur in natural materials. It turns out that the standard case of negative refraction with $\varepsilon = \mu = -1$ is an example of a transformation medium (Leonhardt and Philbin [2006]). As in perfect invisibility devices, the unusual properties of the device are based on an unusual topology of the transformation. For achieving invisibility, virtual space does not cover the entire physical space, whereas for negative refraction virtual space is multi–valued: single points in virtual space are mapped to multiple points in physical space.

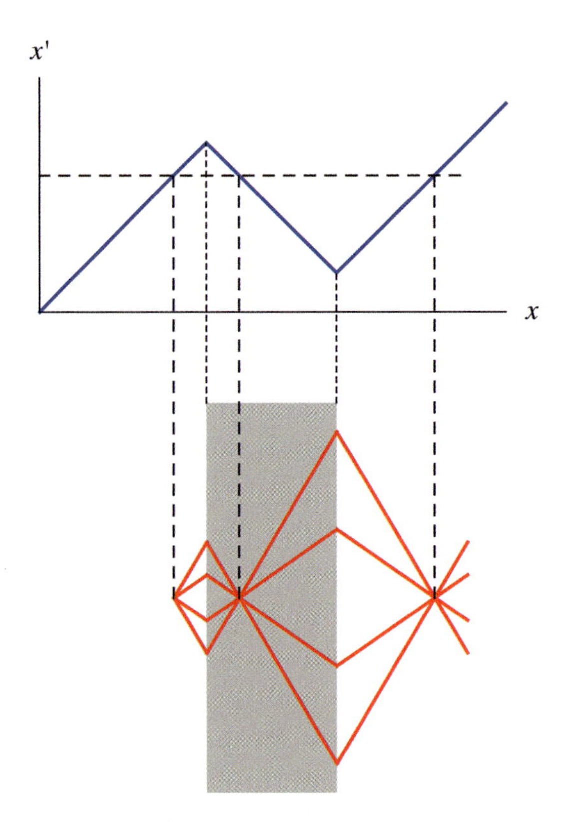

Figure 5.4: Perfect lens. The top figure shows a suitable coordinate transformation from the physical x–axis to the x'–axis in virtual space; the lower figure illustrates the corresponding device. The inverse transformation from x' to x is either triple or single–valued. The triple–valued segment on the physical x–axis corresponds to the focal region of the lens: any source point has two images, one inside the lens and one on the other side. Since the device facilitates an exact coordinate transformation, the images are perfect.

To discuss perfect imaging by negative refraction, consider in Cartesian coordinates the transformation

$$x = x(x') , \quad y = y' , \quad z = z' \tag{34.1}$$

by the folded curve shown in Fig. 5.4; this folding of space is visualized in Fig. 5.5. We see in Fig. 5.4 that in the fold of the function $x(x')$ each point x' in virtual space has three faithful images in physical space, so the electromagnetic field at these three points is the same as that at the point x'. Electromagnetic fields at each of the three points in physical space are therefore perfectly imaged at the other two: the device is a perfect lens (Fig. 5.6). We easily obtain from the theory of transformation media that this lens is a left–handed medium with negative permittivity and permeability: the Cartesian transformation (34.1) has the transformation matrix $\boldsymbol{\Lambda} = \mathrm{diag}\,(\mathrm{d}x/\mathrm{d}x',\ 1,\ 1)$ and we find from the recipe (32.5)

$$\boldsymbol{\varepsilon} = \boldsymbol{\mu} = \mathrm{diag}\left(\frac{\mathrm{d}x}{\mathrm{d}x'}, \frac{\mathrm{d}x'}{\mathrm{d}x}, \frac{\mathrm{d}x'}{\mathrm{d}x}\right). \tag{34.2}$$

Inside the device, i.e. inside the fold in the transformation of Fig. 5.4, the derivative $\mathrm{d}x'/\mathrm{d}x$ becomes negative and the coordinate system changes handedness. The electromagnetic left–handedness of the negative–index material appears through a transformation to a left–handed coordinate system. When the negative slope in the transformation is $\mathrm{d}x'/\mathrm{d}x = -1$, Eq. (34.2) gives the standard perfect lens made of an isotropic material with $\varepsilon = \mu = -1$ (otherwise the material (34.2) is anisotropic). As Fig. 5.4 shows, the imaging range is equal to the thickness of the lens in this case.

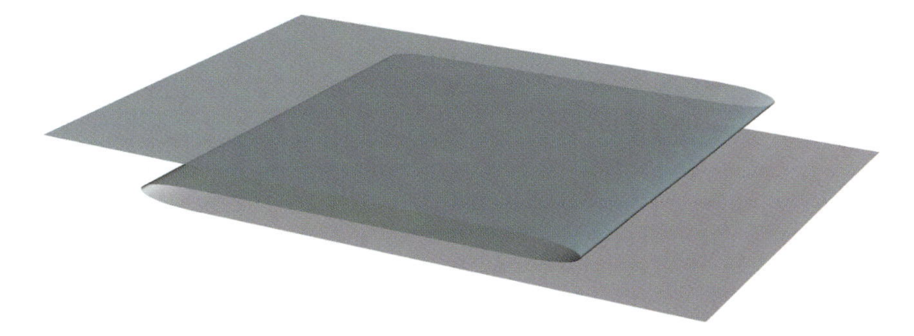

Figure 5.5: Negative refraction and folding of space. In negative refraction, physical space appears to be folded in virtual space (Fig. 5.4). Physical space seems to run backwards inside the negatively–refracting medium.

Perfect lensing was first analyzed through the imaging of evanescent waves in a slab of negatively–refracting material, waves that may carry images finer than the optical resolution limit (Pendry [2000]). Various aspects of this idea have been subject to a considerable theoretical debate (see Minkel [2002]) but experiments

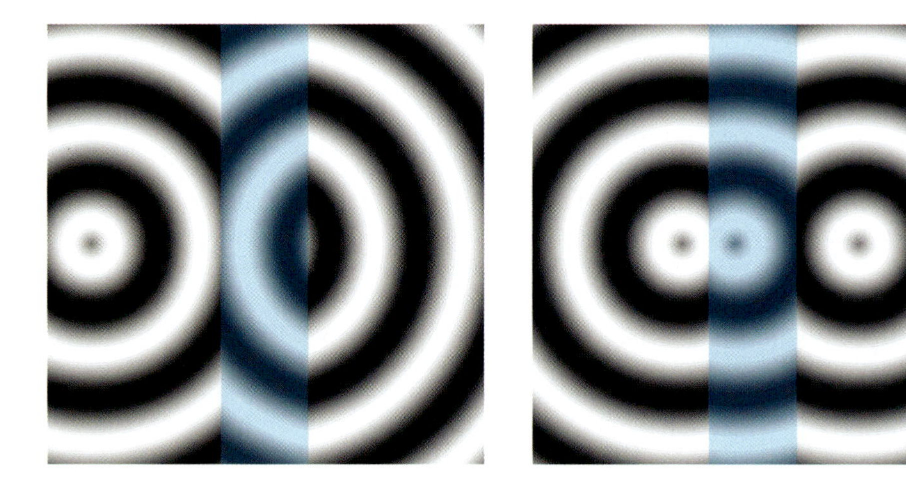

Figure 5.6: Propagation of monochromatic electromagnetic waves in a perfect lens. The lens facilitates the coordinate transformation shown in Fig. 5.4. The wave on the left is emitted from a point source outside the imaging range of the lens; this wave is transformed by the lens, but the device is not sufficiently thick to form an image. The wave on the right is emitted inside the imaging range, creating two images of the source, one inside the device and one outside, corresponding to the image points of Fig. 5.4.

have confirmed negative refraction (see e.g. Valentine et al. [2008]). Sub–resolution imaging was observed for a "poor–man's perfect lens" (Fang et al. [2005]) where the lens is effectively implemented by a sub–wavelength sheet of silver. Our pictorial argument leads to a simple intuitive explanation of why such lenses are indeed perfect. It will also reveal some of the practical limitations of perfect imaging by negative refraction. But this requires a brief intermezzo on dipole radiation.

Exercise 34.1
Consider a vector potential of the form

$$\boldsymbol{A} = A(r,t)\,\boldsymbol{e}_z\,, \quad r = \sqrt{x^2+y^2+z^2}\,, \tag{34.3}$$

where $\boldsymbol{e}_z$ is the Cartesian basis vector of the z–coordinate. Write the magnetic induction $\boldsymbol{B}$ corresponding to this vector potential. Derive a wave equation for general $\boldsymbol{B}$ from the source–free, vacuum Maxwell equations. Use this equation for $\boldsymbol{B}$ to find the wave equation that $A(r,t)$ must obey for expression (34.3) to describe a solution of the source–free, vacuum Maxwell equations. Show that the equation for $A(r,t)$ is satisfied by

$$A(r,t) = \frac{1}{r}\,f(r \pm ct)\,, \tag{34.4}$$

where f is an arbitrary function of either the variable $r+ct$, or the variable $r-ct$.
Solution
From the representation (26.14) we obtain the magnetic induction $\boldsymbol{B}$ corresponding to the vector potential (34.3):

$$\boldsymbol{B} = \boldsymbol{\nabla}\times\boldsymbol{A} = \boldsymbol{\nabla}\times[A(r,t)\boldsymbol{e}_z]\,. \tag{34.5}$$

In empty space (or a spatial geometry) the wave equation of the magnetic field is exactly the same as the wave equation (29.4) of the electric field (§29). In flat space one obtains:

$$\nabla^2 \boldsymbol{B} - \frac{1}{c^2}\frac{\partial^2 \boldsymbol{B}}{\partial t^2} = 0\,. \tag{34.6}$$

The Levi–Civita tensor, used to construct the curl (34.5), has vanishing covariant derivative (§16) and, as we are in flat space, covariant derivatives commute (§21). Moreover, the covariant derivatives of Cartesian basis vectors vanish (§15), so $\nabla \boldsymbol{e}_z = 0$. Hence substitution of the $\boldsymbol{B}$–field (34.5) into the wave equation (34.6) gives

$$\nabla \times \left[\boldsymbol{e}_z \left(\nabla^2 A(r,t) - \frac{1}{c^2}\frac{\partial^2 A(r,t)}{\partial t^2}\right)\right] = 0\,, \tag{34.7}$$

which in turn furnishes a scalar wave equation for the function $A(r,t)$:

$$\nabla^2 A(r,t) - \frac{1}{c^2}\frac{\partial^2 A(r,t)}{\partial t^2} = 0\,. \tag{34.8}$$

This equation is most easily solved in spherical polar coordinates, because of the spatial dependence of $A(r,t)$ on the radial distance r. We recall the expression (17.24) for the Laplacian of a scalar field in spherical polar coordinates and obtain from the wave equation (34.8):

$$\frac{1}{r^2}\frac{\partial}{\partial r}\left(r^2\frac{\partial A(r,t)}{\partial r}\right) - \frac{1}{c^2}\frac{\partial^2 A(r,t)}{\partial t^2} = 0\,. \tag{34.9}$$

The first term can be re–expressed as

$$\frac{1}{r^2}\frac{\partial}{\partial r}\left(r^2\frac{\partial A(r,t)}{\partial r}\right) = \frac{1}{r}\frac{\partial^2}{\partial r^2}\left[rA(r,t)\right]\,, \tag{34.10}$$

so Eq. (34.9) is

$$\frac{\partial^2}{\partial r^2}\left[rA(r,t)\right] - \frac{1}{c^2}\frac{\partial^2}{\partial t^2}\left[rA(r,t)\right] = 0\,. \tag{34.11}$$

Equation (34.11) is the one–dimensional wave equation for the function $rA(r,t)$, one of the most familiar equations in physics. It is a well–known, and easily verified, fact that the one–dimensional wave equation is satisfied by any function of the variables $r \pm ct$; the plus sign gives a wave propagating in the negative r–direction whereas the minus sign gives a wave propagating in the positive r–direction. Taking f to be an arbitrary function, the solution for $A(r,t)$ is therefore Eq. (34.4).

Exercise 34.2

If we choose the minus sign in Eq. (34.4) we obtain a wave emerging from the origin $r = 0$ and propagating out to infinite radial distance:

$$\boldsymbol{A} = A(r,t)\boldsymbol{e}_z\,, \quad A(r,t) = \frac{1}{r}\,f(r-ct)\,, \tag{34.12}$$

where $f(r-ct)$ is an arbitrary function. We obtain formula (34.12) by solving the source–free Maxwell equations, but since the wave (34.12) emerges from the point $r = 0$ there must in fact be a source at $r = 0$ producing the wave. By solving the source–free Maxwell equations we have therefore found a solution for Maxwell's equations with a source! This apparent paradox is explained by the fact that the solution (34.12) diverges at $r = 0$, so it is not valid at the source point; on the other hand, at every point other than $r = 0$ there is no source present and we have correctly solved Maxwell's equations at these points. The solution (34.12) describes a point dipole, an infinitesimal current at the origin $r = 0$, oscillating along the z–axis. An important particular solution is

$$A(r,t) = \frac{d_0}{c}\frac{\mathrm{e}^{\mathrm{i}kr - \mathrm{i}\omega t}}{r}\,, \quad k = \frac{\omega}{c}\,, \tag{34.13}$$

where d_0 is a constant related to the strength of the dipole. The solution (34.13) describes a monochromatic wave, a wave of one frequency ω; it is the spherical version of a plane wave. For the monochromatic solution (34.13) the dipole oscillates continuously at frequency ω, producing a continuous spherical wave. By choosing the function $f(r-ct)$ in expression (34.12) to be peaked at some value of $r-ct$ one obtains a wave packet; in an extreme case where $f(r-ct)$ is sharply peaked at $r-ct=0$, and zero for all other values of $r-ct$, the dipole emits a flash of light at time $t=0$.

Calculate the electric and magnetic fields for the monochromatic dipole solution (34.12) and (34.13). Give a physical explanation of the form of the $\boldsymbol{H}$–field by referring to the nature of the source. Compare the result with the geometrical–optics solution in Exercise 30.2.

Solution

The coordinate system best adapted to the symmetry of the wave (34.12) is the spherical polar system. The Cartesian basis vector $\boldsymbol{e}_z$ in Eq. (34.12) has the following expansion in the spherical polar basis (use the matrix (10.12) and the inverse of the transformation (12.5)):

$$\boldsymbol{e}_z = \cos\theta\,\boldsymbol{e}_r - \frac{\sin\theta}{r}\,\boldsymbol{e}_\theta\,. \tag{34.14}$$

The components of the vector potential (34.12) in spherical polars are therefore

$$A^r = A\cos\theta\,, \quad A^\theta = -\,A\,\frac{\sin\theta}{r}\,, \quad A^\phi = 0\,, \tag{34.15}$$

giving the one–form components (use the spherical polar metric)

$$A_r = A\cos\theta\,, \quad A_\theta = -A\,r\sin\theta\,, \quad A_\phi = 0\,. \tag{34.16}$$

The $\boldsymbol{H}$–field in spherical polars is found from the $\boldsymbol{B}$–field (34.5) to be (use the Levi–Civita tensor (14.8) in spherical polars)

$$H_r = H_\theta = 0\,, \quad H_\phi = -\frac{r\,A_{,r}}{\mu_0}\sin^2\theta\,, \tag{34.17}$$

which for the wave (34.13) is

$$H_r = H_\theta = 0\,, \quad H_\phi = \frac{d_0}{\mu_0 c}(1-\mathrm{i}kr)\frac{\mathrm{e}^{\mathrm{i}kr-\mathrm{i}\omega t}}{r}\sin^2\theta\,. \tag{34.18}$$

The magnetic field thus circulates around the z–axis; this is as expected since a current produces a magnetic field that encircles it, and the dipole source is an (infinitesimal) oscillating current along the z–axis. The magnitude of $\boldsymbol{H}$ (remember to use the spherical polar metric) diverges as r^{-2} as $r\to 0$ and falls off as r^{-1} at large r. To obtain the $\boldsymbol{E}$–field we use the Maxwell equation

$$\nabla\times\boldsymbol{H} = \varepsilon_0\frac{\partial\boldsymbol{E}}{\partial t}\,. \tag{34.19}$$

As the $\boldsymbol{H}$–field (34.18) oscillates with frequency ω, we see from Eq. (34.19) that the $\boldsymbol{E}$–field does likewise. Taking the curl of the $\boldsymbol{H}$–field (34.18) we obtain from (34.19)

$$E_r = 2d_0\left(\frac{\mathrm{i}}{kr^3}+\frac{1}{r^2}\right)\mathrm{e}^{\mathrm{i}kr-\mathrm{i}\omega t}\cos\theta\,, \quad E_\theta = d_0\left(\frac{\mathrm{i}}{kr^2}+\frac{1}{r}-\mathrm{i}k\right)\mathrm{e}^{\mathrm{i}kr-\mathrm{i}\omega t}\sin\theta\,, \quad E_\phi = 0\,. \tag{34.20}$$

The $\boldsymbol{E}$–field is thus everywhere orthogonal to the $\boldsymbol{H}$–field; its magnitude diverges as r^{-3} as $r\to 0$ and falls off as r^{-1} at large r. The solution (34.18) and (34.20), to leading order in r, approaches the geometrical–optics solution (Exercise 30.2) as $r\to\infty$. Thus geometrical optics captures the far field of the radiation from the dipole, but not the near–field behaviour. Note in particular that the exact solution (34.20) for the $\boldsymbol{E}$–field has an r–component and is thus not orthogonal to the wave vector $\boldsymbol{k}=(\omega/c,0,0)$.

Consider the monochromatic point dipole of Exercise 34.2 placed within the imaging range of a perfect lens. The solution derived in Exercise 34.2 is correct in virtual space, because virtual space is empty. By applying the transformation (34.1) of the perfect lens, we find the behaviour of the wave in physical space (Fig. 5.6). The dipole is imaged inside the lens and also on the opposite side of the medium. We also discussed in Exercise 34.2 a wave–packet solution describing a flash of light produced by the dipole at an instant of time. If we consider the behaviour of this flash of light when the dipole is located within the imaging range of a perfect lens we encounter a paradox: the multivaluedness of the transformation defining the medium clearly implies that the flash from the dipole is instantaneously present at the two image points, inside the lens and on the opposite side. How can this flash appear instantaneously at two separated points without violating the fundamental law that signals cannot be sent faster than the speed of light in vacuum? There is no way of resolving this difficulty; a flash of light cannot exhibit such behaviour. If we consider the transformation that defines the perfect lens the problem is clear: single points in virtual space become multiple points in physical space, these points in physical space therefore have the same electromagnetic fields at all times, the situation is therefore necessarily a stationary one where the radiation in physical space is continuously propagating such that the fields at the three transformed points are the same. This implicit assumption of a stationary configuration rules out the flash of radiation as a solution of Maxwell's equations to which we can apply the multivalued transformation. On the other hand, the stationary wave created by the monochromatic source in Fig. 5.6 is a solution to which the transformation of the perfect lens can be applied. The picture of the perfect lens as a transformation medium thus indicates that there must be physical restrictions on materials with negative ε and μ. In particular it must be impossible for any medium to have $\varepsilon = \mu = -1$ for a narrow wave packet of radiation, and this restriction must arise from the requirement that sources radiate forward in time with signal speed equal to c. This is indeed what a detailed analysis of electromagnetism in media shows (Stockman [2007]). It is only possible for a material to have negative ε and μ for a narrow range of frequencies, whereas the flash of radiation considered above necessarily contains a very large range of frequencies and ε and μ will be positive for much of that range. The perfect lens will in reality only operate for approximately monochromatic waves at a frequency where ε and μ are negative. Moreover the same considerations lead to a conclusion that is not apparent from the transformation picture: negative ε and μ is unavoidably accompanied by significant absorption of the radiation by the material and this will degrade the image formation of the perfect lens (Stockman [2007]).

§35. CLOAKING AT A DISTANCE

The simple idea behind the cloaking devices described in §33 is to guide light around an interior region so that the light emerges as if it had propagated through empty space; any object in the interior region is thereby rendered invisible. A far less intuitive design for invisibility is cloaking at a distance (Lai et al. [2009a]), which involves placing the object at a specific position near a different type of cloaking device, but not inside it, one that is tailor made for the object. If, and only if, the object is placed at the required position outside the device, both object and cloak become invisible. This cloaking at a distance uses the idea of negatively refracting transformation media.

The simple transformation in Fig. 5.4 is the recipe for a medium with negative refraction that acts as a perfect lens. For an image to be formed the source must be within a certain range from the lens, the maximum distance being the thickness of the lens for the transformation in Fig. 5.4 (corresponding to a medium with $\varepsilon = \mu = -1$). This maximum distance for imaging can be made greater or less than the lens thickness by changing the value of the negative slope in Fig. 5.4 that determines the negatively refracting medium through Eq. (34.2). But the effect of this transformation medium on electromagnetic waves that originate at large distances beyond the imaging range is also remarkable: the medium cancels out a slab of physical space. To see this, recall that the vertical axis in Fig. 5.4 is virtual space where the light propagates in vacuum. The effective distance experienced by light is that in virtual space where there is no medium to distort its passage. Physical space is the horizontal axis in Fig. 5.4 and as light travels along this axis, traversing the medium defined by the plotted transformation, it clearly covers a distance in physical space that is greater than that in virtual space by twice the thickness of the medium. A slab of material with $\varepsilon = \mu = -1$ therefore cancels out with a slab of physical space of the same thickness, making a slice of physical space disappear that is twice the thickness of the material.

The transformation medium of Fig. 5.4 thus removes a portion of physical space for electromagnetic waves. A continuous stream of light whose wave vector has a non–zero x–component behaves as if a slab of physical space between two x–values were missing. The light is therefore altered compared to propagation through empty space, allowing an observer to detect the presence of the medium. A simple modification of the transformation in Fig. 5.4, however, gives a material configuration in physical space that is invisible—light in physical space traverses the material and emerges as if it had travelled through vacuum. This modified transformation is shown in Fig. 5.7. Note that the distance between two x–values in physical space, with one of the values less than zero and the other greater than x_3, is exactly the same as the distance between the corresponding x'–values in virtual space. Light now experiences the same distance in its traversal of the medium plus the adjacent empty slab $x_2 < x < x_3$ in physical space as it does in its corresponding propagation through (empty) virtual space, so the light in physical space emerges exactly as if the medium were not there: the medium is invisible. The feature of the transformation in Fig. 5.7 that implies invisibility is exactly that already used in §33: the coordinates in virtual space and in physical space agree at large distances. But in Fig. 5.7 there is no region in physical space where an arbitrary object can be placed without destroying the optical illusion; the recipe in Fig. 5.7 is for an invisible material, not for a cloaking device.

In terms of the cancellation property of a negatively refracting transformation medium, the invisibility effect in Fig. 5.7 works as follows. The portion of the plot

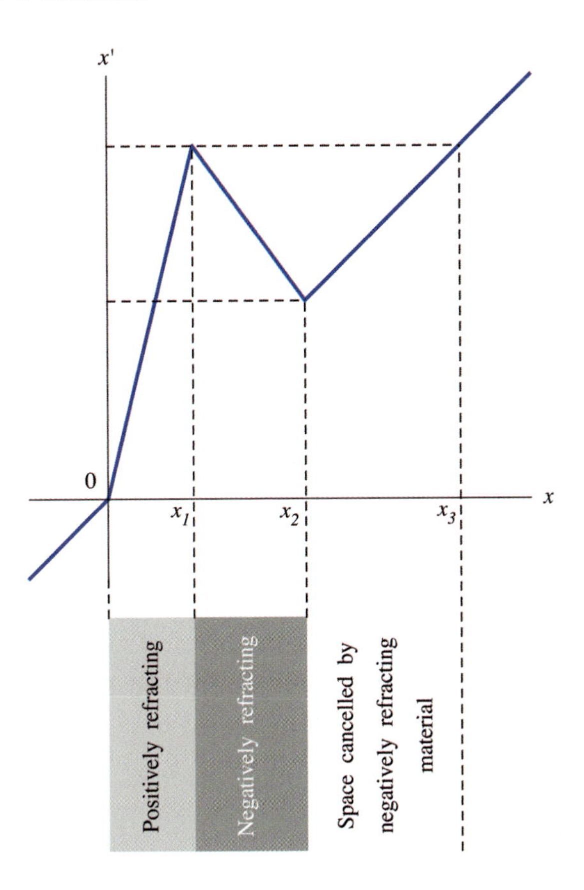

Figure 5.7: An invisible slab. The coordinates in virtual space and in physical space agree for $x = x' < 0$ and $x = x' > x_3$, so the medium in physical space is invisible. The medium consists of a negatively refracting slab, that cancels out with the empty region $x_2 < x < x_3$, and a positively refracting slab that restores the missing space.

with negative slope gives a negatively refracting slab in the region $x_1 < x < x_2$ that cancels out with the slab of empty space between $x = x_2$ and $x = x_3$, the result of which is that the entire region $x_1 < x < x_3$ in physical space is effectively removed. In order to achieve invisibility this cancelled space must be restored and this task is performed by the positively refracting slab in the region $0 < x < x_1$. The figure shows that the positively refracting material of thickness x_1 in physical space corresponds to a thickness x_3 in (empty) virtual space, so the material of thickness x_1 has an effective thickness x_3 for electromagnetic waves, thus restoring the missing space between $x = x_1$ and $x = x_3$.

The foregoing considerations were for idealized slabs of infinite extension in the (y, z)–plane, but the same ideas can be implemented in a finite volume by performing the same transformations on the radial coordinate of a spherical polar system. Figure 5.8 shows the transformation for an invisible medium in the spherical case,

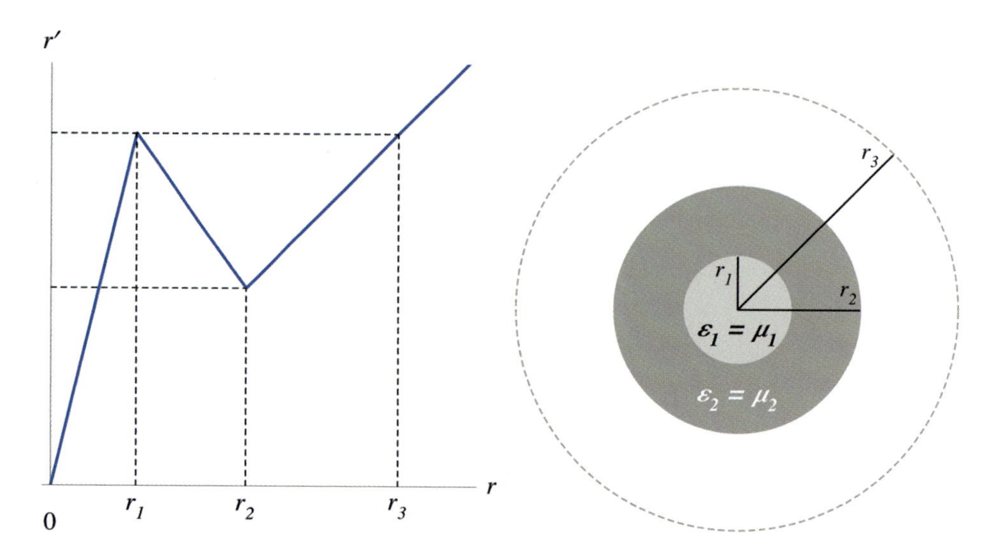

Figure 5.8: An invisible sphere. The plot on the left shows the transformation of Fig. 5.7 applied to the radial coordinate of a spherical polar system. Just as in Fig. 5.7, the result in physical space is an invisible medium, this time a sphere, shown on the right. The permittivities and permeabilities are given in Eqs. (35.2) and (35.3).

which produces an invisible sphere in physical space. As electromagnetic waves traverse the sphere in physical space they move through r–values down to zero and up again to large r–values; the agreement, at large values, of the r coordinate with the r' coordinate of virtual space shows that the result of this propagation in physical space is exactly that of propagating through vacuum. The shell of material between $r = r_1$ and $r = r_2$ is negatively refracting and cancels out with the vacuum shell $r_2 < r < r_3$, so that the entire shell $r_1 < r < r_3$ in physical space is removed. The core sphere of radius r_1 is positively refracting and has an effective radius r_3 for electromagnetic waves, so it restores the removed space. Hence the spherical region of radius r_3 in physical space is exactly equivalent to an empty sphere of the same radius. (The permittivities and permeabilities in Fig. 5.8 will be given below.)

We encountered an invisible sphere in §9, but that recipe was based on ray optics; no negative refraction was required and the invisibility was exact only for rays, not for waves. In contrast, the invisibility of the sphere in Fig. 5.8 is perfect for electromagnetic waves (and therefore also for rays).

All of the ingredients for cloaking at a distance have now been described. The final step is to modify the cancellation part of the invisible–sphere recipe. Recall that by taking virtual space to contain a material with arbitrary permittivity ε' and permeability μ' we obtain transformation media that perform effective coordinate transformations in this material. Consider placing an arbitrary object in virtual space in the shell $r_2 < r' < r_3$ and performing the transformation of Fig. 5.8. The resulting configuration in physical space is equivalent to virtual space, but now virtual space contains an object. Therefore light in physical space traversing the spherical region of radius r_3 emerges as if it had scattered off the object in virtual space. There is nothing dramatic about this overall effect in physical space, but consider the way in which it is achieved by cancellation and restoration. In physical space the shell $r_2 < r < r_3$ now contains the arbitrary object just as in the corresponding shell in virtual space (because the transformation is the identity in this range), but a transformed version of the object also appears in the shell $r_1 < r < r_2$ and in the core sphere $0 \leq r < r_1$ because of the multi–valuedness of the transformation. Just as before, the shell $r_1 < r < r_2$ cancels out with the shell $r_2 < r < r_3$ containing the object, so the version of the object produced in the shell $r_1 < r < r_2$ by the transformation is an "anti–object" that cancels out with the object. Meanwhile, the core sphere $0 \leq r < r_1$ restores the removed shell $r_1 < r < r_3$, but now the core includes a version of the object and is equivalent to the presence of the original object in an otherwise empty sphere of radius r_3, precisely the configuration in virtual space. Note that the cancellation part of this process removes all trace of the object, indeed it removes the entire region $r_1 < r < r_3$; it is the restoration performed by the core sphere, containing its version of the object, that effectively places the original object in the shell $r_1 < r < r_2$, thus reproducing the configuration in virtual space. Now, if we combine the cancellation part of the recipe just described, based on virtual space containing an object in the shell $r_2 < r' < r_3$, with the restoration part of the invisible–sphere recipe, in which virtual space is empty, we obtain a recipe for cancelling out the object in physical space and replacing it by empty space: the object will be cloaked.

Figure 5.9 shows the arrangement in physical space resulting from this cloaking prescription. For simplicity we will assume that the object is isotropic, with scalar permittivity ε' and permeability μ', although it is straightforward to treat a general anisotropic object. The object is placed in the shell $r_2 < r < r_3$, and the transformation of Fig. 5.8 is performed in the range $r_1 < r < r_3$ with the object also present in virtual space. This produces a shell $r_1 < r < r_2$ of material in physical space with the parameters required for it to cancel out with the shell $r_2 < r < r_3$; these parameters consist of an impedance–matched part ($\boldsymbol{\varepsilon}_2 = \boldsymbol{\mu}_2$) that cancels out with the empty space in the shell $r_2 < r < r_3$ (the same $\boldsymbol{\varepsilon}_2 = \boldsymbol{\mu}_2$ as in Fig. 5.8), and a part with permittivity $\boldsymbol{\varepsilon}_a$ and permeability $\boldsymbol{\mu}_a$ that cancels out with the object. The region with permittivity $\boldsymbol{\varepsilon}_a$ and permeability $\boldsymbol{\mu}_a$ is the anti–object that removes the object. To obtain the core sphere $0 \leq r < r_1$ the transformation of Fig. 5.8 is performed in empty virtual space, giving the same impedance–matched core ($\boldsymbol{\varepsilon}_1 = \boldsymbol{\mu}_1$) as for the invisible sphere. The core sphere is equivalent to an empty sphere of radius r_3, so the content of the dotted sphere in Fig. 5.9 is invisible.

It remains to compute the permittivities and permeabilities in Fig. 5.9 using the transformation recipe. First, we write the formula for the transformation of the radial coordinate shown in Fig. 5.8; it is easy to see that this piece–wise linear transformation is

$$r' = \begin{cases} \dfrac{r_3}{r_1} r & \text{for} \quad 0 < r < r_1 \,, \\[2ex] \dfrac{r_3 - r_2}{r_1 - r_2} (r - r_2) + r_2 & \text{for} \quad r_1 < r < r_2 \,, \\[2ex] r & \text{for} \quad r_2 < r \,. \end{cases} \tag{35.1}$$

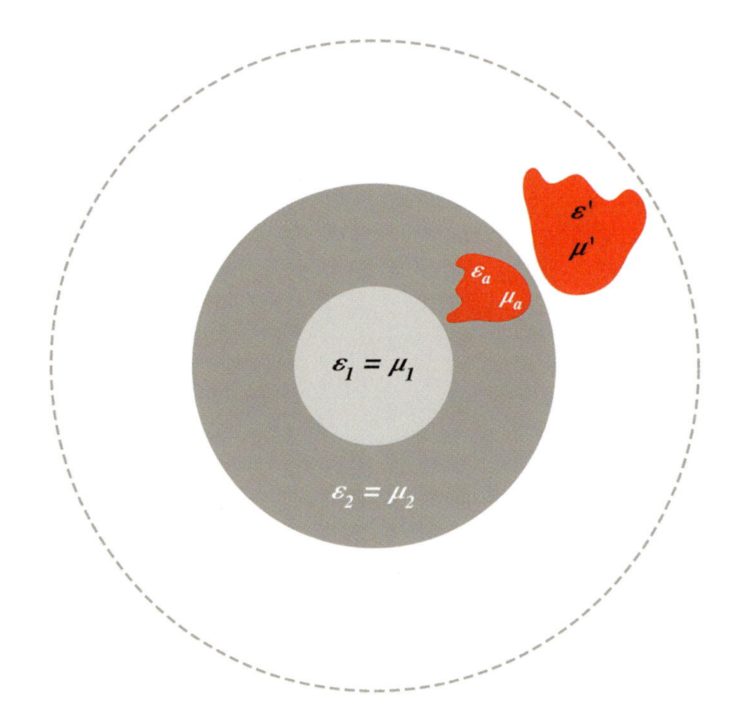

Figure 5.9: Cloaking at a distance. The shell containing an arbitrary object with permittivity ε' and permeability μ' cancels out with the outer shell of the cloaking device containing an anti–object with permittivity $\boldsymbol{\varepsilon}_a$ and permeability $\boldsymbol{\mu}_a$. The core of the cloak restores the removed region. The material properties are given in Eqs. (35.2)–(35.4)

The core sphere in Fig. 5.9 corresponds to the range $0 \leq r < r_1$ of the transformation (35.1), with empty virtual space. Applying the rule (32.5) and lowering an index with the physical–space metric (11.17) we obtain the following expression for the dielectric tensors of the core in spherical polar coordinates $x^i = \{r, \theta, \phi\}$:

$$\varepsilon_1{}^i{}_j = \mu_1{}^i{}_j = \frac{r_3}{r_1}\,\mathrm{diag}\,(1,\ 1,\ 1)\,. \tag{35.2}$$

In the range $r_1 < r < r_3$ the transformation (35.1) is applied to virtual space containing an object in the shell $r_2 < r' < r_3$. For those points in the shell that are not points of the object, the transformation is that of empty virtual space; this gives the dielectric tensors $\boldsymbol{\varepsilon}_2 = \boldsymbol{\mu}_2$ of Fig. 5.9 and the result is

$$\varepsilon_2{}^i{}_j = \mu_2{}^i{}_j = -\frac{r_3 - r_2}{(r_2 - r_1)}\,\mathrm{diag}\left(\left[\frac{r_1 r_2 + r_3 r - r_2(r + r_3)}{r(r_3 - r_2)}\right]^2,\ 1,\ 1\right). \tag{35.3}$$

For those points in the range $r_1 < r' < r_2$ that are points of the object the transformation is that of virtual space with permittivity ε and permeability μ. The material properties $\boldsymbol{\varepsilon}_a$ and $\boldsymbol{\mu}_a$ produced by the transformation are those of the (anisotropic) anti–object in Fig. 5.9; for the isotropic object we are considering, $\boldsymbol{\varepsilon}_a$ and $\boldsymbol{\mu}_a$ differ from $\boldsymbol{\varepsilon}_2$ and $\boldsymbol{\mu}_2$ by the factors ε' and μ', respectively:

$$\varepsilon_{a\ j}^{\ i} = \varepsilon' \, \varepsilon_{2\ j}^{\ i} \,, \quad \mu_{a\ j}^{\ i} = \mu' \, \mu_{2\ j}^{\ i} \,. \tag{35.4}$$

The material of Eq. (35.3) is negatively refracting; the anti–object (35.4) is negatively refracting if the object is positively refracting and vice versa. The shape of the anti–object is the transformed shape of the object; there is a flip of the shape in the radial direction and a distortion due to the compression of the distance between radial lines as one moves towards the origin, as is seen in the example in Fig. 5.9.

The cloaking device in Fig. 5.9 is tailor made for the object, and invisibility will occur only when the object has the position and orientation determined by the transformation that produces the anti–object in the cloak. We also assumed that the object does not absorb light, it only scatters light—the cloaking device reverses the scattering. In addition, the cloaking only works in a steady state where a continuous stream of light has adjusted to the combination of cloak plus object—cloaking at a distance is not an example of spooky action at a distance. Although in its entirety the spherical region bounded by the dotted curve in Fig. 5.9 is equivalent to empty space, any observer inside this sphere would detect both the object and the cloak.

In terms of the scattering of light, cloaking at a distance accomplishes an extraordinary feat: the scattered light from the cloaking device interferes destructively with that from the object, removing all evidence of the presence of both. Needless to say, it would be an enormous challenge to try to deduce the properties of such a cloaking device by thinking about the interference of scattered light. In contrast, geometry supplies the answer through the almost trivial transformation of Fig. 5.8.

The idea of this Section can be pursued further. Instead of restoring empty space in place of the cancelled–out object, one can "restore" a completely different object, thus replacing the object by an illusion (Lai et al. [2009b]).

§36. PERFECT IMAGING WITH POSITIVE REFRACTION

The perfect imaging by negative refraction we discussed in §34 is a beautiful example for the theory of transformation optics, but it does not come without problems in practice. We have seen that negatively refracting materials can only operate in a narrow range of the spectrum. It also turns out that such materials are inevitably absorptive (Stockman [2007]) and that absorption thoroughly ruins the superresolution of the (theoretically) perfect lens. But there are alternatives to perfect imaging by negative refraction. In §9 we already encountered one example of such perfect optical instruments: Maxwell's fish–eye with refractive–index profile

$$n = \frac{2}{1 + r^2/a^2} \,, \tag{36.1}$$

where r denotes the distance from the centre of the lens and a its characteristic length scale. The simple ray optics we had at our disposal at the stage of §9 did not allow us to prove that the fish–eye lens perfectly images light waves. In this Section we solve Maxwell's equations for Maxwell's fish–eye (Leonhardt and Philbin [2010]), which combines electromagnetism with non–Euclidean geometry.

At the heart of our theory lies Luneburg's visualization of Maxwell's fish–eye (Luneburg [1964]): the fish–eye lens implements the stereographic projection from a virtual sphere (Fig. 2.39) that is familiar from the Mercator projection of cartography (Fig. 2.40 and Exercise 9.10). In 2D, Maxwell's fish–eye projects the surface of a sphere of a radius a to the plane; in 3D, the case we consider here, the fish–eye projects the 3–dimensional surface of the 4–dimensional hypersphere to 3–dimensional space. Although hyperspace is not exactly easy to visualize (Fig. 3.25), intuition from ordinary spheres in 3–dimensional space will guide us on the way of our calculations, supplemented by minor modifications in 4D we worked out in §24.

Luneburg's visualization allows us to understand the focusing of light rays in Maxwell's fish–eye without much calculation. Consider a light ray in the refractive–index profile (36.1). As the profile is spherically symmetric, we know from §7 that the ray trajectory must lie in a plane (due to the conservation of angular momentum). The 3–dimensional problem is therefore reduced to the 2–dimensional case we can easily picture and have discussed in detail in §9. Let us recap the essentials: light follows the geodesics on the virtual sphere that are the great circles (Exercise 19.1). Light rays emitted from one point on the sphere reunite at the antipodal point. As the rays in physical space are faithfully represented by the rays on the virtual sphere, all rays that are emitted from one point $\boldsymbol{r}_0$ in physical space must focus at the stereographic projection (24.6) of the antipodal point:

$$\boldsymbol{r}_0' = -\frac{a^2}{|\boldsymbol{r}_0|^2}\,\boldsymbol{r}_0\,, \tag{36.2}$$

provided the rays all lie in the same plane. Note, however, that we can rotate this plane around the line between source $\boldsymbol{r}_0$ and image $\boldsymbol{r}_0'$ and get all ray trajectories emitted from $\boldsymbol{r}_0$ (Fig. 5.10), because this line intersects the origin and so defines an axis of symmetry in the spherically symmetric profile (36.1). Maxwell's fish–eye thus focuses all light rays from any point $\boldsymbol{r}_0$ at the corresponding image (36.2).

Exercise 36.1
The antipode to the point (X_0, Y_0, Z_0, W_0) on the hypersphere is $(-X_0, -Y_0, -Z_0, -W_0)$. Show that the stereographic projection (24.6) gives relation (36.2) between the projected points.

Maxwell's fish–eye perfectly focuses light rays, but the resolution of optical instruments is usually limited by the wave nature of light. Conventional lenses cannot focus much better than the wavelength (Born and Wolf [1999]), which in microscopy prevents us from seeing subwavelength detail—nanostructures, molecules, atoms are all blurred. Let us investigate whether Maxwell's fish–eye resolves better than conventional lenses. We anticipate that if the geometry of light is as perfect for electromagnetic waves as it is for rays, light waves are as perfectly focused as rays. We know from §26 that we need to require impedance matching (26.8) for having a perfect geometry in 3D:

$$\varepsilon = \mu = n\,. \tag{36.3}$$

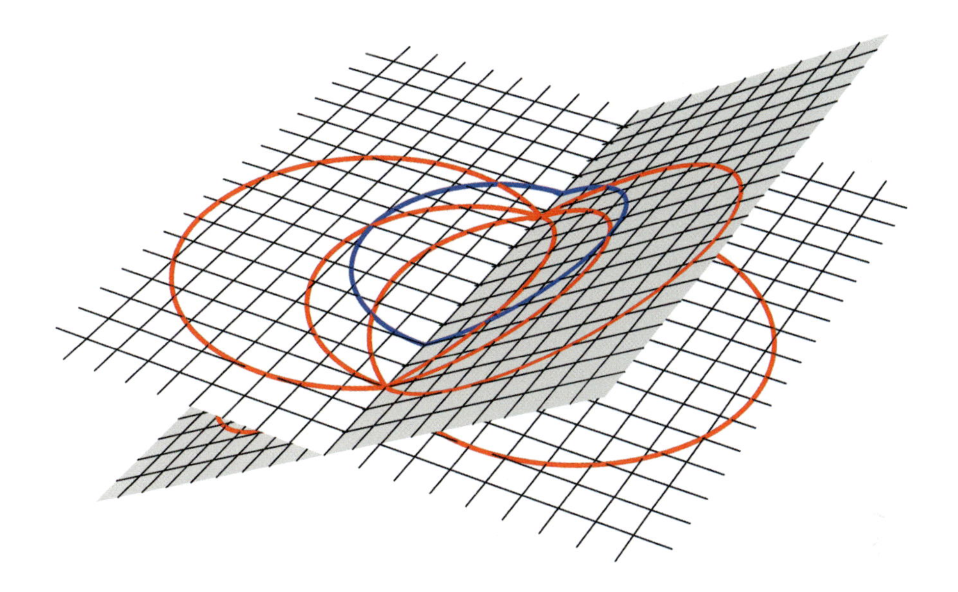

Figure 5.10: Light rays (red) in the 3–dimensional fish–eye. Each ray trajectory lies in a plane and is the stereographic projection of a great circle on a sphere that intersects the plane at its equator (blue circle). The rays shown are all emitted from one point (source) and perfectly focused at an image point.

Consider the simplest case first: electromagnetic waves emitted from a point dipole placed at the centre of the fish–eye lens. In this case we can exploit the spherical symmetry of the profile (36.1) and proceed in a similar way as for deriving dipole radiation in empty space (Exercise 34.2). In analogy to empty space, we conjecture that the vector potential $\boldsymbol{A}$ of the radiation points in the same direction as the dipole, say in the z–direction, and that the magnitude A only depends on the radius:

$$\boldsymbol{A} = A(r)\,\boldsymbol{e}_z\,. \tag{36.4}$$

In order to take advantage of the spherical symmetry, we use spherical polar coordinates $\{r, \theta, \phi\}$. There, as we have seen in Eq. (34.16):

$$A_i = (A\cos\theta,\ -A\,r\sin\theta,\ 0)\,. \tag{36.5}$$

The optical geometry of the fish–eye is conformally flat in Cartesian coordinates (29.17) but we need to express it in spherical polars, which is easily done by multiplying the metric tensor (11.17) in spherical polar coordinates by n^2 and calculating its inverse and determinant:

$$\begin{aligned} g_{ij} &= n^2\,\mathrm{diag}\left(1,\ r^2,\ r^2\sin^2\theta\right), \\ g^{ij} &= \frac{1}{n^2}\,\mathrm{diag}\left(1,\ \frac{1}{r^2},\ \frac{1}{r^2\sin^2\theta}\right), \\ g &= n^6\,r^4\,\sin^2\theta\,. \end{aligned} \tag{36.6}$$

After these preliminaries we are well–equipped to address Maxwell's equations. We calculate the magnetic field (26.15):

$$H^i = \frac{1}{\mu_0}\,\epsilon^{ijk} A_{k,j} = \frac{1}{\mu_0\sqrt{g}}\,[ijk] A_{k,j}\,, \tag{36.7}$$

where we applied formula (14.7) for the Levi–Civita tensor ϵ^{ijk}. As A_i of our ansatz (36.4) has no ϕ–component and does not depend on ϕ either, the permutation symbol $[ijk]$ applied to $A_{k,j}$ vanishes unless $i = \phi$. Consequently, the magnetic field points in the ϕ–direction, and we obtain from Eqs. (36.6) and (36.7)

$$H^\phi = \frac{1}{\mu_0 n^3 r^2}\,(-\partial_r r A + A) = \frac{H}{\mu_0 n^2 r^2} \tag{36.8}$$

with the abbreviation

$$H = -\frac{r\,\partial_r A}{\mu_0 n}\,. \tag{36.9}$$

Lowering the index of H^ϕ with the geometry (36.6) we get

$$H_\phi = H \sin^2\theta\,. \tag{36.10}$$

Having expressed the $\boldsymbol{H}$ field in terms of the vector potential, we insert H_i in Maxwell's equation (26.1) in order to obtain the derivative of the electric field:

$$\varepsilon_0 \frac{\partial E^i}{\partial t} = \epsilon^{ij\phi} H_{\phi,j} = \frac{c^2}{n^3 r^2}\big(2H\cos\theta,\ -\partial_r H \sin\theta,\ 0\big)\,, \tag{36.11}$$

where we used Eqs. (14.7), (36.6) and expression (36.7). We lower the index of the electric field with metric (36.6),

$$\varepsilon_0 \frac{\partial E_i}{\partial t} = \left(\frac{2H}{n\,r^2}\cos\theta,\ -\frac{\partial_r H}{n}\sin\theta,\ 0\right), \tag{36.12}$$

and obtain from Faraday's law (26.1) and $\varepsilon_0\mu_0 = c^{-2}$ for the magnetic field:

$$\frac{\partial^2 H^\phi}{\partial t^2} = -\frac{1}{\mu_0}\frac{\partial}{\partial t}\,\epsilon^{\phi jk} E_{k,j} = \frac{c^2}{n^3 r^2}\left(\partial_r \frac{1}{n}\,\partial_r H - \frac{2H}{n\,r^2}\right). \tag{36.13}$$

All other components of $\epsilon^{ijk} E_{k,j}$ vanish for the same reason as they did for the magnetic field (36.7); which proves *a–posteriori* that our ansatz (36.4) is justified. Finally, from expression (36.8) follows the wave equation of the magnetic field:

$$\frac{1}{n}\partial_r \frac{1}{n}\,\partial_r H - \frac{2H}{n^2 r^2} = \frac{1}{c^2}\frac{\partial^2 H}{\partial t^2}\,. \tag{36.14}$$

For finding an instructive solution of the wave equation (36.14) we express the magnetic–field amplitude H as

$$H = -2r\,\partial_r D\,, \tag{36.15}$$

where D is a function of r and t we determine later (note that D is not related to the dielectric replacement here). One easily verifies the following relationship for the magnetic field (36.15) in the fish–eye profile (36.1):

$$\frac{1}{n}\partial_r \frac{1}{n}\,\partial_r H - \frac{2H}{n^2 r^2} = -2r\,\partial_r\left(\frac{1}{n^3 r^2}\,\partial_r\, n\, r^2 \partial_r D - \frac{D}{a^2}\right). \tag{36.16}$$

We thus realize that the wave equation (36.14) with relation (36.15) is satisfied if we require

$$\frac{1}{n^3 r^2}\,\partial_r\, n\, r^2 \partial_r D - \frac{D}{a^2} = \frac{1}{c^2}\,\frac{\partial^2 D}{\partial t^2}\,. \tag{36.17}$$

Notice that the left–hand side of Eq. (36.17) is the Laplacian (17.23) in the optical geometry (36.6) applied on a function D that depends only on the radius r and time t, as we have in our case. We thus realize that D obeys

$$\nabla_j \nabla^j D - a^{-2} D = \frac{1}{c^2}\,\frac{\partial^2 D}{\partial t^2}\,, \tag{36.18}$$

the scalar wave equation in the geometry of light established by Maxwell's fish–eye, i.e. the scalar wave equation on a hypersphere of radius a. Therefore D describes a scalar wave on the hypersphere. Furthermore, we can express the a^{-2} term in the wave equation (36.18) in terms (24.23) of the curvature scalar R on the hypersphere:

$$\nabla_j \nabla^j D - \frac{R}{6} D = \frac{1}{c^2}\,\frac{\partial^2 D}{\partial t^2}\,. \tag{36.19}$$

Equation (36.19) resembles the wave equation (29.4) of electromagnetic vector waves. In §31 we showed that electromagnetism is conformally invariant in space–time. One can prove that the scalar wave equation (36.19) also is conformally invariant in space–time, due to the curvature term (Birrell and Davies [1982]). Our scalar wave D is said to be conformally coupled on the hypersphere.

We transformed the problem of the radiating dipole in the centre of Maxwell's fish–eye to the propagation of a scalar wave on the hypersphere. The wave is emitted at the origin in physical space that, by inverse stereographic projection (24.7), corresponds to the South Pole on the virtual hypersphere. Let us pause here and ponder about waves on the hypersphere, regardless of where they are emitted. The high degree of symmetry on the hypersphere allows us to deduce the principal structure of these waves with little algebra. Imagine a wave is emitted at point (X_0, Y_0, Z_0, W_0). As the hypersphere is hyperspherically symmetric, we can construct a coordinate system such that (X_0, Y_0, Z_0, Y_0) is the new South Pole. The new and the old systems are related to each other by rotation, so the scalar wave emitted at (X_0, Y_0, Z_0, W_0) simply is the wave from the old South Pole rotated in its entirety to the actual point of emission (Fig. 5.11). Now, the waves from the South Pole must be spherically symmetric, they can only depend on the hyperspherical angle χ, but not on θ and ϕ (Fig. 3.25). Indeed, χ directly corresponds to the radius r of physical space in stereographic projection (24.9) and we know already that the wave D emitted at the origin, the projection of the South Pole, is a function of r, but not of θ and ϕ.

As the scalar waves on the hypersphere are the rotated waves from the South Pole, they can only depend on the radius r' in physical space obtained by rotation on the hypersphere to the point of emission. To work out r' and double–check our ideas, we recall some results from §24. There we have seen that the rotations on the virtual hypersphere correspond to the Möbius transformations (24.12) in physical space. Exercise 24.2 showed that the transformations (24.12) do not change the refractive–index profile (36.1) that establishes the optical geometry and hence determines the wave equation (36.18). Therefore, Möbius transformations do not change the wave equation (36.18); a transformed solution still is a solution, i.e. a proper scalar wave, which rigorously justifies our geometrical picture (Fig. 5.11). The transformed wave is a function $D(r')$ of the Möbius–transformed radius $r' = |\boldsymbol{r}'|$, and we obtain from the Möbius transformation (24.12):

$$r' = \frac{|\boldsymbol{r} - \boldsymbol{r}_0|}{\sqrt{1 + 2\,a^{-2}\,\boldsymbol{r}\cdot\boldsymbol{r}_0 + a^{-4}\,r^2\,r_0^2}} \,. \tag{36.20}$$

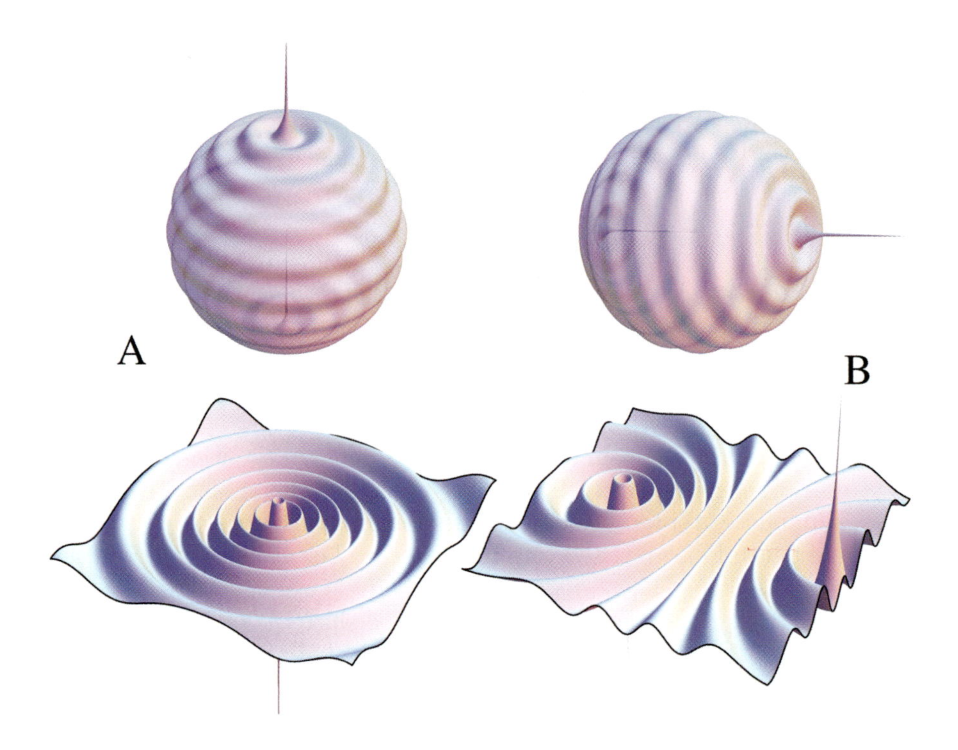

Figure 5.11: Waves on a sphere and their stereographic projection to the plane. The pictures show a sphere, not a hypersphere, and so the scalar wave shown is that for the 2D fish–eye (Leonhardt [2009]). (A) The wave is emitted at the South Pole of the sphere; the image at the North Pole is at infinity upon projection to the plane. (B) A wave emitted at an arbitrary point is obtained by rotating the wave in (A) on the sphere; the source and image points are now visible in the projection to the plane.

The point of emission lies at the Möbius–transformed origin, where $r' = 0$. Equation (36.20) shows that r' vanishes when $\boldsymbol{r} = \boldsymbol{r}_0$, so the parameter $\boldsymbol{r}_0$ in the Möbius transformation (24.12) corresponds to the source point:

$$r' = 0 \quad \Longleftrightarrow \quad \boldsymbol{r} = \boldsymbol{r}_0 \qquad \text{(source)}. \tag{36.21}$$

We expect the image point to lie at the rotated antipodal point to the South Pole. The antipode to the South Pole at $W = -a$ resides at the North Pole at $W = a$ (Fig. 3.25), where, in stereographic projection (24.6) $r = \infty$. Hence the image of the source (36.21) should be formed at the point in physical space where $r' = \infty$. Indeed, the denominator of r' in formula (36.20) goes to zero at the image point (36.2) we already obtained in ray optics, confirming our reasoning:

$$r' = \infty \quad \Longleftrightarrow \quad \boldsymbol{r} = \boldsymbol{r}_0' = -\frac{a^2}{|\boldsymbol{r}_0|^2}\,\boldsymbol{r}_0 \qquad \text{(image)}. \tag{36.22}$$

We have seen that Möbius–transforming a particular solution of the wave equation, i.e. rotating it on the hypersphere, has given us simplifications and insights.

Let us return from the scalar waves (36.19) on the lofty hypersphere to full electromagnetic waves in physical space. So far we considered a special case only, radiation emitted by a dipole pointing in the z–direction. Could we follow a similar procedure as for scalar waves and obtain all the physically meaningful solutions by Möbius transformation? In this way we could circumvent calculation by imagination and simply write down the solution without further ado. However, we hit a little snag in the vector nature of electromagnetic waves. Unlike the scalar wave D, the electromagnetic field is a vector field (or a field of one–forms, depending on representation); the electric and magnetic field strengths point in directions that change under Möbius transformations. We have also started with a special solution that is too special, because we have distinguished the z–axis, requiring that the dipole points in the z–direction. We could equally well choose dipoles in the x– and y–directions where the vector potential (36.4) would be $A\,\boldsymbol{e}_x$ and $A\,\boldsymbol{e}_y$, respectively. In order to take these possibilities into account, we combine the electromagnetic waves for the three principal cases, dipoles in the x, y, z–directions, in one object, the matrix of the electric field strengths

$$\mathbf{E} = \left({}_x\boldsymbol{E},\ {}_y\boldsymbol{E},\ {}_z\boldsymbol{E} \right). \tag{36.23}$$

The indices to the left of E indicate the dipole directions. In the same way we also introduce the matrix $\mathbf{H}$ of the magnetic field. In formula (36.15) we expressed the magnetic field $\boldsymbol{H}$ for the dipole pointing in z–direction in terms of the scalar wave D. This formula resembles the expression (36.9) of the magnetic field in terms of the vector potential, apart from a factor of n and a constant prefactor. According to relation (26.15) the one–form H_i of the magnetic field strength corresponds to the curl $\epsilon_i^{\ jk}\partial_j A_k$ of the vector potential with respect to the optical geometry (29.17). We see from Eqs. (14.7) and (29.17) that this curl differs from the curl in physical space by exactly a factor of n,

$$\epsilon_i^{\ jk} = \frac{1}{n}\,[ijk] \quad \text{in Cartesian coordinates}. \tag{36.24}$$

Comparing expressions (36.9) and (36.15) we thus realize that the $\boldsymbol{H}$ field must be the curl of $2D\boldsymbol{e}_z$ in physical space for the dipole in z–direction. For the magnetic–field matrix $\mathbf{H}$ we replace $\boldsymbol{e}_z$ by the three dipole directions $\boldsymbol{e}_i$ and get

$$ {}_jH_i\,\big|_{\boldsymbol{r}_0=\boldsymbol{0}} = 2[ijk]\partial_k D \quad \text{in Cartesian coordinates.} \tag{36.25}$$

To obtain the matrix of the electric field we apply Maxwell's equation (26.1) with optical geometry (29.17) in Cartesian coordinates

$$ \varepsilon_0 \frac{\partial}{\partial t}\, {}_jE_i \bigg|_{\boldsymbol{r}_0=\boldsymbol{0}} = -\frac{2}{n}\,[ilm]\,\partial_l\,[mpj]\,\partial_p D(r)\,. \tag{36.26}$$

The matrix (36.23) with explicit expression (36.26) describes the radiation emitted by dipoles in the three principal directions of space. To obtain the radiation for a source with arbitrary dipole moment $\boldsymbol{p}$, we decompose $\boldsymbol{p}$ into Cartesian components (p_x, p_y, p_z), multiply each component with the corresponding electric field in the matrix (36.23) and vector–sum the result. In short, we perform the projection

$$ E^i = p^j\,{}_jE^i \tag{36.27}$$

and get the electric field of the wave emitted by the dipole $\boldsymbol{p}$. As we can represent the dipole vector in general coordinates, the left indices should not only refer to Cartesian systems. Furthermore, if we intend to move the field to new source locations $\boldsymbol{r}_0$, the source components will change in general: the field matrix (36.23) thus is a one–form with respect to the source locations $\boldsymbol{r}_0$. The electric field is also a one–form in physical space $\{\boldsymbol{r}\}$. Geometrical objects that are vectors or one–forms on two different spaces are called *bi–tensors*; the electric–field matrix (36.23) thus represents a bi–tensor.* The distinction between the two spaces, $\{\boldsymbol{r}\}$ and $\{\boldsymbol{r}_0\}$, is important in our case, because the two vectors $\boldsymbol{r}$ and $\boldsymbol{r}_0$ are independent; $\boldsymbol{r}$ describes the spectator point and $\boldsymbol{r}_0$ the source point. Formula (36.26) thus describes a bi–tensor in disguise evaluated at $\boldsymbol{r}_0 = 0$. What do we know about this elusive bi–tensor? Like the scalar wave D on the hypersphere, it can only be a function of r', but it also must be a bi–tensor of course. One obvious candidate is the derivative of $D(r')$ with respect to $\boldsymbol{r}$ and $\boldsymbol{r}_0$,

$$ \partial_i\partial_j^0 D(r') = \frac{\partial^2 D(r')}{\partial x^i \partial x_0^j}\,, \tag{36.28}$$

because it is made of the scalar D that depends only on r' and the derivatives ∂_i and ∂_j^0 that are one–forms on $\{\boldsymbol{r}\}$ and $\{\boldsymbol{r}_0\}$, respectively. Expression (36.28) makes a perfect Möbius–invariant bi–tensor constructed from the scalar $D(r')$. Another physically prominent scalar is the refractive index n with profile (36.1), so we could consider

$$ \partial_i\partial_j^0\, n(r') = \frac{\partial^2\, n(r')}{\partial x^i \partial x_0^j}\,. \tag{36.29}$$

*The matrix (36.23) describes the diadic Green function of the electromagnetic field, also known as the Green tensor (Tai [1994]).

As both bi–tensors (36.28) and (36.29) represent independent one–forms on $\{\boldsymbol{r}\}$ and $\{\boldsymbol{r}_0\}$, we can combine them by forming a vector product on $\boldsymbol{r}$ space and a vector product on $\boldsymbol{r}_0$ space; the result is a vector in $\{\boldsymbol{r}\}$ and a vector in $\{\boldsymbol{r}_0\}$ that constitutes a bi–tensor. Let us try out this bi–tensor product:

$$\varepsilon_0 \frac{\partial}{\partial t}\, {}_jE_i = a^2 \epsilon_i^{\ lm}(\boldsymbol{r})\, \epsilon_j^{\ pq}(\boldsymbol{r}_0)\, \frac{\partial^2 n(r')}{\partial x^l \partial x_0^p}\, \frac{\partial^2 D(r')}{\partial x^m \partial x_0^q}\,. \tag{36.30}$$

The $\boldsymbol{r}$ and $\boldsymbol{r}_0$ variables in the Levi–Civita tensors shall indicate that they belong to two different spaces;

$$\epsilon_i^{\ lm}(\boldsymbol{r}) = \frac{[ilm]}{n(r)}\,, \quad \epsilon_j^{\ pq}(\boldsymbol{r}_0) = \frac{[jpq]}{n(r_0)} \quad \text{(in Cartesian coordinates)} \tag{36.31}$$

for the Levi–Civita tensor (14.7) with lowered first index in the optical geometry (29.17). The bi–tensor (36.30) evaluated at $\boldsymbol{r}_0 = \boldsymbol{0}$ turns out to be exactly the same as expression (36.26), because for any $f(r')$:

$$\epsilon_i^{\ lm}(\boldsymbol{r})\, \epsilon_j^{\ pq}(\boldsymbol{r}_0)\, \frac{\partial^2 n(r')}{\partial x^l \partial x_0^p}\, \frac{\partial^2 f(r')}{\partial x^m \partial x_0^q}\bigg|_{\boldsymbol{r}_0=\boldsymbol{0}} = -\frac{2}{a^2 n}\,[ilm]\,\partial_l\,[mpj]\,\partial_p\, f(r)\,, \tag{36.32}$$

as one can verify by straightforward, mechanical but tedious calculation that is best left to computer algebra. Formula (36.30) thus fullfills all our requirements: it describes an r'–dependent bi–tensor that agrees with the special solution (36.26) of Maxwell's equations for $\boldsymbol{r}_0 = \boldsymbol{0}$. Consequently, formula (36.30) must describe the electromagnetic wave emitted at an arbitrary source point $\boldsymbol{r}_0$ with arbitrary dipole direction. Note that it is sufficient to know the time–derivative (36.30) of $\boldsymbol{E}$, because we can get $\boldsymbol{E}$ itself by time integration and $\boldsymbol{B}$ (and hence $\boldsymbol{H}$) by a curl and another time integration according to Faraday's law (26.1). Finally, for stationary electromagnetic waves oscillating at frequency ω the electric field is directly given by

$${}_jE_i = \mathrm{i}\frac{a^2}{\varepsilon_0 \omega}\, \epsilon_i^{\ lm}(\boldsymbol{r})\, \epsilon_j^{\ pq}(\boldsymbol{r}_0)\, \frac{\partial^2 n(r')}{\partial x^l \partial x_0^p}\, \frac{\partial^2 D(r')}{\partial x^m \partial x_0^q}\,. \tag{36.33}$$

We have thus liberated ourselves from the technicalities of electromagnetic waves in Maxwell's fish–eye by completely reducing the problem to the propagation of scalar waves on the hypersphere. We already know a great deal about these scalar waves: we can rotate them around on the hypersphere and they only depend on the hypersphere distance from the point of emission to the point of observation. In physical space, the scalar waves are functions $D(r')$ of the Möbius–transformed radius (36.20). So we only need to consider the scalar waves emitted at one point, say the South Pole that corresponds to the origin in physical space, and rotate them on the hypersphere or, equivalently, Möbius–transform them in physical space. Consider a stationary wave oscillating at frequency ω. Let us motivate a formula for the wave function $D(r)$ by analogy to spherical scalar waves in empty, flat space and then verify that this guess is indeed a solution of the radial wave equation (36.17).

A spherical wave emitted at $\boldsymbol{r}_0 = \mathbf{0}$ in empty space oscillates with a phase (4.5) that is the product of the distance travelled and the wave number

$$k = \frac{\omega}{c}\,. \tag{36.34}$$

The amplitude of the wave is determined from spherical symmetry and energy conservation: in order to maintain a stationary flux through spherical surfaces around the origin, the intensity $|D(r)|^2$ must fall in accord with the increasing surface area: $|D(r)|^2$ must be inversely proportional to the area $2\pi r^2$, so the amplitude is proportional to r^{-1}. We thus obtain

$$D = \frac{\mathrm{e}^{ikr-\mathrm{i}\omega t}}{4\pi r} \qquad \text{(in free space)}, \tag{36.35}$$

where the 4π is a matter of normalization and convention. On the hypersphere, the distance (24.27) from the South Pole is $a\alpha$ with

$$\alpha = \pi - \chi\,. \tag{36.36}$$

Imagine light confined to the surface of an ordinary sphere in 3–dimensional space. There radiation emitted at the South Pole passes through circles of length $2\pi a \sin\alpha$. On the hypersphere (Fig. 3.25), these circles are replaced by spherical surfaces of radius $a\sin\alpha$ with area $4\pi a^2 \sin^2\alpha$. We thus conjecture that the amplitude of the hyperspherical wave should be inversely proportional to $a\sin\alpha$ and write, in analogy to the spherical wave (36.35) in free space,

$$D = \frac{\mathrm{e}^{ika\alpha-\mathrm{i}\omega t}}{4\pi a \sin\alpha}\,. \tag{36.37}$$

Let us express this formula in Cartesian coordinates in physical space by stereographic projection. We obtain from Eqs. (24.9) and (36.36)

$$\tan\left(\frac{\alpha}{2}\right) = \cot\left(\frac{\chi}{2}\right) = \frac{r}{a}\,, \tag{36.38}$$

$$\frac{a}{r} + \frac{r}{a} = \cot\left(\frac{\chi}{2}\right) + \tan\left(\frac{\chi}{2}\right) = \frac{2}{\sin\chi}\,, \tag{36.39}$$

and therefore from formula (36.37)

$$\begin{aligned} D &= \frac{1}{8\pi a}\left(\frac{a}{r}+\frac{r}{a}\right)\exp\left[2ika\arctan(r/a) - \mathrm{i}\omega t\right] \\ &= \frac{1}{8\pi a}\left(\frac{a}{r}+\frac{r}{a}\right)\left(\frac{a+\mathrm{i}r}{a-\mathrm{i}r}\right)^{k}\mathrm{e}^{-\mathrm{i}\omega t}\,. \end{aligned} \tag{36.40}$$

Exercise 36.2

Verify that expression (36.40) is a solution of the radial wave equation (36.17).

Exercise 36.2 proves that our guess has been correct; formula (36.40) describes a stationary radial wave. We obtain all the physically relevant stationary scalar waves in Maxwell's fish–eye by hypersphere rotation, i.e., in physical space, by replacing r by the Möbius–transformed radius (36.20):

$$D = \frac{1}{8\pi a}\left(\frac{a}{r'} + \frac{r'}{a}\right) \exp\left[2\mathrm{i}ka \arctan(r'/a) - \mathrm{i}\omega t\right]. \tag{36.41}$$

At two positions the amplitude of the wave (36.41) diverges; the wave (36.41) has two singularities, one at $r' = 0$ and the other when $r' = \infty$, one at the source point (36.21) and the other at the image point (36.22). At the source, the entire energy of the wave is concentrated in one point with infinite intensity; at the image, the wave focuses into one point with the original, infinite intensity. This perfect focusing of waves indicates already that Maxwell's fish–eye has unlimited resolution. At the image point the wave must be phase–delayed, because it takes time for light to travel from the source to the image. We can easily work out the phase delay φ_∞ from formula (36.41) by taking the limit $r' \to \infty$ in $\arctan(r'/a)$, and obtain

$$\varphi_\infty = k\pi a\,. \tag{36.42}$$

This result is easy to understand on the virtual hypersphere, because there the distance traveled from one point to its antipode is πa and the travel time is

$$t_\infty = \pi\frac{a}{c}\,, \tag{36.43}$$

so the phase should be $\varphi_\infty = \omega t_\infty = \pi a\,\omega/c$, as formula (36.42) with wave number (36.34) confirms. Our result (36.41) is consistent and makes sense, but it also contains a feature that may appear strange. The wave is emitted at the source point and approaches the image in a single point, but it never returns: the wave disappears at the image.

In assuming a stationary state of radiation continuously emitted at the source, we implicitely assumed a drain absorbing the radiation at the image. Otherwise, a stationary state is impossible. If we were in empty space, the emitted light would radiate out to ∞ and disappear. But Maxwell's fish–eye represents a finite virtual space with finite volume (24.26) where light is confined. Hence a drain is required to remove the radiation at the rate it is produced. The drain at the image could be a photoactive substance or a detector. However, instead of an artificially stationary situation, we can also describe the evolution of non–stationary wavepackets.

Consider the opposite of the stationary case: a flash of light emitted at the source $\boldsymbol{r}_0$ at time t_0. We describe the flash by a wavefunction δ that is infinitely short but infinitely intense such that the integrated amplitude is finite, a *delta function.* Any wavepacket $D(\boldsymbol{r}, t)$, including the flash, can be understood as a superposition of stationary waves (36.37) with different frequencies ω and amplitudes $\widetilde{\mathcal{A}}(\omega)$ (by Fourier analysis, see Riley, Hobson and Bence [2006]):

$$D = \int_{-\infty}^{+\infty} \frac{\widetilde{\mathcal{A}}(\omega)}{4\pi a \sin\alpha} \exp\left[\mathrm{i}\omega\left(\frac{a\alpha}{c} - t\right)\right] \mathrm{d}\omega = \frac{D_0(\alpha - \varphi)}{\sin\alpha} \tag{36.44}$$

in terms of the function D_0, the wave form

$$D_0 = \int_{-\infty}^{+\infty} \frac{\widetilde{A}(\omega)}{4\pi a} \exp\left(\mathrm{i}\frac{\omega a}{c} - \mathrm{i}\omega t_0\right) \mathrm{d}\omega\,, \tag{36.45}$$

and the dynamical phase

$$\varphi = \frac{c}{a}(t - t_0)\,. \tag{36.46}$$

The light flash has a delta function as wave form:

$$D = \frac{\delta(\alpha - \varphi)}{\sin\alpha}\,. \tag{36.47}$$

It is advantageous to transform the delta function in expression (36.47). The δ function is a generalized function $\delta(x)$ that is defined to be different from zero only for $x = 0$, but the integral over $\delta(x)$ is 1 (Riley, Hobson and Bence [2006]). In formulae:

$$\delta(x) = 0 \quad \text{for } x \neq 0\,, \quad \int_{-\infty}^{+\infty} \delta(x)\,\mathrm{d}x = 1\,. \tag{36.48}$$

The only contribution to the integral comes from $\delta(x)\mathrm{d}x$ at $x = 0$ and so $\delta(x)\mathrm{d}x = 1$ here. The same applies to $\delta(y)\mathrm{d}y$ where y is a function of x, so $\delta(y)\mathrm{d}y = 1$ for $\mathrm{d}y > 0$ at $y = 0$. Consequently,

$$\delta(y) = \left|\frac{\mathrm{d}x}{\mathrm{d}y}\right| \delta(x) \tag{36.49}$$

where we take the modulus for allowing also for $\mathrm{d}y < 0$, i.e. for decreasing functions $y(x)$. Writing $x = \alpha - \varphi$ and $y = \cos\varphi - \cos\alpha$ we obtain according to the transformation rule (36.49)

$$D = \delta(\cos\varphi + \zeta) \tag{36.50}$$

with the abbreviation

$$\zeta = -\cos\alpha = \cos\chi = \frac{\cos^2(\chi/2) - \sin^2(\chi/2)}{\cos^2(\chi/2) + \sin^2(\chi/2)} = \frac{r^2 - a^2}{r^2 + a^2} \tag{36.51}$$

where we applied Eqs. (36.36) and (36.38) for the special case of light emission from the South Pole (or the origin in physical space). As we know, for the general case we only need to replace r by the Möbius–transformed radius (36.20):

$$\zeta = \frac{r'^2 - a^2}{r'^2 + a^2}\,. \tag{36.52}$$

Equations (36.50), (36.46) and (36.52) show that the light pulse reaches the image as a perfect flash. Without the drain the flash (36.50) is reflected there (changing sign upon reflection) and bounces back and forth, as it remains located at

$$\cos\varphi + \zeta = 0 \tag{36.53}$$

and $\cos\varphi$ is periodic. At the source (36.21) $\zeta = -1$, which, according to definition (36.46), is consistent with $t = t_0$, but also with $t = t_0 + 2m\,t_\infty$, where t_∞ is the travel time (36.43) from source to image and m an integer. At the image (36.22) $\zeta = +1$ and so $t = t_0 + (2m+1)\,t_\infty$. The period of the bouncing flash is twice the travel time from source to image, as it should. (Provided of course, the light is not absorbed in the material.) Now, any light field carrying an image we can imagine as made of light flashes emitted from various positions with various amplitudes, as a linear superposition of the elementary waves (36.50). As the fish–eye lens perfectly focuses individual light flashes, it must perfectly image any wave, at least in principle.

Having solved Maxwell's equations for Maxwell's fish–eye, we proved that light waves are as perfectly imaged as light rays. Maxwell's fish–eye makes a perfect lens, but it is a rather peculiar lens, because both source and image are inside it. The lens rather resembles a mirror surrounding the source and focusing its light at the image. An elliptical mirror focuses light, but only at the two focal points of the ellipse, whereas in Maxwell's fish–eye, all points are focal points. It turns out (Leonhardt [2009]) that the two–dimensional version of Maxwell's fish–eye is also a perfect imaging device for waves with the right polarization (when the planar medium of the lens establishes a geometry according to §27). The wave functions, however, are expressed in terms of higher transcendental functions (Erdelyi [1953]); so we preferred the 3–dimensional version here where the solutions are elementary functions. On the other hand, the 2–dimensional case seems to be much more useful and easier in practice, because in 2D impedance matching is not required for establishing a geometry, as we have seen in §27. Both in 2D and 3D one could apply a circular or spherical mirror to reduce the index profile to a finite range and still get perfect imaging (Fig. 5.12). In practical applications, one could imagine the lens as a thin film placed over or under the object to be perfectly transferred as an image with, in principle, unlimited resolution. But at present it remains to be seen whether such perfect lenses become practically useful.

§37. Moving media

The devices considered so far have exploited the spatial part of the geometry of light, but we have seen in §31 that the geometry of light extends to 4–dimensional space–time. Here we show that a moving medium with isotropic $\varepsilon = \mu = n$ creates an effective space–time geometry for light. For this purpose we generalize Fermat's principle for a medium with $\varepsilon = \mu = n$ from 3–dimensional space to space–time.

Imagine the medium consists of infinitesimally small cells moving with velocities $\boldsymbol{u}$. The velocity profile may vary in space and time, but for a given space–time point we can always construct a locally co–moving inertial frame in Cartesian coordinates. The co–moving frames of different points will differ in general, but for our argument it is sufficient to consider only one, arbitrary point with its associated co–moving frame. We will formulate our results as tensor equations such that they are valid in all frames, including the co–moving frames of all other points. In this way we shall establish the space–time geometry of light in moving media.

In the co–moving frame, the infinitesimal cell does not move, by definition, and so we can apply to this cell the standard theory of electromagnetism in media at rest. In particular, Fermat's principle is exact for non–moving media with $\varepsilon = \mu = n$

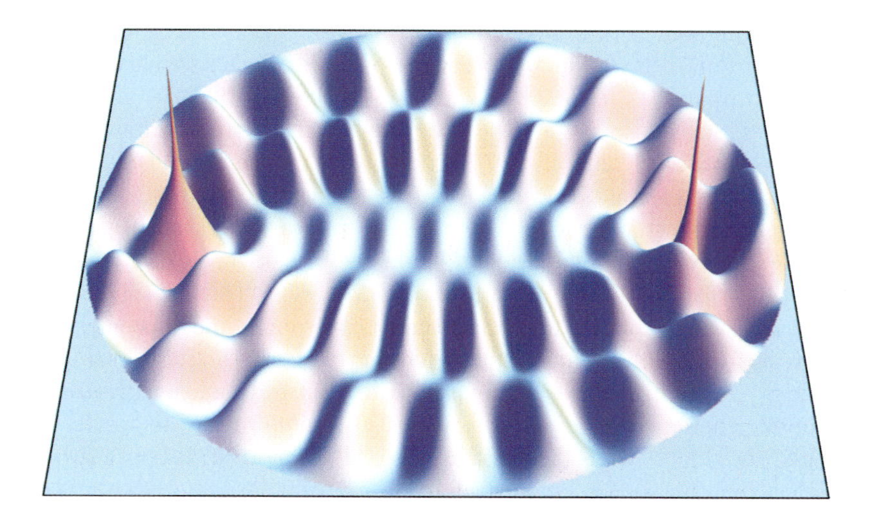

Figure 5.12: Imaging in the fish–eye mirror (Fig. 2.38). The infinitely sharp field emitted at the source point (left tip) propagates as an electromagnetic wave until it focuses at the image point (right tip) with infinite resolution. The focusing is done by Maxwell's fish eye in 2D constrained by the mirror around the Equator (Leonhardt [2009]).

(§26) so in the cell of interest the electromagnetic field perceives a spatial geometry with the line element

$$\mathrm{d}l^2 = n^2(\mathrm{d}x^2 + \mathrm{d}y^2 + \mathrm{d}z^2)\,. \tag{37.1}$$

The line element of the space–time geometry with spatial part (37.1) is

$$\mathrm{d}s^2 = -(\mathrm{d}x^0)^2 + \mathrm{d}l^2\,, \quad x^0 = ct\,, \tag{37.2}$$

and hence we obtain the space–time metric tensor (31.6)

$$g_{\alpha\beta} = \mathrm{diag}\,(-1,\ n^2,\ n^2,\ n^2)\,. \tag{37.3}$$

In the description so far, the medium creates a purely spatial geometry: it influences only the spatial part of the metric (37.3), not the measure of time. We can change this by making use of the conformal invariance of Maxwell's equations (in the absence of external charges and currents) that was described in §31. We may multiply the space–time metric by an arbitrary conformal factor and still obtain exactly the same constitutive equations (31.2) as before, and therefore exactly the same electromagnetic waves. Multiplying the metric (37.3) by n^{-2} we obtain the new metric tensor (31.7) with determinant g and inverse $g^{\alpha\beta}$ as

$$g_{\alpha\beta} = \mathrm{diag}\,(-n^{-2},\ 1,\ 1,\ 1)\,, \tag{37.4}$$

$$g = -n^{-2}\,, \quad g^{\alpha\beta} = \mathrm{diag}\,(-n^2,\ 1,\ 1,\ 1)\,. \tag{37.5}$$

In this geometry the medium influences the measure of time (through the metric element $g_{00} = -n^2$) but not the spatial part of the metric. We can rewrite the metric tensor (37.4) as

$$g_{\alpha\beta} = \eta_{\alpha\beta} + (1 - n^{-2})\, u_\alpha\, u_\beta\,, \tag{37.6}$$

where $\eta_{\alpha\beta}$ is the Minkowski metric (25.2) and u_α is a 4–dimensional one–form in space–time with components

$$u_\alpha = (-1,\ 0,\ 0,\ 0) \quad \text{in the co–moving frame.} \tag{37.7}$$

Equations (37.6) and (37.7) describe the effective space–time geometry created by the infinitesimal cell in an inertial frame co–moving with this cell at one particular time. If the cell is accelerating then a new inertial frame is required at each moment of time. Furthermore, other infinitesimal cells may move at different velocities and so their co–moving frames may differ from each other. We wish to write the effective metric in the global inertial frame in physical space–time in a form that describes every cell of the moving medium. For this, we must distinguish the effective metric (37.6) created by the cell of material from the Minkowski metric $\eta_{\alpha\beta}$ that describes the flat background space–time in which the cell is moving (physical space–time). The theory of special relativity shows that the components of the metric tensor in flat space–time are given by the Minkowski metric in every inertial frame (Landau and Lifshitz [1975], Misner, Thorne and Wheeler [1973]). This property is the only ingredient from special relativity we need. In particular, the transformation of the effective metric (37.6) from the co–moving inertial frame to the laboratory frame does not change the Minkowski term $\eta_{\alpha\beta}$. But we need to find how the one–form (37.7) changes when transformed to the global frame of physical space–time. Consider the associated vector u^α in the flat background space–time with Minkowski metric $\eta_{\alpha\beta}$; this vector is related to the one–form u_α by

$$u^\alpha = \eta^{\alpha\beta} u_\beta\,, \quad u_\alpha = \eta_{\alpha\beta} u^\beta\,, \tag{37.8}$$

where

$$\eta^{\alpha\beta} = \text{diag}(-1,\ 1,\ 1,\ 1) \tag{37.9}$$

is the inverse Minkowski metric:

$$\eta^{\alpha\gamma}\eta_{\gamma\beta} = \delta^\alpha_\beta\,. \tag{37.10}$$

From Eqs. (37.7) and (37.9) follows that the vector (37.8) has components

$$u^\alpha = (1,\ 0,\ 0,\ 0) \quad \text{in the co–moving frame.} \tag{37.11}$$

Consider the space–time trajectory $x^\alpha(s_0)$ of the moving infinitesimal cell of the medium. We parametrize this curve by the distance s_0 in physical space–time with Minkowski line element (25.1). In the co–moving frame, the cell is not moving in space, only in time; the curve $x^\alpha(s)$ runs in the time direction only:

$$x^\alpha = (s_0,\ 0,\ 0,\ 0) \quad \text{in the co–moving frame.} \tag{37.12}$$

We see that the components for the space–time tangent vector $\mathrm{d}x^\alpha(s)/\mathrm{d}s_0$ are equal to those of the vector (37.11) in the co–moving frame:

$$u^\alpha = \frac{\mathrm{d}x^\alpha(s_0)}{\mathrm{d}s_0} . \tag{37.13}$$

Equation (37.13) is a tensor equation and so holds in all frames, not just the co–moving frame; u^α is always the space–time tangent vector. The length of u^α in the co–moving frame is easily found from the metric (25.2) and Eq. (37.11):

$$\eta_{\alpha\beta} u^\alpha u^\beta = -1 , \tag{37.14}$$

and this is a scalar, so it has the same value in all frames. Equation (37.13) and the Minkowski line element (25.1) applied to $\mathrm{d}x^\alpha(s_0)$ can be used to express the length $\eta_{\alpha\beta} u^\alpha u^\beta = -1$ in an arbitrary frame:

$$\begin{aligned} -1 = \eta_{\alpha\beta} u^\alpha u^\beta &= \frac{1}{\mathrm{d}s_0^2}\, \eta_{\alpha\beta}\, \mathrm{d}x^\alpha(s_0)\, \mathrm{d}x^\beta(s_0) \\ &= \frac{1}{\mathrm{d}s_0^2} \left\{ -c^2 \left[\mathrm{d}t(s_0)\right]^2 + \left[\mathrm{d}x(s_0)\right]^2 + \left[\mathrm{d}y(s_0)\right]^2 + \left[\mathrm{d}z(s_0)\right]^2 \right\} \\ &= -\frac{\left[\mathrm{d}t(s_0)\right]^2}{\mathrm{d}s_0^2} \left[c^2 - \left(\frac{\mathrm{d}x}{\mathrm{d}t}\right)^2 - \left(\frac{\mathrm{d}y}{\mathrm{d}t}\right)^2 - \left(\frac{\mathrm{d}z}{\mathrm{d}t}\right)^2 \right] . \end{aligned} \tag{37.15}$$

As expression (37.15) holds in an arbitrary frame, it also holds in a frame that is co–moving with velocity

$$\boldsymbol{u} = \frac{\mathrm{d}\boldsymbol{r}}{\mathrm{d}t} . \tag{37.16}$$

Writing Eq. (37.15) in terms of the velocity (37.16) we obtain

$$\mathrm{d}s_0 = c\, \mathrm{d}t(s_0) \left(1 - \frac{u^2}{c^2}\right) . \tag{37.17}$$

where $u^2 = \boldsymbol{u} \cdot \boldsymbol{u}$ is the square of the speed. To find the components u^α and u_α in an arbitrary frame we apply Eq. (37.17) to the tensor equation (37.13) and use the relationship (37.8):

$$u^\alpha = \frac{(1, \boldsymbol{u}/c)}{\sqrt{1 - u^2/c^2}} , \quad u_\alpha = \frac{(-1, \boldsymbol{u}/c)}{\sqrt{1 - u^2/c^2}} . \tag{37.18}$$

The space–time vector u^α is called the *four–velocity* of the cell; it is the space–time generalization of the ordinary (three–velocity) $\boldsymbol{u}$ and in a frame co–moving with the cell (in which $\boldsymbol{u} = \boldsymbol{0}$) it reduces to the simple time–oriented vector (37.11).

Having determined the four–velocity we have gathered all ingredients for the space–time geometry of light. The metric (37.6) created by the infinitesimal cell of the medium depends on the four–velocity one–form u_α given by Eqs. (37.18) in terms of the velocity $\boldsymbol{u}$ of the cell in physical space. But the derivation just given applies to every cell of the moving medium at every moment of time, where $\boldsymbol{u}$ is the velocity of that cell at that moment. In this way, we have established the space–time geometry of moving media discovered by Walter Gordon [1923]. The inverse of the metric (37.6) is

$$g^{\alpha\beta} = \eta^{\alpha\beta} + (1 - n^2)\, u^\alpha\, u^\beta\,. \tag{37.19}$$

Exercise 37.1
Show that the tensor (37.19) is the matrix–inverse of the metric tensor (37.6) using Eqs. (37.14) and (37.8).

In matrix form, the metric $g_{\alpha\beta}$, its determinant g and inverse $g^{\alpha\beta}$ appear as

$$g_{\alpha\beta} = \begin{pmatrix} \dfrac{u^2 - c^2 n^{-2}}{c^2 - u^2} & \dfrac{n^{-2} - 1}{c^2 - u^2}\, c\, \boldsymbol{u} \\ \dfrac{n^{-2} - 1}{c^2 - u^2}\, c\, \boldsymbol{u} & \mathbb{1} - \dfrac{n^{-2} - 1}{c^2 - u^2}\, \boldsymbol{u} \otimes \boldsymbol{u} \end{pmatrix}, \tag{37.20}$$

$$g = \det\left(g_{\alpha\beta}\right) = -n^{-2}, \tag{37.21}$$

$$g^{\alpha\beta} = \begin{pmatrix} \dfrac{u^2 - c^2 n^2}{c^2 - u^2} & \dfrac{1 - n^2}{c^2 - u^2}\, c\, \boldsymbol{u} \\ \dfrac{1 - n^2}{c^2 - u^2}\, c\, \boldsymbol{u} & \mathbb{1} - \dfrac{n^2 - 1}{c^2 - u^2}\, \boldsymbol{u} \otimes \boldsymbol{u} \end{pmatrix}. \tag{37.22}$$

Exercise 37.2
Verify that expression (37.21) is the determinant of $g_{\alpha\beta}$.

We obtain from Eq. (31.2) the dielectric tensors of moving media (that are locally isotropic and impedance matched):

$$\boldsymbol{\varepsilon} = \boldsymbol{\mu} = \frac{n}{1 - u^2 n^2/c^2} \left(\left(1 - \frac{u^2}{c^2}\right) \mathbb{1} + \left(1 - n^{-2}\right) \frac{\boldsymbol{u} \otimes \boldsymbol{u}}{c^2} \right) \approx n\, \mathbb{1}\,, \tag{37.23}$$

where the approximation refers to the case of low velocity, $u/c \ll 1$. Furthermore, since for moving media $g_{\alpha\beta}$ contains mixed space–time elements g_{0i}, we also get the vector (31.2):

$$\boldsymbol{w} = \frac{n^2 - 1}{1 - u^2 n^2/c^2}\, \boldsymbol{u} \approx (n^2 - 1)\, \boldsymbol{u}\,. \tag{37.24}$$

In this way we have found one possible physical interpretation for the bi–anisotropy vector $\boldsymbol{w}$ of electromagnetism in space–time geometries: $\boldsymbol{w}$ may appear as the velocity of a moving medium.

The spatial geometry of light in media at rest plays a prominent role in geometrical optics (§30). We expect that the space–time geometry established by moving media is equally instructive. Geometrical optics is governed by the phase φ of electromagnetic waves with wave vector $\boldsymbol{k}$, a one–form, the phase gradient (30.3), and the frequency ω, the negative time derivative (30.2) of the phase. In space–time, frequency and wave vector are combined in one object, the 4–dimensional wave vector (one–form):

$$k_\alpha = \varphi_{,\alpha} = \left(-\frac{\omega}{c}, \boldsymbol{k}\right). \tag{37.25}$$

In a locally co–moving frame, the frequency ω and 3–dimensional wave vector $\boldsymbol{k}$ are connected by the dispersion relation (4.2) of an optically isotropic medium at rest:

$$n^2\frac{\omega^2}{c^2} - k^2 = 0 \quad \text{in the co–moving frame}, \tag{37.26}$$

or, expressed in terms of the 4–dimensional wave vector (37.25) and the inverse metric tensor (37.5),

$$g^{\alpha\beta}k_\alpha k_\beta = 0\,. \tag{37.27}$$

Relation (37.27) is a scalar equation obtained from the tensor $g^{\alpha\beta}$ by contraction with the one forms k_α, so it must hold in all frames, not only the co–moving frame attached to an infinitesimal cell of the medium. The dispersion relation (37.27) with tensor (37.22) thus describes geometrical optics in moving media.

In most cases, the speed u of the medium is small in comparison with the speed of light in the medium, c/n, so it is wise to approximate the dispersion relation (37.27) in the limit of low velocities. One finds in this case:

$$\frac{\omega}{c} = \frac{|\boldsymbol{k}|}{n} + \left(1 - n^{-2}\right)\frac{\boldsymbol{u}\cdot\boldsymbol{k}}{c}\,. \tag{37.28}$$

Exercise 37.3
Show that relation (37.28) approximates the dispersion relation (37.27) to first order in u/c.
Solution
By first–order Taylor expansion in u/c of expression (37.22) for the metric, we obtain from the dispersion relation (37.27) with 4–dimensional wave vector (37.25):

$$k^2 = n^2\frac{\omega^2}{c^2} + 2(1-n^2)\frac{\omega}{c}\frac{\boldsymbol{u}\cdot\boldsymbol{k}}{c} \approx n^2\left(\frac{\omega}{c} - \left(1-n^{-2}\right)\frac{\boldsymbol{u}\cdot\boldsymbol{k}}{c}\right)^2, \tag{37.29}$$

adding, in the last step, a term quadratic in u/c that we can neglect. Equation (37.29) gives the dispersion relation (37.28).

In geometrical optics, the frequency ω plays the role of the Hamiltonian and the ray trajectories are given by Hamilton's equations (4.8). We have seen in §30 that the Hamiltonian time t is the physical travel time of light, so $\mathrm{d}\boldsymbol{r}/\mathrm{d}t$ gives the velocity of light. We obtain from the dispersion relation (37.28):

$$\frac{\mathrm{d}\boldsymbol{r}}{\mathrm{d}t} = \frac{\partial\omega}{\partial\boldsymbol{k}} = \frac{c}{n}\frac{\boldsymbol{k}}{|\boldsymbol{k}|} + \left(1 - n^{-2}\right)\boldsymbol{u}\,. \tag{37.30}$$

We see that in a medium at rest (or in a co–moving frame) where $\boldsymbol{u} = \boldsymbol{0}$ the speed of light $|\mathrm{d}\boldsymbol{r}/\mathrm{d}t|$ is c/n, as it should be. We also see that in a moving medium the velocity $\boldsymbol{u}$ adds to the speed of light; the medium drags light with it. But not all of the velocity $\boldsymbol{u}$ is added, only a fraction of $(1-n^{-2})$, as if only part of the medium is able to drag light. Fresnel [1818] derived the *dragging coefficient* of $(1-n^{-2})$. Only in the limit $n \to \infty$, for very slow light, do the speed of light in the medium and the speed of the medium add directly. For $n = 1$, when light travels at the speed of light in vacuum, light is not dragged at all. The *Fresnel drag* of light in moving media turns out to be an example of Einstein's relativistic addition of velocities (Van Bladel [1984]). It is thus quite remarkable that Fresnel deduced the correct dragging coefficient in 1818, long before special relativity was discovered (Einstein [1905]). It is equally remarkable that Fizeau [1851] was able to observe the Fresnel drag of light in an experiment (Fig. 5.13) more than a century before lasers were invented.

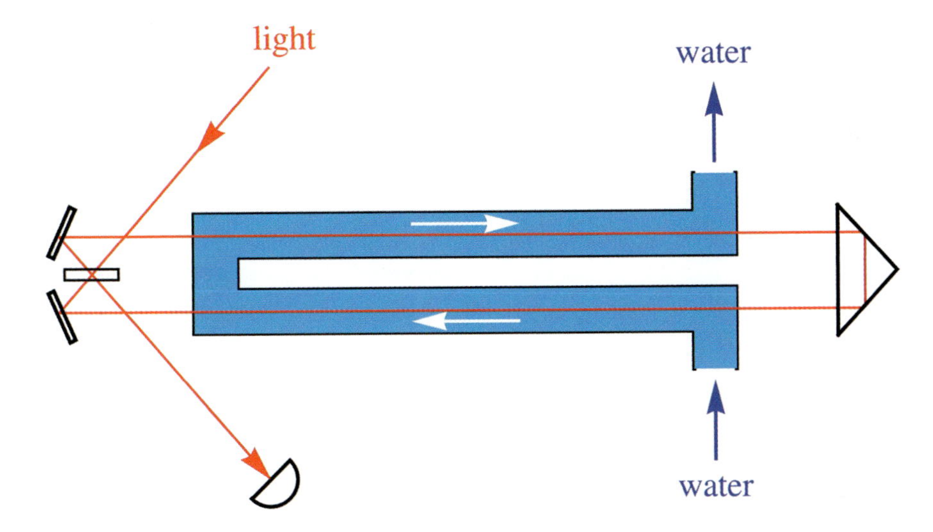

Figure 5.13: Fizeau's experimental demonstration of the dragging of light by a moving medium (Fresnel drag). Light is passed through two glass tubes in which water is flowing in opposite directions. The light path is arranged so that the two tubes form sections of the arms of an interferometer. If the water drags light, as predicted by Fresnel, the interference fringes will shift when the speed of the water is changed.

Exercise 37.4
The velocity vector $\boldsymbol{v}$ in the laboratory frame is given in terms of the velocity $\boldsymbol{v}'$ in the moving frame by Einstein's addition theorem (Van Bladel [1984]) in full vectorial form:

$$\boldsymbol{v} = \frac{\boldsymbol{v}' + \boldsymbol{u} + (\gamma - 1)\boldsymbol{u}\left(u^2 + \boldsymbol{u}\cdot\boldsymbol{v}'\right)/u^2}{\gamma\left(1 + \boldsymbol{u}\cdot\boldsymbol{v}'/c^2\right)}\,, \qquad \gamma = \frac{1}{\sqrt{1-u^2/c^2}}\,. \tag{37.31}$$

Derive from this expression formula (37.30) for the Fresnel drag in the limit $u \ll c$.

Exercise 37.5
Deduce the Hamiltonian ω of light rays in moving media by solving for ω in the dispersion relation (37.27) with metric (37.22) and the 4–dimensional wave vector (37.25).

Solution
Written in explicit form,

$$\omega^2 - c^2k^2 + (n^2-1)\frac{(\omega - \boldsymbol{u}\cdot\boldsymbol{k})^2}{1-u^2/c^2} = 0\,, \tag{37.32}$$

the dispersion relation defines a quadratic equation for ω with the solution

$$\omega = \left(\frac{c^2-u^2}{n^2c^2-u^2}\right)^{1/2}\left(c^2k^2 - \frac{n^2c^2-c^2}{n^2c^2-u^2}(\boldsymbol{u}\cdot\boldsymbol{k})^2\right)^{1/2} + \frac{n^2c^2-c^2}{n^2c^2-u^2}\boldsymbol{u}\cdot\boldsymbol{k}\,. \tag{37.33}$$

Exercise 37.6
According to Hamilton's equations (4.8) the velocity $\boldsymbol{v}$ of light in the moving medium is given by $\partial\omega/\partial\boldsymbol{k}$ with Hamiltonian (37.33). Show that $\boldsymbol{v}$ is the relativistic sum (37.31) of the velocity $\boldsymbol{u}$ of the medium and a velocity $\boldsymbol{v}'$ of light in the medium with $v'^2 = c^2/n^2$.
Solution
From the inverse of the velocity addition (37.31)

$$\boldsymbol{v}' = \frac{\boldsymbol{v} - \boldsymbol{u} - (\gamma-1)\boldsymbol{u}\left(u^2 - \boldsymbol{u}\cdot\boldsymbol{v}\right)/u^2}{\gamma\left(1-\boldsymbol{u}\cdot\boldsymbol{v}/c^2\right)}\,, \qquad \gamma = \frac{1}{\sqrt{1-u^2/c^2}}\,, \tag{37.34}$$

follows the relation

$$1 - \frac{v'^2}{c^2} = \frac{(1-u^2/c^2)(1-v^2/c^2)}{(1-\boldsymbol{u}\cdot\boldsymbol{v}/c^2)^2}\,. \tag{37.35}$$

Calculating the derivative $\partial\omega/\partial\boldsymbol{k}$ and substituting the result for $\boldsymbol{v}$ in the right–hand side of expression (37.35) gives after some algebra $1-n^{-2}$ and hence $v'^2 = c^2/n^2$.

§38. Optical Aharonov–Bohm effect

Perfect invisibility devices and perfect lenses with negative refraction exploit transformations to non–trivial topologies in space—excluded regions in physical space or folds in virtual space. Here we study a simple example, the propagation of light through a vortex, that turns out to represent a transformation medium with a multi–valued space–time geometry.

Consider a vortex in a fluid, say in water flowing down a plug hole (Fig. 5.14). A vortex concentrates the vorticity $\boldsymbol{\nabla}\times\boldsymbol{u}$ of the swirling velocity flow $\boldsymbol{u}$ in one line, the vortex core. In Cartesian coordinates the velocity profile of a straight vortex is given by the expression

$$\boldsymbol{u} = \frac{\mathcal{W}}{x^2+y^2}\begin{pmatrix} -y \\ x \\ 0 \end{pmatrix}, \tag{38.1}$$

because for this profile $\boldsymbol{\nabla}\times\boldsymbol{u}$ vanishes except at the z axis where $\boldsymbol{u}$ diverges, but the circulation $\oint \boldsymbol{u}\cdot\mathrm{d}\boldsymbol{r}$ has the finite value $2\pi\mathcal{W}$. All vorticity is concentrated along the vortex line.

It is useful to express the velocity profile (38.1) of the vortex in cylindrical coordinates, the most natural coordinates for this situation. Transforming the Cartesian components (38.1) to cylindrical polar coordinates using the rule (12.1) and the

Figure 5.14: Vortex in flowing water. (St Andrews duck pond; photo: Tomáš Tyc.)

transformation matrix (10.15) one finds the cylindrical components

$$u^i = \frac{\mathcal{W}}{r^2}\begin{pmatrix}0\\1\\0\end{pmatrix}, \quad u_i = (0,\ \mathcal{W},\ 0)\,, \tag{38.2}$$

where the index is lowered using the cylindrical–polar metric $\gamma_{ij} = \mathrm{diag}(1, r^2, 1)$ of physical space (derived in Exercise 11.3). We immediately see from the expressions (17.19) of the curl in arbitrary coordinates that $\boldsymbol{\nabla} \times \boldsymbol{u}$ vanishes, because u_i is constant. The vorticity is concentrated in the z axis that is excluded from the cylindrical polar system because points on this axis do not have unique $\{r, \phi, z\}$ coordinates (the angle ϕ is undefined on the axis $z = 0$).

Imagine the vortex is illuminated from the side. A moving medium Fresnel–drags light (§37): light propagating with the flow is advanced, whereas light propagating against the flow lags behind. Dragged by the moving medium, light rays experience phase shifts and, if the phase fronts are deformed, light rays are deflected. For flow speeds u much smaller than the speed of light in the medium, c/n, we expect that the phase shift is proportional to the integral of $\boldsymbol{u}$ along the propagation of light. For regions with vanishing vorticity, we can deform the contours of the phase integrals $\int \boldsymbol{u}\cdot \mathrm{d}\boldsymbol{r}$ and so the phase difference between light waves that have passed the vortex on different sides is proportional to the circulation $\oint \boldsymbol{u} \cdot \mathrm{d}\boldsymbol{r}$, a constant. Hence we expect that the water vortex does not deflect light, but imprints a characteristic phase slip, in analogy to the Aharonov–Bohm effect (Aharonov and Bohm [1959], Tonomura [1998]). In the Aharonov–Bohm effect charged matter waves, electrons for example, are not deflected by a vortex in the magnetic vector potential, but experience a characteristic phase shift. The optical analogue of the Aharonov–Bohm

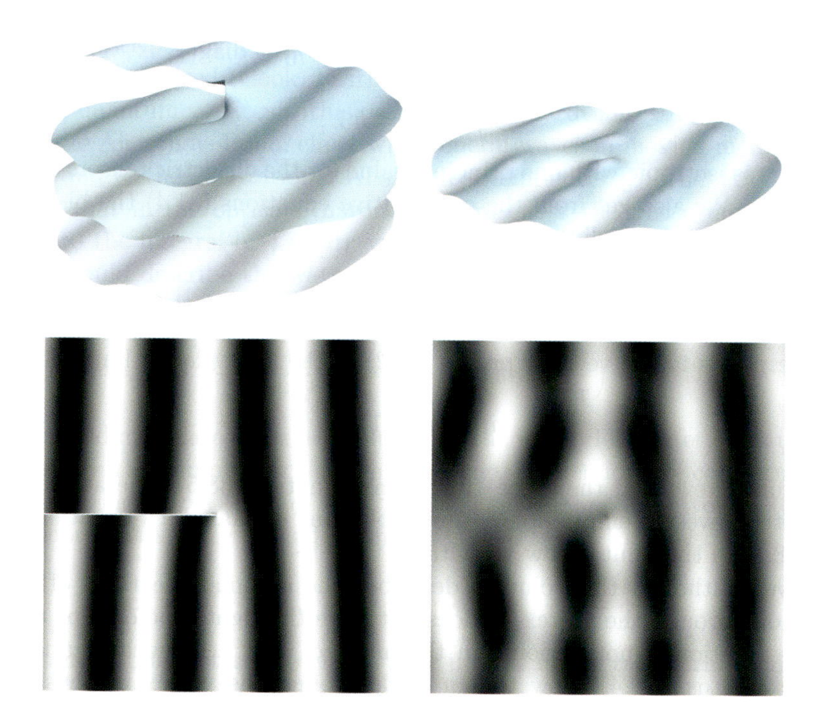

Figure 5.15: Aharonov–Bohm effect. A fluid vortex generates the optical Aharonov–Bohm effect described by the coordinate transformation (38.4). Light, incident from the right, is Fresnel–dragged by the moving medium: light propagating with the flow is advanced, whereas light propagating against the current is retarded. The wave should develop the phase slip shown in the left figures. However, although the transformation (38.4) is exact, it gives a physical space–time that is described by multi–valued coordinates (top left). Instead of the simple phase slip, the light turns out to exhibit the characteristic interference pattern illustrated in the right figures (Aharonov and Bohm [1959]).

Solution
One of Hamilton's equations (4.8) gives the Hamiltonian velocity (37.30) that we multiply by nk/ω in order to get the Newtonian velocity $\mathrm{d}\boldsymbol{r}/\mathrm{d}\xi$. Note that the wave number k depends on the velocity of the medium according to the dispersion relation (37.28), but we obtain to first order in u/c:

$$
\begin{aligned}
\frac{\mathrm{d}\boldsymbol{r}}{\mathrm{d}\xi} &= \frac{c}{\omega}\,\boldsymbol{k} + (n^2-1)\frac{\boldsymbol{u}}{c} \quad \text{and from Eqs. (4.8) and (37.28)} \\
\frac{\mathrm{d}^2\boldsymbol{r}}{\mathrm{d}\xi^2} &= -\frac{cnk}{\omega}\left(1-n^{-2}\right)\boldsymbol{\nabla}(\boldsymbol{u}\cdot\boldsymbol{k}) + \frac{n^2-1}{c}\left(\frac{\mathrm{d}\boldsymbol{r}}{\mathrm{d}\xi}\cdot\boldsymbol{\nabla}\right)\boldsymbol{u} \\
&\approx \frac{n^2-1}{c}\left[\left(\frac{\mathrm{d}\boldsymbol{r}}{\mathrm{d}\xi}\cdot\boldsymbol{\nabla}\right)\boldsymbol{u} - \boldsymbol{\nabla}\left(\frac{\mathrm{d}\boldsymbol{r}}{\mathrm{d}\xi}\cdot\boldsymbol{u}\right)\right]
\end{aligned} \tag{38.15}
$$

to first order in u/c, which gives the equation of motion (38.14) according to formula (14.17). The light ray appears to experience a *Lorentz force*, as if it were a charged particle in the magnetic field $\boldsymbol{\nabla}\times\boldsymbol{u}$. The mass of the particle corresponds to n^{-2} and the charge to Fresnel's dragging coefficient $(1-n^{-2})$. The velocity $\boldsymbol{u}$ of the medium appears like the vector potential $\boldsymbol{A}$ according to

Figure 5.14: Vortex in flowing water. (St Andrews duck pond; photo: Tomáš Tyc.)

transformation matrix (10.15) one finds the cylindrical components

$$u^i = \frac{\mathcal{W}}{r^2}\begin{pmatrix}0\\1\\0\end{pmatrix}, \quad u_i = (0,\ \mathcal{W},\ 0)\,, \tag{38.2}$$

where the index is lowered using the cylindrical–polar metric $\gamma_{ij} = \mathrm{diag}(1, r^2, 1)$ of physical space (derived in Exercise 11.3). We immediately see from the expressions (17.19) of the curl in arbitrary coordinates that $\boldsymbol{\nabla} \times \boldsymbol{u}$ vanishes, because u_i is constant. The vorticity is concentrated in the z axis that is excluded from the cylindrical polar system because points on this axis do not have unique $\{r, \phi, z\}$ coordinates (the angle ϕ is undefined on the axis $z = 0$).

Imagine the vortex is illuminated from the side. A moving medium Fresnel–drags light (§37): light propagating with the flow is advanced, whereas light propagating against the flow lags behind. Dragged by the moving medium, light rays experience phase shifts and, if the phase fronts are deformed, light rays are deflected. For flow speeds u much smaller than the speed of light in the medium, c/n, we expect that the phase shift is proportional to the integral of $\boldsymbol{u}$ along the propagation of light. For regions with vanishing vorticity, we can deform the contours of the phase integrals $\int \boldsymbol{u} \cdot \mathrm{d}\boldsymbol{r}$ and so the phase difference between light waves that have passed the vortex on different sides is proportional to the circulation $\oint \boldsymbol{u} \cdot \mathrm{d}\boldsymbol{r}$, a constant. Hence we expect that the water vortex does not deflect light, but imprints a characteristic phase slip, in analogy to the Aharonov–Bohm effect (Aharonov and Bohm [1959], Tonomura [1998]). In the Aharonov–Bohm effect charged matter waves, electrons for example, are not deflected by a vortex in the magnetic vector potential, but experience a characteristic phase shift. The optical analogue of the Aharonov–Bohm effect was first described by Jon Hannay [1976] in his PhD thesis and has also been independently rediscovered (Cook, Fearn and Milonni [1995]). The effect could be used to detect quantum vortices with slow light (Leonhardt and Piwnicki [2000]); it is related to the Aharonov–Bohm effect with general waves in moving media (Berry et al. [1980]) and to the gravitational Aharonov–Bohm effect (Stachel [1982]).

Assuming that in the optical Aharonov–Bohm effect the phase modulation is proportional to the integral of $\int \boldsymbol{u}\cdot \mathrm{d}\boldsymbol{r}$ we guess a space–time coordinate transformation that should describe the wave propagation. Then we use our formalism to verify that this guess is correct. The integral of the velocity profile (38.1) gives

$$\int \boldsymbol{u}\cdot \mathrm{d}\boldsymbol{r} = \mathcal{W}\int \frac{-y\,\mathrm{d}x + x\,\mathrm{d}y}{x^2+y^2} = \mathcal{W}\int \mathrm{d}\left(\arctan\frac{y}{x}\right) = \mathcal{W}\phi\,. \tag{38.3}$$

This phase should modulate the time evolution of the wave. The water has a uniform refractive index n that simply rescales the wavelength. Therefore we conjecture that the cylindrical coordinates $\{ct, r, \phi, z\}$ of physical space–time are given in terms of cylindrical coordinates $\{ct', r', \phi', z'\}$ in empty virtual space–time by

$$ct = ct' - a\,\phi'\,,\quad r = \frac{r'}{n}\,,\quad \phi = \phi'\,,\quad z = \frac{z'}{n} \tag{38.4}$$

with a constant a to be determined later. The inverse of the transformation is

$$ct' = ct + a\,\phi\,,\quad r' = n\,r\,,\quad \phi' = \phi\,,\quad z' = n\,z\,. \tag{38.5}$$

Virtual space–time is flat and empty with the Minkowski line element (25.1) written in the primed coordinates; expressed in physical coordinates this line element reads

$$\mathrm{d}s^2 = (c\,\mathrm{d}t + a\,\mathrm{d}\phi)^2 - n^2(\mathrm{d}r^2 + r^2\mathrm{d}\phi^2) - n^2\,\mathrm{d}z^2\,. \tag{38.6}$$

Hence we obtain for the metric tensor of virtual space–time transformed to physical coordinates

$$g_{\alpha\beta} = \begin{pmatrix} 1 & 0 & a & 0 \\ 0 & -n^2 & 0 & 0 \\ a & 0 & a^2 - n^2 r^2 & 0 \\ 0 & 0 & 0 & -n^2 \end{pmatrix}, \tag{38.7}$$

$$g = \det\left(g_{\alpha\beta}\right) = -r^2 n^6\,, \tag{38.8}$$

$$g^{\alpha\beta} = \begin{pmatrix} 1 - \dfrac{a^2}{n^2r^2} & 0 & \dfrac{a}{n^2r^2} & 0 \\ 0 & -\dfrac{1}{n^2} & 0 & 0 \\ \dfrac{a}{n^2r^2} & 0 & -\dfrac{1}{n^2r^2} & 0 \\ 0 & 0 & 0 & -\dfrac{1}{n^2} \end{pmatrix}. \tag{38.9}$$

Using this metric in the constitutive equations (31.4), together with the determinant $\gamma = r^2$ of the spatial cylindrical metric of physical space, we obtain the dielectric tensors

$$\varepsilon^{ij} = \mu^{ij} = n\,\mathrm{diag}\left(1,\ r^{-2},\ 1\right)\,. \tag{38.10}$$

Lowering an index with the physical–space metric $\gamma_{ij} = \text{diag}(1,\ r^2,\ 1)$ produces the simple result

$$\varepsilon^i{}_j = \mu^i{}_j = n\,\mathbb{1}\,, \tag{38.11}$$

the medium is isotropic with one principle value for the dielectric tensors, the refractive index n. We also get from the constitutive equations (31.4) the bi–anisotropy vector in cylindrical coordinates

$$w_i = (0,\ a,\ 0) \tag{38.12}$$

that corresponds to the velocity profile (38.2) in the low–velocity limit (37.24) for

$$a = (n^2 - 1)\,\mathcal{W}. \tag{38.13}$$

These results prove that the space–time transformation (38.5) generates, from electromagnetic waves in empty space–time, solutions of Maxwell's equations for the vortex (38.1) in the limit of low velocities.

However, although this solution describes the dominant features of the wave propagation through the vortex, it does not have the right topology for a physical wave; the solution is not periodic in the angle ϕ. For a monochromatic wave of frequency ω, Eqs. (38.4) show that a circle around the vortex in physical space changes the phase by $2\pi a\omega$, which is not an integer multiple of 2π in general. If we take the coordinate transformation (38.4) literally, physical space has become multi–valued, with a branch cut in the direction of incidence (Fig. 5.15). This branch cut describes the phase slip of light after passing through the vortex, but it cannot be mathematically rigorous. Figure 5.15 also shows the correct solution due to Aharonov and Bohm [1959] when additional scattering resolves the topological dilemma of the coordinate transformation (38.4).* But this is a subject beyond the scope of this Section, as is the theory of light propagation in rapidly spinning media that no longer act merely as transformation media: they may suck light into the vortex (Leonhardt and Piwnicki [1999]).

Exercise 38.1

Consider a moving medium with uniform refractive index, $n = \text{const}$, but non–uniform velocity profile $\boldsymbol{u}$. Assume that the speed of the medium is small compared with the speed of light in the medium, $u \ll c/n$. Derive from the Hamiltonian ω with dispersion relation (37.28) the following equation of motion in the limit of small u/c:

$$\frac{1}{n^2}\frac{\mathrm{d}^2\boldsymbol{r}}{\mathrm{d}\xi^2} = \left(1 - n^{-2}\right)(\boldsymbol{\nabla}\times\boldsymbol{u})\times\frac{\mathrm{d}\boldsymbol{r}}{\mathrm{d}\xi} \quad \text{with parameterization} \quad \mathrm{d}\xi = \frac{\omega}{nk}\,\mathrm{d}t\,. \tag{38.14}$$

From this equation also follows that light rays are straight lines in the optical Aharonov–Bohm effect, because the curl of the vortex profile (38.1) vanishes—no force is acting on them; the trajectories are straight.

Give a physical interpretation for the equation of motion (38.14) for general $\boldsymbol{u}$ based on the analogy of the light ray with the trajectory of a mechanical particle. Which force does the fictitious particle experience?

---

*See also the "many–whirls interpretation" of Aharonov–Bohm scattering (Berry [1980]).

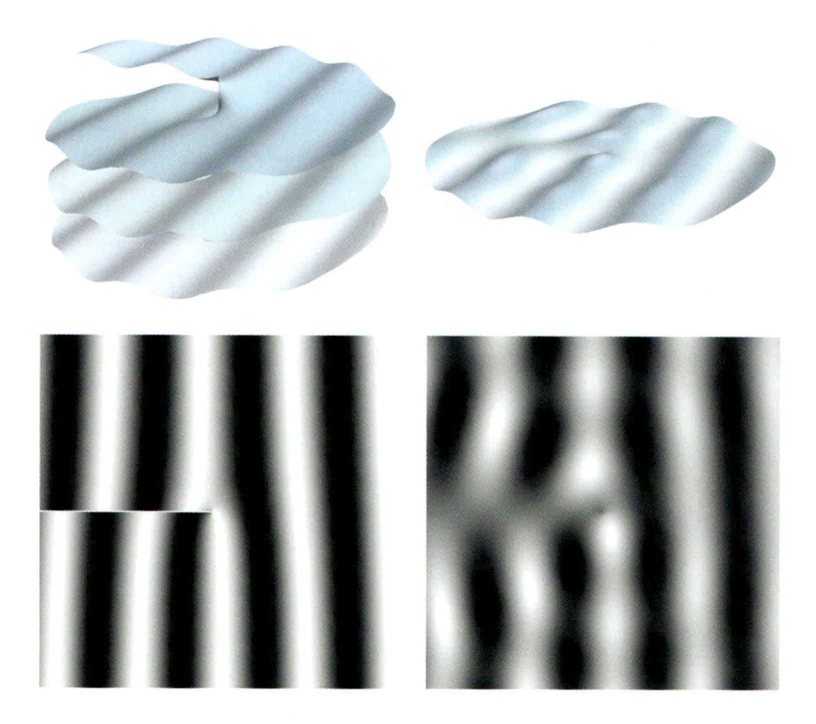

Figure 5.15: Aharonov–Bohm effect. A fluid vortex generates the optical Aharonov–Bohm effect described by the coordinate transformation (38.4). Light, incident from the right, is Fresnel–dragged by the moving medium: light propagating with the flow is advanced, whereas light propagating against the current is retarded. The wave should develop the phase slip shown in the left figures. However, although the transformation (38.4) is exact, it gives a physical space–time that is described by multi–valued coordinates (top left). Instead of the simple phase slip, the light turns out to exhibit the characteristic interference pattern illustrated in the right figures (Aharonov and Bohm [1959]).

Solution
One of Hamilton's equations (4.8) gives the Hamiltonian velocity (37.30) that we multiply by nk/ω in order to get the Newtonian velocity $\mathrm{d}\boldsymbol{r}/\mathrm{d}\xi$. Note that the wave number k depends on the velocity of the medium according to the dispersion relation (37.28), but we obtain to first order in u/c:

$$\begin{aligned} \frac{\mathrm{d}\boldsymbol{r}}{\mathrm{d}\xi} &= \frac{c}{\omega}\boldsymbol{k} + (n^2-1)\frac{\boldsymbol{u}}{c} \quad \text{and from Eqs. (4.8) and (37.28)} \\ \frac{\mathrm{d}^2\boldsymbol{r}}{\mathrm{d}\xi^2} &= -\frac{cnk}{\omega}\left(1-n^{-2}\right)\nabla(\boldsymbol{u}\cdot\boldsymbol{k}) + \frac{n^2-1}{c}\left(\frac{\mathrm{d}\boldsymbol{r}}{\mathrm{d}\xi}\cdot\nabla\right)\boldsymbol{u} \\ &\approx \frac{n^2-1}{c}\left[\left(\frac{\mathrm{d}\boldsymbol{r}}{\mathrm{d}\xi}\cdot\nabla\right)\boldsymbol{u} - \nabla\left(\frac{\mathrm{d}\boldsymbol{r}}{\mathrm{d}\xi}\cdot\boldsymbol{u}\right)\right] \end{aligned} \tag{38.15}$$

to first order in u/c, which gives the equation of motion (38.14) according to formula (14.17). The light ray appears to experience a *Lorentz force*, as if it were a charged particle in the magnetic field $\nabla\times\boldsymbol{u}$. The mass of the particle corresponds to n^{-2} and the charge to Fresnel's dragging coefficient $(1-n^{-2})$. The velocity $\boldsymbol{u}$ of the medium appears like the vector potential $\boldsymbol{A}$ according to

relation (26.14). One can turn this analogy around and visualize the vector potential as a velocity (Rousseaux and Guyon [2002]). Incidentally, this is how Maxwell imagined the vector potential.

Exercise 38.2
One can dramatically slow down the speed of light in media where the refractive index depends strongly on frequency (Milonni [2004]). However, typically only the group velocity $v_g = \partial\omega/\partial k$ is reduced in the slow–light medium, but not the phase velocity c/n; for all practical purposes $n = 1$ in the frequency band of slow light, close to some frequency ω_0. However, the dragging coefficient turns out to depend on the group velocity such that the dispersion relation (37.28) is replaced by (Leonhardt and Piwnicki [2001]):

$$\frac{\omega^2}{c^2} = k^2 + \frac{2}{v_g}\frac{\omega_0}{c}\,\boldsymbol{u}\cdot\boldsymbol{k} = \left(\boldsymbol{k} + \frac{\omega_0}{v_g c}\boldsymbol{u}\right)^2 - \frac{\omega_0^2 u^2}{v_g^2 c^2}\,. \tag{38.16}$$

Deduce the equation of motion

$$\frac{\mathrm{d}^2\boldsymbol{r}}{\mathrm{d}\xi^2} = \frac{1}{v_g}(\boldsymbol{\nabla}\times\boldsymbol{u})\times\frac{\mathrm{d}\boldsymbol{r}}{\mathrm{d}\xi} + \boldsymbol{\nabla}\frac{u^2}{2v_g^2} \quad \text{with} \quad \mathrm{d}\xi = c\,\mathrm{d}t \quad \text{for} \quad \omega = \omega_0\,. \tag{38.17}$$

How does slow light experience a vortex? Relate the equation of motion (38.17) in the vortex profile (38.1) to one of the problems considered in Chapter 2.
Solution
It is advantageous to employ instead of the Hamiltonian ω a new Hamiltonian $\Omega = \omega^2/(2\omega_0)$; Hamilton's equations (4.8) remain the same for trajectories with $\omega = \omega_0$. We obtain from the dispersion relation (38.16):

$$\Omega = \frac{\omega_0}{2}\left(\frac{c}{\omega_0}\boldsymbol{k} + \frac{\boldsymbol{u}}{v_g}\right)^2 - \frac{u^2}{2v_g^2} \tag{38.18}$$

and from Hamilton's equations (4.8) with Ω as Hamiltonian

$$\begin{aligned}\frac{\mathrm{d}\boldsymbol{r}}{\mathrm{d}\xi} &= \frac{1}{c}\frac{\partial\Omega}{\partial\boldsymbol{k}} = \frac{c}{\omega_0}\boldsymbol{k} + \frac{\boldsymbol{u}}{v_g}\,,\\ \frac{\mathrm{d}^2\boldsymbol{r}}{\mathrm{d}\xi^2} &= -\frac{\boldsymbol{\nabla}\Omega}{\omega_0} + \left(\frac{\mathrm{d}\boldsymbol{r}}{\mathrm{d}\xi}\cdot\boldsymbol{\nabla}\right)\frac{\boldsymbol{u}}{v_g} = \boldsymbol{\nabla}\frac{u^2}{2v_g^2} - \boldsymbol{\nabla}\left(\frac{\boldsymbol{u}}{v_g}\cdot\frac{\mathrm{d}\boldsymbol{r}}{\mathrm{d}\xi}\right) + \left(\frac{\mathrm{d}\boldsymbol{r}}{\mathrm{d}\xi}\cdot\boldsymbol{\nabla}\right)\frac{\boldsymbol{u}}{v_g}\,,\end{aligned} \tag{38.19}$$

which gives the equation of motion (38.17). For the vortex (38.1) the curl of $\boldsymbol{u}$ vanishes and the u^2 term acts like the r^{-2} potential of fatal attraction (Exercise 7.8). Slow light is spiraling into the vortex (Fig. 2.29).

---

§39. Analogue of the event horizon

Transformation media are at the heart of perfect invisibility devices, perfect lenses with negative refraction and the optical Aharonov–Bohm effect; here we explain that they also describe much of the essential physics at the event horizon of a black hole.

In 1972 William Unruh invented a simple analogy of the event horizon as an illustration for a colloquium at Oxford: black holes resemble rivers (Unruh [2008]). Imagine a river is flowing towards a waterfall. Suppose that the river carries waves that propagate with speed c' relative to the water, but the water is rapidly moving with velocity u that at some point exceeds c'. Beyond this point, waves trying to propagate upstream are swept away towards the waterfall. The point of no return is the *horizon* of the *black hole*. Imagine another situation: a river flowing out through an expanding channel into the sea, getting slower. Waves coming from the sea are blocked at the line where the flow exceeds the wave speed. Here the river establishes the analogue of a *white hole*, an object that nothing can enter. Such analogies occur also in many other situations and they illustrate some essential features of the event horizon (Fig. 5.16).

Horizons are perfect traps: one would not notice anything suspicious while passing the horizon with the flow, but to return is impossible. Waves propagating against the flow get stuck at the horizon; they freeze with dramatically shrinking wavelengths. These similarities to event horizons are not mere analogies, there is a mathematical equivalence to wave propagation in general relativity (as long as higher–order dispersion is irrelevant). For example, sound waves in moving fluids behave like waves in a certain space–time geometry that depends on the flow and the local velocity of sound (White [1973], Unruh [1981]). Another example has been described in §38; light experiences a moving isotropic impedance–matched medium as the effective space–time geometry (37.6). At the horizon, where the flow u reaches

Figure 5.16: Aquatic analoque of the event horizon. The Point of no return is the analogue of the event horizon for fish in a river. To the left of it, the water flows faster than a fish can swim. If a fish happens to venture beyond this line, it can never get back upstream; it is doomed to be crushed in Singularity Falls. (Reproduced with friendly permission of Yan Nascimbene.)

the speed of light c/n, the measure of time g_{00} in Gordon's metric (37.20) vanishes. Time comes to a standstill. Close to the horizon the wavelength is dramatically reduced (until due to optical dispersion the refractive index changes, tuning the light out of the grip of the horizon.) For short wavelengths the lateral dimensions of the horizon are irrelevant, therefore the optics at the horizon is essentially one–dimensional.

Consider a one–dimensional model: both medium and light move along the z–axis (Fig. 5.17) and the electromagnetic field vectors are pointing orthogonal to z in the xy–plane. In the following we show that this situation corresponds to a space–time transformation medium. We will deduce a coordinate transformation that maps one–dimensional wave propagation in moving media to propagation in empty virtual space–time. In empty 1–dimensional space, i.e. (1+1)–dimensional space–time, light would freely propagate with the speed c as a superposition of wave packets that move either in the positive or negative spatial direction while maintaining their shapes. Hence these two wavepackets depend only on

$$t_\pm = t' \mp \frac{z'}{c}\,. \tag{39.1}$$

In physical space–time, where there is a moving medium, light in an infinitesimal cell of the medium propagates at the velocity $\pm c/n$ in the positive or negative direction in a frame co–moving with this cell. Therefore we expect that in physical space the light propagates at a velocity $v_\pm$ that is the relativistic addition of $\pm c/n$ and the local flow speed u. We obtain from Einstein's velocity addition (37.31) in one

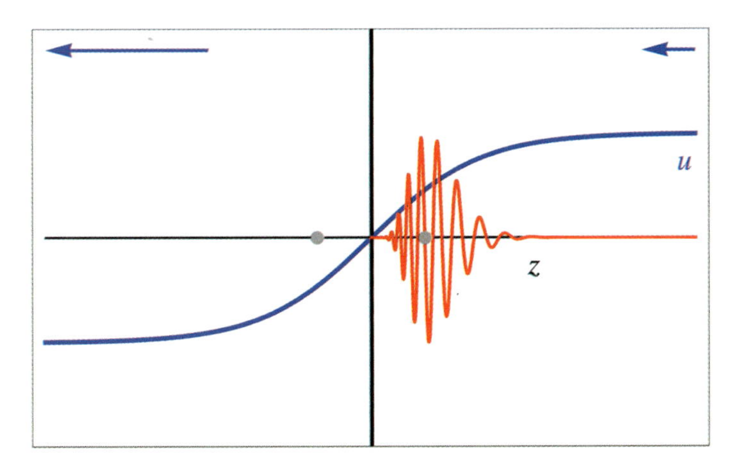

Figure 5.17: Velocity profile with horizon. A wave packet (red) propagates against a medium moving with velocity u (blue arrows). The blue curve shows the velocity profile; u is negative and its magnitude reaches the speed of light c/n at the horizon (black line). The grey dots mark two spatial positions, one in front and one behind the horizon, that appear in the Penrose diagrams 5.18 and 5.19.

dimension:

$$v_\pm = \frac{u \pm c/n}{1 \pm u/(cn)} \,. \tag{39.2}$$

In physical space–time, wave packets therefore depend only on

$$t_\pm = t - \int \frac{\mathrm{d}z}{v_\pm} \,. \tag{39.3}$$

Comparing Eqs. (39.1) and (39.3) we expect that the one–dimensional wave propagation in the moving medium corresponds to waves in empty space–time through the transformation

$$ct' = \frac{c}{2}\,(t_- + t_+)\,, \quad x' = x\,, \quad y' = y\,, \quad z' = \frac{c}{2}\,(t_- - t_+)\,. \tag{39.4}$$

To prove this assertion we apply the theory of transformation media. For $\boldsymbol{u} = (0,\ 0,\ u)$ the effective space–time geometry of the moving medium is described by the metric (37.20) with

$$g^{\alpha\beta} = \begin{pmatrix} \dfrac{u^2 - c^2 n^2}{c^2 - u^2} & 0 & 0 & \dfrac{(1-n^2)\,c\,u}{c^2 - u^2} \\ 0 & 1 & 0 & 0 \\ 0 & 0 & 1 & 0 \\ \dfrac{(1-n^2)\,c\,u}{c^2 - u^2} & 0 & 0 & \dfrac{c^2 - n^2 u^2}{c^2 - u^2} \end{pmatrix} . \tag{39.5}$$

We transform $g^{\alpha\beta}$ to virtual space–time,

$$g^{\alpha'\beta'} = \Lambda^{\alpha'}{}_{\alpha}\, g^{\alpha\beta}\, \Lambda^{\beta'}{}_{\beta} \tag{39.6}$$

with the matrix $\Lambda^{\alpha'}{}_{\alpha}$ that we obtain by differentiating the new coordinates $x^{\alpha'}$ with respect to the coordinates x^{α} of physical space–time

$$\Lambda^{\alpha'}{}_{\alpha} = \begin{pmatrix} 1 & 0 & 0 & \dfrac{(n^2-1)\,c\,u}{c^2 - n^2 u^2} \\ 0 & 1 & 0 & 0 \\ 0 & 0 & 1 & 0 \\ 0 & 0 & 0 & \dfrac{n(c^2 - u^2)}{c^2 - n^2 u^2} \end{pmatrix} . \tag{39.7}$$

The result is the diagonal matrix

$$g^{\alpha'\beta'} = \mathrm{diag}\left(-\frac{n^2(c^2-u^2)}{c^2 - n^2u^2},\ \frac{n^2(c^2-u^2)}{c^2 - n^2u^2},\ 1,\ 1\right) \tag{39.8}$$

with the inverse

$$g_{\alpha'\beta'} = \mathrm{diag}\left(-\frac{c^2 - n^2u^2}{n^2(c^2-u^2)},\ \frac{c^2 - n^2u^2}{n^2(c^2-u^2)},\ 1,\ 1\right) . \tag{39.9}$$

The determinant of the metric is

$$g' = -\frac{(c^2 - n^2u^2)^2}{n^4(c^2 - u^2)^2}\,. \tag{39.10}$$

The metric $g_{\alpha'\beta'}$ describes the geometry in virtual space–time. To find out how this geometry appears as a medium we use the constitutive equations (31.2) in virtual space–time, with primed instead of unprimed metric tensors. Since $g_{\alpha'\beta'}$ is diagonal, the bi–anisotropy vector $\boldsymbol{w}'$ vanishes: in virtual space–time the medium is at rest. For the dielectric tensors we obtain

$$\varepsilon'_x = \mu'_x = \varepsilon'_y = \mu'_y = 1\,. \tag{39.11}$$

Since electromagnetic waves propagating in the z–direction are polarized in the x, y plane their electromagnetic fields only experience the x and y components of the dielectric tensors. Consequently, for one–dimensional wave propagation virtual space–time is empty, waves are free here, whereas in physical space–time they appear as modulated wavepackets according to the transformations (39.3) and (39.4).

We mapped one–dimensional electromagnetic waves in moving media onto waves in empty space–time, but what happens at the horizon? Why are horizons special? Suppose, without loss of generality, that the medium moves in the negative z–direction and develops a black–hole horizon at $z = 0$: for positive z the flow speed $|u|$ lies below c/n and for negative z the flow exceeds c/n. For the waves that propagate against the current in the positive z–direction the transformation (39.3) develops a pole at the horizon. We linearize $u(z)$ here and obtain

$$t_+ = t - \frac{\ln|z|}{\alpha}\,, \qquad \alpha = \frac{1}{1-n^{-2}}\left.\frac{\mathrm{d}u}{\mathrm{d}z}\right|_0\,. \tag{39.12}$$

For any given time t, the entire range of t_+ is filled for $z > 0$ or $z < 0$, for either side of the horizon. So virtual space–time has become multi–valued, as illustrated by the Penrose diagram 5.18. Physical space–time seems to be cut into two distinct regions, because waves confined to either one of the two sides would never interact with each other. But this last assertion is not entirely true (Hawking [1974]).

Picture a wavepacket that, after having barely escaped from the horizon, propagates in a region where the medium moves at uniform speed. The wavepacket is purely forward–propagating, so it consists of a superpostion of plane waves with purely positive wavenumbers k and corresponding positive frequencies ω. On the other hand, according to a result from complex analysis (Ablowitz and Fokas [1997]) Fourier transforms with positive spectra are analytic on the upper half plane, because integrals over k containing the analytic function $\exp(\mathrm{i}kz)$ converge and are continuous for positive k and positive $\mathrm{Im}\, z$. Consequently, if we analytically continue the wave packet on the complex z plane it should be analytic for $\mathrm{Im}\, z > 0$, even if we trace its evolution back to the horizon. Analytic functions are continuous; so the escaping wavepacket must have partially originated from beyond the horizon: some part of the wave must have tunneled through (Fig. 5.19).

In order to calculate the tunnel amplitude we use the following trick (Damour and Ruffini [1976]): we combine two waves $\exp(-\mathrm{i}\omega t_+)$ localized on either side of

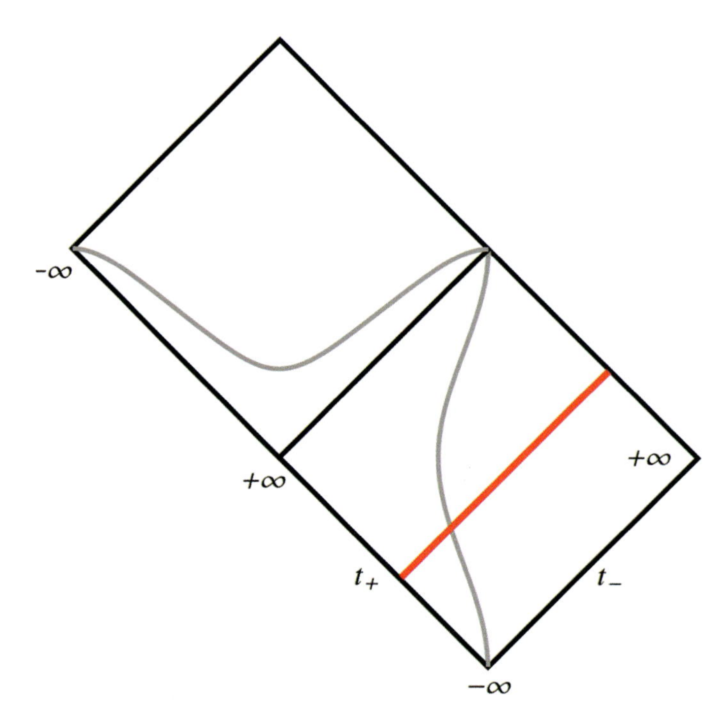

Figure 5.18: Penrose diagram with horizon. One–dimensional moving media are space–time transformation media. The picture represents the virtual space–time of the moving medium of Fig. 5.17 as a Penrose diagram (§31) spanned by the null coordinates $t_\pm$. The horizon cuts space–time into two parts; in the lower part (in front of the horizon) t_+ runs from $-\infty$ to $+\infty$ and in the upper part (behind the horizon) from $+\infty$ to $-\infty$. The grey curves represent the two spatial markers of Fig. 5.17 in space–time. The red line shows the space–time trajectory of a forward–propagating wave packet of light that therefore depends only on t_+. The wave packet begins its journey immediately in front of the horizon, it passes the grey marker and leaves to $t_- = +\infty$ that corresponds to $z = +\infty$.

the horizon such that they seamlessly form an analytic function on the upper–half of the complex z–plane. From

$$\exp\left(-\pi\frac{\omega}{\alpha} + \mathrm{i}\ln(-z)\right) = \exp\left(\mathrm{i}\frac{\omega}{\alpha}\ln(z)\right) \tag{39.13}$$

follows that we should give the waves on the left side of the horizon the prefactor $\exp(-\pi\omega/\alpha)$. Then the two waves combined make an analytic function. On the left side the medium moves faster than the speed of light in the medium, so here the phase of the wave moves backwards; in locally co–moving frames the wave would oscillate with negative frequencies. We discussed in §29 (pp. 177) that the complex waves with positive frequencies are always "shadowed" by negative–frequency waves. We have seen here that the horizon partially mixes the positive– and negative frequency parts. As the negative–frequency wave makes a negative contribution to the

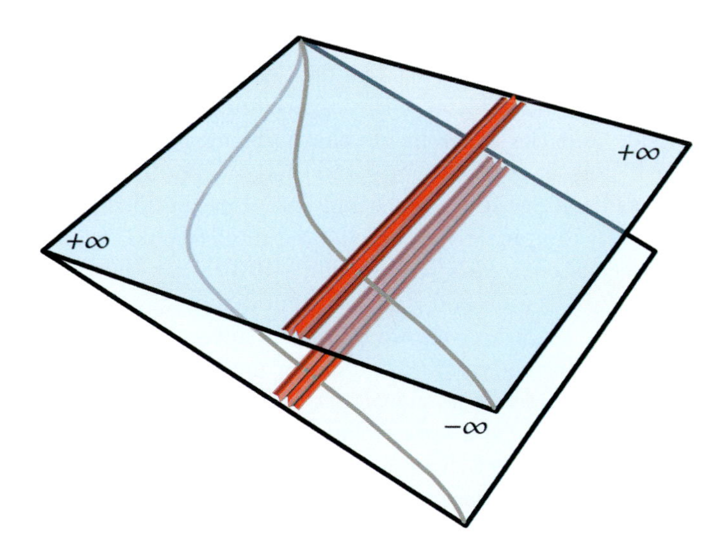

Figure 5.19: Folded Penrose diagram with horizon. Figure 5.18 shows the Penrose diagram of an outgoing wave that is either in front or behind the horizon. In contrast, this diagram visualizes the tunnelling of an ingoing wave. The Penrose diagram is folded such that the two sides of the horizon are connected and the wave can tunnel from one sheet to the other, say from the top (in front of the horizon) to the bottom sheet (beyond the horizon). There the wave appears as a mirror image, in time, of the wave in front of the horizon, as a negative–frequency wave. The conversion from positive– to negative–frequency waves is the classical cause behind the spontaneous creation of quanta at horizons, Hawking radiation.

energy, we ought to subtract the modulus square of its amplitude in the total intensity of the wave (a technically more involved argument (Leonhardt [2010]) confirms this reasoning). Therefore the ratio N of the tunneled to the total spectral intensity is

$$N = \frac{\exp(-2\pi\omega/\alpha)}{1 - \exp(-2\pi\omega/\alpha)} = \left[\exp\left(\frac{2\pi\omega}{\alpha}\right) - 1\right]^{-1} . \tag{39.14}$$

According to quantum field theory, positive– and negative–frequency photon pairs are spontaneously created from the quantum vacuum at horizons, because they do not cost any energy (Birrell and Davies [1984], Brout et al. [1995]). They are emitted at the tunneling rate (39.14) that, remarkably, one can also interpret as the Planck spectrum

$$N = \left[\exp\left(\frac{\hbar\omega}{k_B T}\right) - 1\right]^{-1} \tag{39.15}$$

with temperature (k_B denotes Boltzmann's constant)

$$k_B T = \frac{\hbar\alpha}{2\pi} . \tag{39.16}$$

Black holes are not black after all. The horizon emits light with a Planck spectrum (39.15) and temperature (39.16). The light consists of entangled photon pairs where each pair contains one photon with positive and one with negative frequency, one emitted outside and the other inside the horizon. Seen from one side of the horizon, the black hole is a black–body radiator. This insight, due to Stephen Hawking [1974], has been one of the most influential predictions of modern theoretical physics, because it illuminated an unexplored intellectual landscape: Hawking's theory supports Bekenstein's idea (Bekenstein [1973]) that the horizon carries an entropy proportional to its area and it gives a tantalizing glimpse into the quantum physics of a space–time geometry. Hawking's effect thus connects three vastly different areas of physics—thermodynamics, quantum mechanics and general relativity. In modern attempts for finding a quantum theory of gravity, such as loop quantum gravity (Rovelli [1998]) and superstring theory (Green, Schwarz and Witten [1987]), the correct prediction of the Bekenstein–Hawking entropy has been used as a benchmark. However, the Hawking temperature of solar–mass black holes lies about eight orders of magnitude below the temperature of the cosmic microwave background, so there is probably no chance of observing Hawking radiation in astrophysics. The benchmark of some of the most advanced theories of physics seems destined to remain theory.

On the other hand, laboratory analogues of the event horizon may demonstrate the physics behind Hawking radiation (Novello, Visser and Volovik [2002], Volovik [2003], Schützhold and Unruh [2007]). For example, one could perhaps generate a detectable amount of Hawking radiation using few–cycle light pulses in fibres or waveguides (Philbin et al. [2008]). According to nonlinear fibre optics (Agrawal [2001]) the pulses behave, for all practical purposes, like one–dimensional moving media (Philbin et al. [2008]): due to the optical Kerr effect they create additional contributions to the refractive index that move with the light pulses—media that move at the speed of light. Apart from optical dispersion, such moving media resemble the case considered in this Section. How Hawking radiation emerges from elementary processes in gravity has remained largely a mystery, but laboratory analogues have a chance to, quite literally, shed light on one of the most fascinating areas of physics, the creation of light at horizons.

Further reading

Applications of geometry in optics, in particular in transformation optics have become an active field of research with many papers published. We recommend the review article Chen, Chan and Sheng [2010] for further reading.

Tonomura [1998] is a beautiful book on quantum physics revealed by electron waves that explains the Aharonov–Bohm effect. Novello, Visser and Volovik [2002] and Schützhold and Unruh [2007] contain various ideas on analogues of gravity; Volovik [2003] focuses on the rich physics of Helium–3 that contains the entire standard model of physics; he calls it, half in jest and half in earnest, the "Helium–centric world view". Birrell and Davies [1984] discusses quantum fields in curved space; Brout et al. published a readable primer [1995] on quantum black holes. We also recommend Paul and Jex [2004] and Leonhardt [2010] on quantum optics to complement this book.

Appendix

We show here that the free–space Maxwell equations in an arbitrary space–time geometry are equivalent to the macroscopic Maxwell equations in Cartesian coordinates with the constitutive equations (31.1)–(31.2). To start, we must write the free–space Maxwell equations as tensor equations in space–time. As mentioned in §25, the fact that Maxwell's equations can be cast in such a form reveals that they satisfy Einstein's requirement that the laws of physics should be the same for all observers. The space–time tensor form of the free–space Maxwell equations are written in terms of the electromagnetic field tensor $F_{\mu\nu}$, constructed from the $\boldsymbol{E}$ and $\boldsymbol{B}$ fields; in a right–handed coordinate system

$$F_{\mu\nu} = \begin{pmatrix} 0 & -E_1 & -E_2 & -E_3 \\ E_1 & 0 & cB_3 & -cB_2 \\ E_2 & -cB_3 & 0 & cB_1 \\ E_3 & cB_2 & -cB_1 & 0 \end{pmatrix} . \tag{A1}$$

The free–space Maxwell equations are (Misner, Thorne and Wheeler [1973]):

$$F_{[\mu\nu;\lambda]} = F_{[\mu\nu,\lambda]} = 0 \, , \quad \varepsilon_0 F^{\mu\nu}{}_{;\nu} = \frac{\varepsilon_0}{\sqrt{-g}} \left(\sqrt{-g} F^{\mu\nu}\right)_{,\nu} = j^{\mu} \, , \tag{A2}$$

where $j^{\mu} = (\rho, j^i/c)$ is the four–current, and the square brackets denote antisymmetrization. This last operation produces a completely antisymmetric tensor and in (A2) it is explicitly

$$F_{[\mu\nu,\lambda]} = F_{\mu\nu,\lambda} + F_{\nu\lambda,\mu} + F_{\lambda\mu,\nu} \, , \tag{A3}$$

since the electromagnetic field tensor (A1) is antisymmetric: $F_{\mu\nu} = -F_{\nu\mu}$. The forms of (A2) containing partial rather than covariant derivatives are obtained by using the antisymmetry of $F_{\mu\nu}$ and $F^{\mu\nu}$ (the latter is of course obtained from $F_{\mu\nu}$ by raising the indices using the space–time metric tensor). In a left–handed coordinate system the relation between $F_{\mu\nu}$ and the magnetic field differs by a sign from that in (A1). Here we confine ourselves to right–handed systems; the issue of handedness is dealt with in §31. In flat space–time, with Minkowski metric (25.2), it is straightforward to show that Eqs. (A2) are the familiar Maxwell equations on p. 165. As in the consideration of spatial curvature on p. 166, the local flatness of space–time leads to Eqs. (A2) as the free–space Maxwell equations in arbitrary curved space–time.

We define a quantity $H^{\mu\nu}$ by

$$H^{\mu\nu} = \varepsilon_0\sqrt{-g}F^{\mu\nu} = \varepsilon_0\sqrt{-g}\, g^{\mu\lambda}g^{\nu\rho}F_{\lambda\rho} \tag{A4}$$

$$\Longrightarrow \qquad F_{\mu\nu} = \frac{1}{\varepsilon_0\sqrt{-g}}\, g_{\mu\lambda}g_{\nu\rho}\, H^{\lambda\rho} \tag{A5}$$

and regard $H^{\mu\nu}$ as being constructed from $\boldsymbol{D}$ and $\boldsymbol{H}$ fields as follows:

$$H^{\mu\nu} = \begin{pmatrix} 0 & D^1 & D^2 & D^3 \\ -D^1 & 0 & H^3/c & -H^2/c \\ -D^2 & -H^3/c & 0 & H^1/c \\ -D^3 & H^2/c & -H^1/c & 0 \end{pmatrix}. \tag{A6}$$

Then, introducing a new four–current $J^\mu = \sqrt{-g}\, j^\mu$, the free–space equations (A2) can be written

$$F_{[\mu\nu,\lambda]} = 0\,, \qquad H^{\mu\nu}{}_{,\nu} = J^\mu\,, \tag{A7}$$

which are easily shown to be the macroscopic Maxwell equations (26.4) in right–handed Cartesian coordinates. The constitutive equations are given by (A1) and (A4)–(A6).

To obtain relations between the vector fields $\boldsymbol{D}$, $\boldsymbol{H}$ and $\boldsymbol{E}$, $\boldsymbol{B}$ consider first the components F_{0i}; from Eqs. (A1), (A5) and (A6) one obtains

$$E_i = \frac{1}{\varepsilon_0\sqrt{-g}}\left(g_{i0}g_{j0} - g_{ij}g_{00}\right) D^j - \frac{1}{\varepsilon_0 c\sqrt{-g}}\,[jkl]g_{0j}g_{ik}\, H^l\,. \tag{A8}$$

We simplify this result as follows. The identity

$$g_{\mu\lambda}g^{\lambda\nu} = \delta^\nu_\mu \tag{A9}$$

gives

$$g_{i\lambda}g^{\lambda 0} = 0 \quad \Longrightarrow \quad g_{i0} = -\frac{1}{g^{00}}\, g_{ij}g^{j0}\,, \tag{A10}$$

$$g_{0\lambda}g^{\lambda i} = 0 \quad \Longrightarrow \quad g^{i0} = -\frac{1}{g_{00}}\, g^{ij}g_{j0}\,, \tag{A11}$$

$$g_{j\lambda}g^{\lambda i} = g_{j0}g^{0i} + g_{jk}g^{ki} = \delta^i_j\,. \tag{A12}$$

Use of Eqs. (A10) or (A11) in Eq. (A12) produces the two relations

$$\left(g^{ij} - \frac{1}{g^{00}}g^{i0}g^{k0}\right) g_{kj} = \delta^i_j\,, \quad g^{ik}\left(g_{kj} - \frac{1}{g_{00}}g_{k0}g_{j0}\right) = \delta^i_j\,, \tag{A13}$$

which reveal inverse–related 3×3 matrices. In view of Eqs. (A12) and (A13), multiplying (A8) by g^{li} and contracting on the index i yields

$$D_i = -\frac{\varepsilon_0\sqrt{-g}}{g_{00}}g^{ij}E^j + \frac{1}{cg_{00}}\,[ijk]g_{j0}\, H^k\,, \tag{A14}$$

the first of the constitutive equations (31.1)–(31.2).

To obtain the second constitutive relation, we employ the tensors dual to $F_{\mu\nu}$ and $H^{\mu\nu}$. This requires use of the space–time Levi–Civita tensor, which in a right–handed system is given by

$$\epsilon_{\mu\nu\lambda\rho} = \sqrt{-g}\,[\mu\nu\lambda\rho]\,, \quad \epsilon^{\mu\nu\lambda\rho} = -\frac{1}{\sqrt{-g}}\,[\mu\nu\lambda\rho]\,, \quad [0123] = +1\,. \tag{A15}$$

The dual tensors ${}^*F^{\mu\nu}$ and ${}^*H_{\mu\nu}$ are defined by (Misner, Thorne and Wheeler [1973])

$$\begin{aligned} {}^*F^{\mu\nu} &= \frac{1}{2}\epsilon^{\mu\nu\lambda\rho}F_{\lambda\rho} &&\Longrightarrow && F_{\mu\nu} = \frac{1}{2}\epsilon_{\mu\nu\lambda\rho}\,{}^*F^{\lambda\rho}\,, \\ {}^*H_{\mu\nu} &= \frac{1}{2}\epsilon_{\mu\nu\lambda\rho}H^{\lambda\rho} &&\Longrightarrow && H^{\mu\nu} = \frac{1}{2}\epsilon_{\mu\nu\lambda\rho}\,{}^*H^{\lambda\rho}\,, \end{aligned} \tag{A16}$$

so they have components

$${}^*F^{\mu\nu} = \frac{1}{\sqrt{-g}}\begin{pmatrix} 0 & cB_1 & cB_2 & cB_3 \\ -cB_1 & 0 & -E_3 & E_2 \\ -cB_2 & E_3 & 0 & -E_1 \\ -cB_3 & -E_2 & E_1 & 0 \end{pmatrix}, \tag{A17}$$

$${}^*H_{\mu\nu} = \sqrt{-g}\begin{pmatrix} 0 & -H^1/c & -H^2/c & -H^3/c \\ H^1/c & 0 & -D^3 & D^2 \\ H^2/c & D^3 & 0 & -D^1 \\ H^3/c & -D^2 & D^1 & 0 \end{pmatrix}. \tag{A18}$$

Re–expressed in terms of the dual tensors, the constitutive equations (A4)–(A5) read

$${}^*H_{\mu\nu} = \varepsilon_0\sqrt{-g}\,g_{\mu\lambda}g_{\nu\rho}\,{}^*F^{\lambda\rho}\,, \tag{A19}$$

$${}^*F^{\mu\nu} = \frac{1}{\varepsilon_0\sqrt{-g}}\,g^{\mu\lambda}g^{\nu\rho}\,{}^*H_{\lambda\rho}\,, \tag{A20}$$

where use has been made of the identity

$$\epsilon^{\mu\nu\lambda\rho}\epsilon_{\lambda\rho\sigma\tau} = -2(\delta^\mu{}_\sigma\delta^\nu{}_\tau - \delta^\mu{}_\tau\delta^\nu{}_\sigma)\,. \tag{A21}$$

Writing out ${}^*H_{0i}$ using Eqs. (A17)–(A19) one finds

$$H^i = -\frac{\varepsilon_0 c^2}{\sqrt{-g}}\,(g_{00}g_{ij} - g_{i0}g_{j0})\,B_j + \frac{\varepsilon_0 c}{\sqrt{-g}}\,[jkl]g_{j0}g_{ik}\,E_l\,. \tag{A22}$$

Comparison of this with Eqs. (A8) and (A14) shows that

$$B_i = -\frac{\sqrt{-g}}{\varepsilon_0 c^2 g_{00}}\,g^{ij}H^j - \frac{1}{cg_{00}}\,[ijk]g_{j0}\,E_k\,, \tag{A23}$$

which is the second of the constitutive equations (31.1)–(31.2).

In this way we derived Plebanski's constitutive equations (31.1)–(31.2) (Plebanski [1960]). Note that several other relations between $\boldsymbol{D}$, $\boldsymbol{H}$, $\boldsymbol{E}$ and $\boldsymbol{B}$ are contained in (A4)–(A5) and (A19)–(A20). For example, to express $\boldsymbol{D}$ and $\boldsymbol{H}$ in terms of $\boldsymbol{E}$ and $\boldsymbol{B}$ we need only take the time–space components of (A4), obtaining

$$D^i = \varepsilon_0\sqrt{-g}\left(g^{i0}g^{j0} - g^{ij}g^{00}\right)E_j - \varepsilon_0 c\sqrt{-g}\,[jkl]g^{k0}g^{ij}B_l\,, \tag{A24}$$

and the required formulae are Eqs. (A22) and (A24).

Bibliography

Ablowitz, M.J. and Fokas, A.S. [1997] *Complex Variables* (Cambridge University Press, Cambridge).

Agrawal, G. [2006] *Nonlinear Fiber Optics*, 4th Ed. (Academic Press, San Diego).

Aharonov, Y. and Bohm, D. [1959] *Phys. Rev.* **115**, 485.

Alu, A. and Engheta, N. [2005] *Phys. Rev. E* **72**, 016623.

Arnol'd, V.I. [1990] *Huygens & Barrow, Newton & Hooke* (Birkhäuser, Basel).

Ashkin, A. and Dziedzic, J.M. [1973] *Phys. Rev. Lett.* **30**, 139.

Barnett, S.M. and Loudon, R. [2010] *Phil. Trans. R. Soc. A* **368**, 927.

Batz, S. and Peschel, U. [2008] *Phys. Rev. A* **78**, 043821.

Bekenstein, J.D. [1973] *Phys. Rev. D* **7**, 2333.

Berry, M.V. [1980] *Eur. J. Phys* **1**, 240.

Berry, M.V. [1984] *Proc. R. Soc. A* **392**, 45.

Berry, M.V., Chambers, R.G., Large, M.D., Upstill, C. and Walmsley, J.C. [1980] *Eur. J. Phys.* **1**, 154.

Birrell, N.D. and Davies, P.C.W. [1984] *Quantum Fields in Curved Space* (Cambridge University Press, Cambridge).

Born, M. and Wolf, E. [1999] *Principles of Optics*, 7th Ed. (Cambridge University Press, Cambridge).

Bortolotti, E. [1926] *Rend. R. Acc. Naz. Linc.* **4**, 552.

Boyd, R.W. [1992] *Nonlinear Optics* (Academic Press, San Diego).

Brout, R., Massar, S., Parentani, R. and Spindel, P. [1995] *Phys. Rep.* **260**, 329.

Cai, W. and Shalaev, V. [2009] *Optical Metamaterials: Fundamentals and Applications* (Springer, Berlin).

Capolino, F. (ed.) [2009] *Theory and Phenomena of Metamaterials (Metamaterials Handbook)* (CRC Press, Boca Raton).

Chen, H., Chan, C.T. and Sheng, P. [2010] *Nature Materials* **9**, 387.

Cook, R.J., Fearn, H. and Milonni, P.W. [1995] *Am J. Phys.* **63**, 705.

Damour, T. and Ruffini, R. [1976] *Phys. Rev. D* **14**, 332.

Demkov, Y.N. and Osherov, V.I. [1968] *Sov. Phys. JETP* **26**, 916.

Dolin, L.S. [1961] *Isvestiya Vusov* **4**, 964.

Eaton, J.E. [1952] *Trans. IRE Ant. Prop.* **4**, 66.

Einstein, A. [1905] *Ann. Phys.* **17**, 891.

Erdlyi, A., Magnus, W., Oberhettinger, F. and Tricomi, F.G. [1981] *Higher Transcendental Functions, Volume I* (McGraw–Hill, New York).

Erdlyi, A., Magnus, W., Oberhettinger, F. and Tricomi, F.G. [1981] *Higher Transcendental Functions, Volume III* (McGraw–Hill, New York).

Fang, N., Lee, H., Sun, C. and Zhang, X. [2005] *Science* **308**, 534.

Feynman, R.P. [1948] *Rev. Mod. Phys.* **20**, 367.

Feynman, R.P. and Hibbs, A.R. [1965] *Quantum Mechanics and Path Integrals* (McGraw–Hill, New York).

Feynman, R.P., Leighton, R.B. and Sands, M. [1963] *The Feynman Lectures on Physics, Volume I* (Addison Wesley, Reading, Mass.).

Feynman, R.P., Leighton, R.B. and Sands, M. [1964] *The Feynman Lectures on Physics, Volume II* (Addison Wesley, Reading, Mass.).

Firsov, O.B. [1953] *Zh. Eksp. Teor. Fiz.* **24**, 279.

Fizeau, H. [1851] *C. R. Acad. Sci.* (Paris) **33**, 349.

Fresnel, A.J. [1818] *Ann. Chim. Phys.* **9**, 57.

Gbur, G. [2003] *Prog. Opt.* **45**, 273.

Goldstein, H., Poole, C.P. and Safko, J.L. [2001] *Classical Mechanics*, 3rd Ed. (Pearson, Upper Saddle River).

Gordon, W. [1923] *Ann. Phys.* (Leipzig) **72**, 421.

Green, M.B., Schwarz, J.H. and Witten, E. [1987] *Superstring Theory* (Cambridge University Press, Cambridge).

Greenleaf, A., Lassas, M. and Uhlmann, G. [2003a] *Physiol. Meas.* **24**, S1.

Greenleaf, A., Lassas, M. and Uhlmann, G. [2003b] *Math. Res. Lett.* **10**, 1.

Hannay, J.H. [1976] Cambridge University Hamilton prize essay.

Hao, Y. and Mittra, R. [2008] *FDTD Modeling of Metamaterials: Theory and Applications* (Artech House, Norwood).

Hart, J.B., Miller, R.E. and Mills, R.L. [1987] *Am. J. Phys.* **55**, 67.

Hawking, S.W. [1974] *Nature* **248**, 30.

Hawking, S.W. and Ellis, G.F.R. [1973] *The Large Scale Structure of Space–Time* (Cambridge University Press, Cambridge).

Hendi, A., Henn, J. and Leonhardt, U. [2006] *Phys. Rev. Lett.* **97**, 073902.

Jackson, J.D. [1999] *Classical Electrodynamics*, 3rd Ed. (Wiley, New York).

Jammer, M. [1989] *The Conceptual Development of Quantum Mechanics* (McGraw–Hill, New York).

Kleinert, H. [2009] *Path Integrals in Quantum Mechanics, Statistics, Polymer Physics, and Financial Markets* (World Scientific, Singapore).

Lai, Y., Chen, H., Zhang, Z.Q. and Chan, C.T. [2009a] *Phys. Rev. Lett.* **102**, 093901.

Lai, Y., Ng, J., Chen, H., Han, D., Xiao, J., Zhang, Z.Q. and Chan, C.T. [2009b] *Phys. Rev. Lett.* **102**, 253902.

Landau, L.D. and Lifshitz, E.M. [1982] *Mechanics* 3rd Ed. (Butterworth–Heinemann, Oxford).

Landau, L.D. and Lifshitz, E.M. [1975] *The Classical Theory of Fields*, 4th Ed. (Butterworth-Heinemann, Oxford).

Landau, L.D., Lifshitz, E.M. and Pitaevskii, L.P. [1984] *Electrodynamics of Continuous Media*, 2nd Ed. (Butterworth-Heinemann, Oxford).

Leonhardt, U. [2006a] *Science* **312**, 1777.

Leonhardt, U. [2006b] *New. J. Phys.* **8**, 118.

Leonhardt, U. [2006c] *Nature* **444**, 823.

Leonhardt, U. [2009] *New J. Phys.* **11**, 093040.

Leonhardt, U. [2010] *Essential Quantum Optics: From Quantum Measurements to Black Holes* (Cambridge University Press, Cambridge).

Leonhardt, U. and Philbin, T.G. [2006] *New J. Phys.* **8**, 247.

Leonhardt, U. and Philbin, T.G. [2009] *Prog. Opt.* **53**, 69.

Leonhardt, U. and Philbin, T.G. [2010] *Phys. Rev. A* **81**, 011804(R).

Leonhardt, U. and Piwnicki, P. [1999] *Phys. Rev. A* **60**, 4301.

Leonhardt, U. and Piwnicki, P. [2000] *Phys. Rev. Lett.* **84**, 822.

Leonhardt, U. and Piwnicki, P. [2001] *J. Mod. Optics* **48**, 977.

Leonhardt, U. and Tyc, T. [2009] *Science* **323**, 110.

Li, J. and Pendry, J.B. [2008] *Phys. Rev. Lett.* **101**, 203901.

Luneburg, R.K. [1964] *Mathematical Theory of Optics* (University of California Press, Berkeley and Los Angeles).

Mahoney, M.S. [1994] *The Mathematical Career of Pierre de Fermat, 1601-1665* (Princeton University Press, Princeton).

Maxwell, J.C. [1854] *Cambridge and Dublin Math. J.* **8**, 188.

Milonni, P.W. [2004] *Fast Light, Slow Light and Left–Handed Light* (Taylor & Francis, London).

Milton, G.W. [2002] *The Theory of Composites* (Cambridge University Press, Cambridge).

Milton, G.W. and Nicorovici, N.–A.P. [2006] *Proc. Roy. Soc. London A* **462**, 3027.

Miñano, J.C. [2006] *Opt. Express* **14**, 9627.

Miñano, J.C., Bentez, P. and Santamaria, A. [2006] *Opt. Express* **14**, 9083.

Minkel, J.R. [2002] *Phys. Rev. Focus* **9**, 23.

Misner, C.W., Thorne, K.S. and Wheeler, J.A. [1973] *Gravitation* (Freeman, San Francisco).

Nachman, A.I. [1988] *Ann. Math.* **128**, 531.

Nakahara, M. [2003] *Geometry, Topology and Physics*, 2nd Ed. (IOP Publishing, Bristol).

Needham, T. [1993] *Am. Math. Mon.* **100**, 119.

Needham, T. [2002] *Visual Complex Analysis* (Clarendon Press, Oxford).

Nehari, Z. [1952] *Conformal Mapping* (McGraw–Hill, New York).

Newton, I. [1687] *Philosophiae Naturalis Principia Mathematica* (Cambridge University Press, Cambridge).

Novello, M., Visser, M. and Volovik, G.E. (editors) [2002] *Artificial Black Holes* (World Scientific, Singapore).

Paul, H. and Jex, I. [2004] *Introduction to Quantum Optics: From Light Quanta to Quantum Teleportation* (Cambridge University Press, Cambridge).

Peacock, J.A. [1999] *Cosmological Physics* (Cambridge University Press, Cambridge).

Pendry, J.B. [2000] *Phys. Rev. Lett.* **85**, 3966.

Pendry, J.B., Schurig, D. and Smith, D.R. [2006] *Science* **312**, 1780.

Penrose, R. [2004] *The Road to Reality* (Vintage, London).

Philbin, T.G., Kuklewicz, C., Robertson, S., Hill, S., König, F. and Leonhardt, U. [2008] *Science* **319**, 1367.

Plebanski, J. [1960] *Phys. Rev.* **118**, 1396.

Post, E.J. [1962] *Formal Structure of Electromagnetics* (Dover, New York).

Rashed, R. [1990] *Isis* **81**, 464.

Riley, K.F., Hobson, M.P. and Bence, S.J. [1998] *Mathematical Methods for Physics and Engineering* (Cambridge University Press, Cambridge).

Röntgen, W.C. [1888] *Ann. Phys. Chem.* **35**, 264.

Rousseaux, G. and Guyon, E. [2002] *Bulletin de l'Union des Physiciens Français* **841**, 107.

Rovelli, C. [1998] *Living Rev. Rel.* **1**, 1.

Rytov, S.M. [1938] *Compt. Rend. (Doklady) Acad. Sci. URSS* **18**, 263.

Sarychev, A.K. and Shalaev, V.M. [2007] *Electrodynamics of Metamaterials* (World Scientific, Singapore).

Schleich, W. and Scully, M.O. [1984] *General relativity and modern optics*, in *Les Houches Session XXXVIII New trends in atomic physics* (Elsevier, Amsterdam).

Schützhold, R. and Unruh, W.G. (editors) [2007] *Quantum Analogues: From Phase Transitions to Black Holes and Cosmology* (Springer, Berlin).

Schurig, D., Pendry, J.B. and Smith, D.R. [2007] *Opt. Express* **15**, 14772.

Schurig, D., Mock, J.J., Justice, B.J., Cummer, S.A., Pendry, J.B., Starr, A.F. and

Schutz, B.F. [2009] *A First Course in General Relativity*, 2nd Ed. (Cambridge University Press, Cambridge).

Serdyukov, A., Semchenko, I., Tretyakov, S. and Sihvola, A. [2001] *Electromagnetics of Bi–anisotropic Materials* (Gordon and Breach, Amsterdam).

She, W., Yu, J. and Feng, R. [2008] *Phys. Rev. Lett.* **101**, 243601.

Shen, Y.R. [1984] *The Principles of Nonlinear Optics* (Wiley, New York).

Singh, S. [1997] *Fermat's Last Theorem: The story of a riddle that confounded the world's greatest minds for 358 years* (Walker, New York).

Sihvola, A.H., Viitanen, A.J., Lindell, I.V. and Tretyakov, S.A. [1994] *Electromagnetic Waves in Chiral and Bi–Isotropic Media* (Artech House, Norwood).

Smith, D.R. [2006] *Science* **314**, 977.

Snyder, A.W. and Love, J.D. [1983] *Optical Waveguide Theory* (Chapman and Hall, London).

Solymar, L. and Shamonina, E. [2009] *Waves in Metamaterials* (Oxford University Press, Oxford).

Stachel, J. [1982] *Phys. Rev. D* **26**, 1281.

Stavroudis, O.N. [2006] *The Mathematics of Geometrical and Physical Optics* (Wiley–VCH, Weinheim).

Stockman, M.I. [2007] *Phys. Rev. Lett* **98**, 177404.

Stoll, R.R. [1969] *Linear Algebra and Matrix Theory* (Dover, New York).

Stratton, J.A. [1941] *Electromagnetic Theory* (McGraw–Hill, New York).

Tai, C.T. [1994] *Dyadic Green Functions in Electromagnetic Theory* (IEEE Press, Piscataway).

Tamm, I.Y. [1924] *J. Russ. Phys.-Chem. Soc.* **56**, 2-3, 248.

Tamm, I.Y. [1925] *J. Russ. Phys.-Chem. Soc.* **56**, 3-4, 1.

Tonomura, A. [1998] *The Quantum World Unveiled by Electron Waves* (World Scientific, Singapore).

Tyc, T. and Leonhardt, U. [2008] *New J. Phys.* **10**, 115038.

Unruh, W.G. [1981] *Phys. Rev. Lett.* **46**, 1351.

Unruh, W.G. [2008] *Phil. Trans. Roy. Soc. A* **366**, 2905.

Valentine, J., Zhang, S., Zentgraf, T., Ulin–Avila, E., Genov, D.A., Bartal, G. and Zhang, X. [2008] *Nature* **455**, 376.

Van Bladel, J. [1984] *Relativity and Engineering* (Springer, Berlin).

Veselago, V.G. [1968] *Sov. Phys. Usp.* **10**, 509.

Volovik, G.E. [2003] *The Universe in a Helium Droplet* (Clarendon Press, Oxford).

Wagner, F.E., Haslbeck, S., Stievano, L., Calogero, S., Pankhurst, Q.A. and Martinek, K.–P. [2000] *Nature* **407**, 691.

White, R.W. [1973] *J. Acoust. Soc. Amer.* **53**, 1700.

Wiles, A. [1995] *Ann. Math.* **141**, 443.

Index